Political Parties in American Society

SECOND EDITION

Samuel J. Eldersveld
University of Michigan

Hanes Walton, Jr.
University of Michigan

BEDFORD/ST. MARTIN'S Boston ◆ New York

To Alice and Els

ISBN 978-1-349-62492-8 ISBN 978-1-137-11290-3 (eBook)
DOI 10.1007/978-1-137-11290-3

For Bedford/St. Martin's
Political Science Editor: James R. Headley
Senior Editor, Publishing Services: Douglas Bell
Production Supervisor: Cheryl Mamaril
Project Management: Stratford Publishing Services, Inc.
Cover Design: Lucy Krikorian
Cover Photo: Liaison Agency, Inc. Copyright © Rod Rolle
Composition: Stratford Publishing Services, Inc.
Printing and Binding: Haddon Craftsman, an R.R. Donnelley & Sons Company

President: Charles H. Christensen
Editorial Director: Joan E. Feinberg
Director of Editing, Design, and Production: Marcia Cohen
Manager, Publishing Services: Emily Berleth

Library of Congress Catalog Card Number: 99-64268

Softcover reprint of the hardcover 2nd edition 2000 978-0-312-22688-6

5 4 3 2 1 0
f e d c b a

For information write: Bedford/St. Martin's, 75 Arlington Street, Boston, MA 02116 (617-426-7440)

ISBN: 978-0-312-24164-3 (paperback)

Transferred to Digital Printing 2007

Preface

Major changes have taken place in American politics since our earlier book. There has been turnover in party control of the presidency—from the Democrats to the Republicans (1980) and from the Republicans to the Democrats (1992). In addition, Democratic control of Congress gave way to Republican control in 1994. And public disaffection with the major parties led to a vigorous third-party challenge in 1992 by Ross Perot. Public alienation seemed also to be responsible for a continuing decline in voter turnout, reaching a low of 48.8 percent of the eligible adults in 1996. The parties have launched major efforts at reform of their organizations and operations during this period—in Congress, in the national party committees, and at the state and local levels. Scholars have claimed that these reforms have revitalized the parties, arrested their decline, and led to more effective campaigns and more party cohesion in Congress. The leadership selection process, particularly for presidential candidates, although continuously modified during this period still appears flawed and subject to much criticism. Scholars continue to debate the virtues and consequences of party reform in the past twenty years and disagree on the centrality and relevance of American parties today. It is within this context of political change, as well as change in the American society, that this survey of American parties is presented.

This study of American political parties is a radical revision of the 1982 book. It is the result of a reexamination of the working of our system, which led to a new description of how that system has changed, as well as how and why it has endured. It is an update incorporating new data from all recent elections, including 1998, as well as the findings from reports on party development in the eighties and nineties. We focus primarily on the importance of parties as structures of action in American society. We ask a series of critical questions about the character and performance of our parties: How well are they organized, nationally and locally; what changes have occurred in the functions they perform in our political system; in what ways has party leadership—national, state, and local—changed; how have the nature and basis of party competition for political office changed; how has the public's involvement with the parties been altered over the years; how well have the parties met the challenges of political action committees and interest groups; how responsibly do parties represent the interests of the critical groups in the United States, including ethnics, women, and racial groups; how have party conflict and divided government contributed to or inhibited the development of sustained social and economic policy? The emphasis is on adaptation by our parties to the changing needs and demands of the American public.

Our present two-party system has survived for over 130 years, since the Civil War. Ninety percent of American voters still cast ballots for either the Republican or Democratic parties. Two-thirds say they "identify" with one of these two parties. The Republican and Democratic parties win virtually all elections and run all the governments. This is as true today as in the past, despite many challenges over the years from third parties. Perot's Reform party in 1996 is just another example of the brevity of third-party challenges. Our two-party system is thus quite permanent, dominant, and durable. It commands the support of the majority of American citizens who are attentive to politics.

The persistence of our party system presents a puzzle. It has survived despite many third-party challenges, despite great changes in American society, and above all despite the predictions of its demise by scholars and commentators. There has been in fact a considerable controversy over the relevance, the viability, and the state of health of our major parties. Writers have declared at one time or another that the party system was at the point of extinction. One scholar predicted "the gradual disappearance of the political party in the United States."[1] Others warned, "a partyless era . . . may be settling on us."[2] However, another recently stated flatly: "America's political parties are alive and well."[3] One commentator wrote a book entitled *The Party's Over*[4]; another wrote *The Party's Just Begun.*[5] Many scholars are on the fence, uncertain how well parties are doing and what their future will be. One scholar describes three possible models of what is happening to our system: party decline, party stabilization, and party resurgence. He then concludes that "each model captures a part of reality."[6] While there is great disagreement and ambivalence, few political scientists would agree with the extreme position that "the United States actually has no political parties at all."[7] The student of American parties thus faces the task of sorting out the evidence on the state of the parties and interpreting that evidence carefully, aware that American scholars have wildly conflicting images and evaluations of our party system.

How has it happened that the American party system has survived, despite many predictions to the contrary? Our answer is that the parties have adapted successfully by changing significantly in the past 138 years. Party systems are complex institutions, constantly changing, just as the world they work in is constantly changing. The types of parties we have today are very different from those of 1860, or 1900, or 1930, or even 1960. Parties must have the capacity to change, in certain positive and responsive ways, if they are to survive. Not all major parties (to say nothing of minor parties) do survive. A study of European parties revealed that 23 percent of their major parties disappeared and another 22 percent made major structural changes in order to survive.[8] Our system has also undergone considerable structural and functional change and has survived. One important question we will examine in this book is what changes have been made and how successful are they, are they actually linked to survival? A systematic analysis of party change and its effect on party survival is necessary.

All modern societies are in a state of dynamic change, manifested in the constantly altered character of their populations, the emergent political interest groups, the new types of issues which dominate political discourse, the social and economic transformations which occur, the periodic increase in international tensions and pressures, and the fluctuation in the depth of the public's interest in, and involvement with, political matters. Parties everywhere need to respond to these changes, and the American system is no exception. Our society and political system impose an additional set of burdens on our parties: the heterogeneity of our society, the federal and fragmented nature of our political institutions, and the very populist elements of our political culture. We elect more people to public office, we hold elections more frequently, we impose more burdens on citizens to make political choices than any other nation. We confront citizens with more mass media exposure for longer campaign periods, requiring parties to secure and spend more money in elections, than anywhere else in the world. The tasks of the parties are to organize effectively, recruit able leaders, campaign efficiently, finance adequately, communicate with voters convincingly, and govern responsibly—more demanding and challenging today than ever. It is within this context that party performance must be understood and evaluated.

Our aim in this book is to lead students to enlightened, and more confident, interpretations of our party system. We are convinced of the relevance and central importance of the parties for this complex democratic system of ours. Their relevance exists at the individual as well as the system level. Parties have complex meanings for the lives of American citizens. Thousands of Americans are contacted by party workers in each campaign, thousands belong to local party organizations and run for local or state or national offices. Above all, they are exposed daily via the mass media to the Democratic and Republican leaders in Washington, DC, or the legislators in their state capitals, or their county commissioners, or their city councils as these leaders search for solutions to our basic social and economic problems. The meaning of all this exposure to, and involvement with, parties varies of course for the individual. Clearly, however, for a large part of the American public the parties are periodically, if not regularly, salient. They are reference groups for many people, social groups for many activists, political action groups for political careerists, and policy groups affecting all of us. For a few they are career groups. The reality of the political party for the individual citizen is thus highly diverse, but for all of us the actions of the party can have a decisive impact. We therefore need to systematically acquaint ourselves with who the parties are, what they do, and how well they do what they do.

A controversy has persisted for many years among scholars of our party system over what type of system the American party system is. Many writers have seen our parties as nonideological, issueless, and primarily power driven; they see the party system, then, as centrist, divorced of real issue alternatives, a system of benign moderation. Sundquist noted this emphasis early, calling it "the centrist theme in the literature on party behavior."[9] He cited as an example a statement by Scammon and Wattenberg that "the center is the only position of political power."[10] But other great scholars of American parties have pictured our parties somewhat differently. V. O. Key long ago noticed the "policy cleavages" between the Republican and Democratic parties, based on early survey data, and described our system, while moderate, as a "dualism" with significant differences in issues and group support.[11] He quoted Charles Beard's conclusion that "the center of gravity of wealth is on the Republican side while the center of gravity of poverty is on the Democratic side."[12] Later textbooks have also discussed at length the ideological role of parties, but are inclined to diminish its relevance. Sorauf in his early book discussed policy and issue differences at length, but concludes that in the last analysis "the party in the government and its candidates cannot give up their total commitment to electoral victory."[13] Beck, in his revision of Sorauf's book, also describes the "important differences of principle between the parties," but settles for the basic conclusion that "throughout their long histories, the American political parties have been remarkably nonideological."[14] On the other hand, another recent scholar of American parties, Aldrich, arrives at a different conclusion: "the public has perceived the parties as quite distinct," in recent years particularly; "the good reason [for this] is that the parties actually are distinct.[15]

We are inclined in this text to put more emphasis on the nature of ideological divisions in American society, linked to social cleavages, and the way in which parties as conflict structures represent these cleavages as they compete for power. In the first version of this book in 1982, we defined the American parties as "ideological competitors," although "not polarized," and presented evidence of the distinctiveness of party supporters, activists, and leaders in ideology and program. We shall continue to argue here that American parties can be very ideological, programmatic, issue oriented and dualist, rather than "centrist," unprincipled, and exclusively power driven.

Some of the disagreement about the linkage between parties and ideology stems perhaps from the different interpretations of Downs' classic book on democracy, written in 1957. His spatial model of a two-party system assumes, he says, that parties will move closer together to appeal to those voters clustered at the center of the public opinion spectrum, hence becoming more moderate to win middle-of-the-road voters.[16] However, he also reminds us that "political parties are caught in the classic dilemma of all competitive advertisers. Each must differentiate his product."[17] And "any party which is both responsible and reliable will probably have an ideology. . . . parties find ideologies useful in gaining the support of various social groups . . . and, thus, "ideologies develop out of the desire . . . [for] gaining office."[18] This set of theoretical assumptions has guided us in the past and will guide us in this book.

Another goal we seek to pursue in this book is to describe and analyze how our parties have appealed to, and have represented, the interests of minorities, especially racial and ethnic minorities, over the years. Early scholars avoided the race issue or argued it was not important to focus attention on this matter. A renowned early student of parties, Charles Merriam, was an exception. In 1923 he wrote: "among the significant factors in the composition of the political party is that of race affiliation—Sometimes this is the decisive element in determining party allegiance. . . . In a great section of the U.S. it places the question of race above all other issues, for this one problem seems to overshadow them all."[19] In this way Merriam conceptualized "race" as a major variable in party behavior. Yet no one followed up with a systematic analysis for a long time. In fact, there were those who downplayed African American suffrage rights and suggested the defensibility of disenfranchisement.[20] Most early textbooks devoted at most only a few sentences to the subject of race. As Rossiter put it, they "tiptoed around diversity issues like race."[21] There were, however, notable exceptions that gave somewhat more attention to the issue. Sundquist particularly challenged the omission of the racial question in studies of our parties and traced the treatment of it since 1948 by both parties, calling it a "crosscutting" and "realigning" issue.[22] The consensus on avoidance of the subject was shattered in 1989 by Carmines and Stimson in their book *Issue Evolution: Race and the Transformation of American Politics.* They argued that parties had come to terms with the issue of race and had responded to it in their search for political power. They wrote: "the progressive racial tradition in the Republican party gave way to racial conservatism [in the mid-1960s] and the Democratic party firmly embraced racial liberalism. These changes unleashed political forces that permanently reshaped the contours of American politics."[23] Although the importance of this analysis is now acknowledged,[24] American party textbooks have not systematically incorporated data on race in their discussions of party system change. We hope to remedy that omission in this book.

We see our system as essentially one of considerable political conflict and competition, more so today than previously. The divergent coalitional character of our parties reveals this, as do their stands on the critical issues of the day and their basic ideological orientations. We find this conflict healthy and central to democratic belief and practice: Democracies like ours emerged from political struggle, and institutions such as the party system were founded on the principle of the legitimacy and necessity of political opposition. The sharp contest between parties for political office, grounded in programmatic dissent and argument, is a major component of our system. We will explore, describe, and analyze how well our parties function as power contestants and ideological competitors and whether they thus fulfill the requirements for effective party democracy and meaningful social change.

The periodic process of reanalysis and review of the developments in American political parties, an area of intellectual inquiry of great importance for our system, is necessary. The process requires evaluation of developments based on an appraisal of the evidence from important research conducted over the years by able scholars in the field. Only then can useful and valid statements be made about the state of our parties and the impact of changes on party performance. Ad hoc critiques and generalizations not informed by systematic research evidence are too gratuitous and tend to be unreliable. We seek in this book to be as data based in our theoretical conclusions as the available research permits us to be.

Special studies like this of the party system in a particular country are necessary. This study describes and analyzes how the parties in this American system function and how they adapt over time—and thus how the two-party system survives. As a case study of a party system in one country, it identifies the conditions influencing party development and the key factors responsible for successful organizational adaptation. We need such detailed studies of this system and of other party systems in other countries. Our analytical findings help us develop a theory of party system change that we hope will be useful for understanding other party systems, most of which are quite different from ours. Thus, the American party system case study can contribute to a broader knowledge of party performance and change in democratic societies.

ACKNOWLEDGEMENTS

Our hope is that this book will again be useful to students of the American party system. Insofar as that is true, we must give much credit to many others, while accepting the criticisms as ours. We are particularly indebted to the many scholars who over the years have engaged in excellent research about our system, and comparatively have done excellent research comparing our system to others. Since 1982, much research has been completed, on which we have relied heavily. The quality of this scholarship over the entire postwar period has been very high, well informed, and stimulating. It has been a tremendous task to become acquainted with this body of literature, and we are sure we have not adequately covered it, or represented it, in our book. We apologize for omissions. Our students over the years have also contributed to our research and to our insights about American politics, for which we heartily thank them. Several able secretaries helped tremendously in putting this volume together: Holly Bender, Amy Napolitan, Jan Williams, Crystal Williamson, Margaret Mitchell Ilugbo, and Greta Blake. We thank them for their long-suffering assistance. We also were helped a great deal by our graduate students who assisted with preparation of materials: Adam Berinsky, Paul Freedman, Anne M. Bennett, and Volker Krause. Finally, we thank our political friends whom we interviewed in many places. It was through the contacts and discussions with them that we gained a better understanding of the functioning of our parties at the grass roots, in their organizations, and in government. This book, thus, is the product of a tremendous amount of excellent research, the insights of many writers, and experience in the classroom teaching of the parties course, as well as the intellectual interactions with other scholars, students, and politicians. This book also arises in part from some of our mentors: the late Harold Gosnell and Robert Brisbane, and Professors Samuel DuBois Cook and Tobe Johnson to name a few. All were indispensable for this study, whose aims are theoretical relevance, empirical evidence, and practical wisdom about the historical development and contemporary performance of our party system.

Samuel J. Eldersveld
Hanes Walton, Jr.

NOTES

1. Walter Dean Burnham, *Critical Elections and the Mainsprings of American Politics* (New York: Norton, 1970), 133.
2. William J. Crotty and Gary E. Jacobson, *American Parties in Decline* (Boston: Little, Brown, 1980), 255.
3. Joseph A. Schlesinger, "The New American Party," *American Political Science Review* 79 (December 1985): 1152.
4. David Broder, *The Party's Over* (New York: Harper and Row, 1972).
5. Larry J. Sabato, *The Party's Just Begun* (Glenview, IL: Scott Foresman/Little, Brown, 1988).
6. James W. Ceaser, "Political Parties—Declining, Stabilizing, or Resurging?" in Anthony King, ed., *The New American Political System,* 2nd ed. (Washington, DC: AEI Press, 1990), 90.
7. Richard S. Katz and Robin Kolodny, "Party Organization as an Empty Vessel: Parties in American Politics," in Richard S. Katz and Peter Mair, eds., *How Parties Organize* (London: Sage Publications, 1994), 24.
8. Richard Rose and Thomas T. Mackie, "Do Parties Persist or Fail? The Big Trade-Off Facing Organizations," in Kay Lawson and Peter H. Merkl, *When Parties Fail* (Princeton: Princeton University Press, 1988), 543.
9. James L. Sundquist, *Dynamics of the Party System* (Washington, DC: Brookings Institution, 1973), 301.
10. Ibid., 300. Richard M. Scammon and Ben J. Wattenberg, *The Real Majority* (New York: Coward-McCann, 1970), 200.
11. V. O. Key, Jr., *Politics, Parties and Pressure Groups,* 5th ed. (New York: Crowell, 1964), 217-18, 222.
12. Ibid., 215.
13. Frank Sorauf, *Party Politics in America* (Boston: Little Brown, 1968), 360.
14. Paul Allen Beck, *Party Politics in America* (New York: Longman, 1996), 360.
15. John H. Aldrich, *Why Parties?* (Chicago: University of Chicago Press, 1995), 169-70.
16. Anthony Downs, *An Economic Theory of Democracy* (New York: Harper, 1957), 116-17.
17. Ibid., 97.
18. Ibid., 109-13.
19. Charles E. Merriam and Harold F. Gosnell, *The American Party System* (New York: Macmillan, 1922), 12.
20. Edward M. Sait, *American Parties and Elections* (New York: The Century Company, 1927), 31-32.
21. Clinton Rossiter, *Party and Politics in America* (Ithaca: Cornell University Press, 1960), 58.
22. Sundquist, *Dynamics,* 314-19, 355-69.
23. Edward Carmines and James Stimson, *Issue Evolution: Race and the Transformation of American Politics* (Princeton: Princeton University Press, 1989), 58.
24. Alan Abramowitz, "Issue Evolution Reconsidered: Racial Attributes and Partisanship in the American Electorate," *American Journal of Political Science* 38 (February 1994): 1-24.

Contents

About the Authors

Samuel J. Eldersveld is Professor Emeritus at the University of Michigan. He has specialized in American and comparative politics and, though retired, still teaches and works with students in these fields. He received his Ph.D. at the University of Michigan where he was Department Chair and Murfin Professor of Political Science. He was also mayor of Ann Arbor, Michigan, from 1957 to 1959. He has held many fellowships abroad and has taught in the Netherlands, France, India, and China. In 1964, he received the APSA Woodrow Wilson Award for his book *Political Parties: A Behavioral Analysis.* In 1982, he published the first edition of *Political Parties in American Society.* Among his other books are *Citizens and Politics: Mass Political Behavior in India; Elite Images of Dutch Politics; Political Elites in Modern Societies; Party Conflict and Community Development;* and *Local Elites in Western Democracies.*

Hanes Walton, Jr., is Professor of Political Science at the University of Michigan. He holds a Ph.D. from Howard University and specializes in American politics, race and politics, parties and elections, and state and local government. During his career, he has held Guggenheim, Ford, Rockefeller, and APSA congressional fellowships. Over the years, he has frequently taught a course on political parties and his research on the topic has produced several books. In *The Negro in Third Party Politics* and *Black Political Parties,* he has explored African-American third party behavior. In his books *The Native-Son Presidential Candidate: The Carter Vote in Georgia* and *Reelection: William J. Clinton as a Native-Son Presidential Candidate,* he analyzed the rise and impact of southern Democrats as party leaders and influential strategists. His current research on President Lyndon B. Johnson continues this innovative work on how southerners are transforming party behavior in the Democratic and Republican organizations at the elite and mass levels.

CHAPTER 1

Parties in Society: A Theoretical Overview

Political parties are major structures of politics in modern societies. In developing democracies they are universal phenomena. They are objects of intensive study, primarily because, as apparatuses of political action and social power, they engage in activities that may be of great consequence for the individual citizen and for the world in which that citizen lives. In order to understand parties and arrive at a useful way of thinking about them, we will discuss parties from three standpoints: their place in the political system, their nature as a special group, and their importance to democracy.

THE PLACE OF PARTIES IN THE POLITICAL SYSTEM

Modern political systems are highly complex, the American system particularly so. The *institutions* of government in Washington and at the state and local levels consist of huge bureaucracies, large bicameral legislatures with elaborate committee systems, and many layers of courts with different jurisdictions — structures so separated and fragmented as to make coordination of governmental action extremely difficult. In addition to these institutions, a multiplicity of *interest groups* in modern systems, seemingly always increasing in number and diversity and constantly pressing their conflicting claims and demanding governmental attention, makes the representation of these interests and the mediation of group conflicts more involved and trying than ever before. And finally, there is in modern systems the *mass public,* larger than ever because of the extension of the right to vote and the greater opportunities for participation in politics than in the past. The task of political leadership to communicate with this public and mobilize its support is ever more complicated at the same time that it is ever more important.

If such systems are to survive and to be governed effectively, means must be developed to bring the scattered parts together, to lubricate the system so that, at least minimally, leaders, groups, and citizens can work toward the achievement of certain goals for the society. Specifically, leaders must be recruited, policy objectives defined, citizens heard, group conflicts negotiated, and elections conducted. Leaders have to mobilize support, competition has to be organized, and the public has to be educated. These activities must be accomplished in such a way as to make sense to citizens as well as to lead to important policy decisions by elites.

What emerges to facilitate government in modern systems under these complex conditions are linkage structures, intermediary organizations that help produce positive action and effective decisions in the face of fragmentation, conflict, and mass involvement. These structures are groups that engage in activities and organize initiatives that make cooperative behavior possible. The political party is one major type of linkage structure (see Figure 1.1); some would say it is the central one. In what ways is such a structure crucial for the system? First, it provides a basis for interaction and cohesion within legislatures, such as Congress, and often, but not always, between legislative and executive leaders, such as between Congress and the president. Further, a party provides some basis for cooperation between national, state, and local institutions and leadership. Second, the party is a forum within which interest groups can (but not all do) present their views about governmental policies as well as press for particular types of candidates for offices, both elective and appointive. The party is, therefore, an arena for the development of compromises by interest groups as well as the agent in creating interest group coalitions working for particular goals. Third, a party constitutes a medium or chan-

FIGURE 1.1 ■ Parties as Linkage Structures

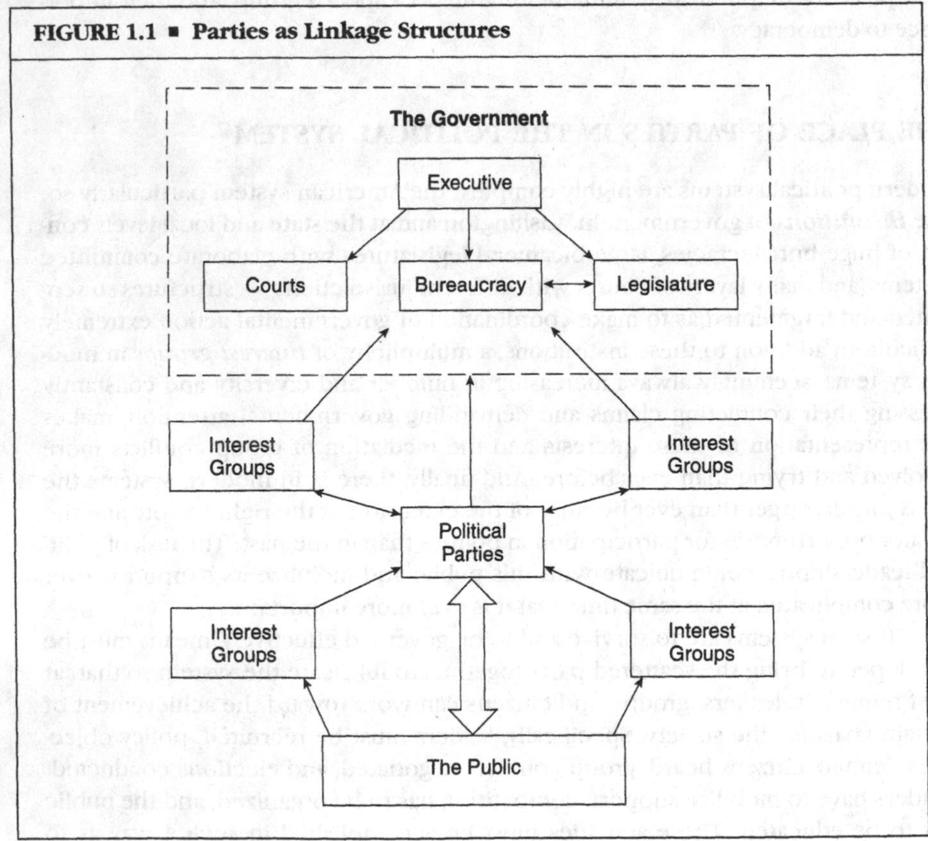

nel for communication linking citizens, organizational leaders, and governmental officials. Parties bring the citizen into contact with government and, conversely, are used by leaders to communicate with the public. Thus, the involvement of many citizens with politics occurs through the parties.

Parties, however, are not the only linkage apparatuses in modern systems. Certainly some of the large-scale interest groups, labor unions, business associations, farm organizations, and civic clubs, as well as single-issue groups, are important in providing linkages between citizens and political leaders. In some societies, indeed, such groups play a role as central as that of parties. Further, the linkage contributions of the mass media, particularly of newspapers and television, are considerable in bringing citizens to a continuing awareness of government. Although parties are not the only linkage structures, to many scholars they are more central to the functioning of the American system than are other structures. An early scholar of American parties, Schattschneider, wrote almost forty years ago that "the parties are not merely . . . appendages of modern government; they are in the center of it and play a determinative and creative role in it."[1] More recently, Samuel Huntington has argued that the party "is the distinctive institution of the modern polity. . . . the function of the party is to organize participation, to aggregate interests, to serve as the link between social forces and the government."[2] Avery Leiserson states the position even more explicitly: "The political party is a strategically critical concept. . . . The political party, or party system, *provides* the major connective linkage between people and government. . . ."[3]

There are scholars who do not precisely share this view of the centrality of parties. What has been presented here is the classic rationale for parties. Other voices in recent years have expressed reservations. One group of scholars, discussed in the Preface, has denied the relevance and even the existence of parties. Others take a more moderate view, de-emphasizing parties while still recognizing their (major) roles. Epstein, in his excellent interpretation of our system, sees parties as "organizationally desirable and probably essential . . . but I do not expect as much of parties as do most of their champions."[4] Fiorina states the importance of parties differently: "The only way collective responsibility has ever existed, and can exist, given our institutions, is through the agency of party."[5] For Aldrich, parties are structures created by politicians "because it has been in their interest to do so." But parties serve more than the narrow self-interest of politicians, he says; "they turn to parties because of their durability as institutionalized solutions" to problems of governance.[6] Scholars thus differ in their conception of how parties are relevant and important.

A concluding point should be made about parties in their context as linkage structures. We live in a society of considerable conflict between individuals and social groups. These cleavages are manifest in the nature and workings of our political institutions as well as in other aspects of our society. The disagreements occur at all levels of our system over a variety of immediate political issues: Should property taxes be cut, should the police force be strengthened, should streets be paved, should pollution standards be raised, should a new bomber be built, should welfare programs be cut, and so forth? Our positions on these issues reflect fundamental philosophical differences concerning the priorities for our society and the proper role of government in solving social problems. It was ever thus. The conflicts among humans about the use of their resources by political leaders and about the proper exercise of political authority have been recurrent and universal.

We also live in a society that is committed to cooperative behavior. That is, we are aware that some resolution of our conflicts must be undertaken if we are to be saved from intolerable turmoil or even civil war. The proper handling of social conflict is crucial for any society. In modern systems we have developed institutions and practices to "manage" conflict. We do this by (1) providing channels for the expression of conflicting viewpoints; (2) providing forums in which cooperative discussions for the resolution of conflict can occur; and (3) making decisions that are responsive to social conflicts and perceived as legitimate and authoritative, even if not happily acceptable to all those involved in a particular controversy.

Parties are groups that are part of this conflict articulation, mediation, and resolution process — not the only groups or institutions involved with it, but usually very much involved at some stage, if not at all stages in the process. This process, in large, is what "party politics" is all about: the struggle for power and influence among individuals and groups, which reflects conflicting objectives and viewpoints and leads eventually to policy making as at least a temporary resolution of the particular conflict.

The way parties provide the organization and leadership to participate in the conflict resolution process varies greatly from one society to another. Parties may be centralized power structures in one system or very decentralized in another. They may be ideological or personalistic and pragmatic. One party may capture majority control of government in one system, while in another system several parties may work in coalition toward social harmony. The parties in one system may be engaged in violent, hostile combat, while in another system the competition may be peaceful. Social groups may work completely within and through the party system; in another society each major interest group may organize its own party. Thus a wide variety of form, style, leadership, and activity may be found. But in all systems in which parties exist, they are groups deeply involved with the expression and mediation of human conflict.

WHAT TYPE OF GROUP IS A PARTY?

It is not enough to visualize the party only as a major linkage structure in the system. The key definition questions that remain are: What kind of group is it? What does it do as a group? What are its distinctive activities?

The search for an acceptable definition of a political party is quite complicated and controversial. The problem is that because there are such strange varieties of groups that are called parties or are considered to be parties, arriving at a universally applicable definition is difficult.

In some democratic systems parties can be easily formed to promote the interests of a small group of people or to advance a particular cause. Most of these parties have a flash-in-the-pan existence. When should they be taken seriously? Sometimes groups behave as parties are thought to behave but they have different labels, such as appeal, movement, rally, league, congress, or front. Some examples are the Christian Democratic Appeal (Holland), the Popular Republican Movement (France), the Awami League (Bangladesh), the Indian National Congress, Rally of the French People (Gaullist), and the Popular Front (France before World War II). What makes these groups parties? Sometimes parties exist for a long time but rarely win elections, as with the Prohibition party in the United States. Are they to be considered genuine parties? And what about factions within parties, such as Reagan's Citizens for the Republic — why are they not political parties? The variability in the phenomena taxes our ability at clear conceptualization. Yet it is possible to develop a clear definition that identifies the essence of party and distinguishes it from other groups.

The basic starting point is to recognize that a party is a group competing for political office through the election process. On this most scholars would agree. They may want to insert language to be more precise, however. They may want to describe what is meant by "competing" and when it is "genuine" competition, or they may want to insist that the election is "free," and even that the outcome for the party, at least occasionally, is a success (for the party). One scholar restricts the definition to parties that compete "successfully" (or for which at least there are "expectations of winning"). These qualifications can be argued. But the key initial way of defining a political party is to say that it contests elections in order to win public office.

In addition, students of parties may add other requirements: that the group has an organization that does the competing, that this organization seeks to mobilize votes, and that it does this in part at least by offering policies and programs to the voters that they think will distinguish the party from other parties. Finally, one could add the expectation that the party, if it succeeds, will use public office to maximize its chances of winning again.

Scholars differ in what they emphasize about parties in their definitions. To us, three aspects or "images" emerge that are particularly important. All three are interrelated, in our opinion, and are derived from, and linked to, the idea of the party as a group seeking to get votes from the public in order to get public office and, hence, political power. We present each of these interconnected images.

The First Image of Party: A Group Seeking Power by Winning Elections

The first image of a party is a group seeking political power by winning elections. The party consists of men and women who either select or endorse candidates for a public office and then work to secure enough votes to put them in office. Thus one scholar says parties are "organizations that pursue the goal of placing their avowed representatives in government positions."[7] And another writes that a party

Variations in Scholars' Conceptions of a Political Party, from 1770 to Today

E. Burke: "[A] party is a body of men united, for promoting by their joint endeavors the national interest, upon some particular principle in which they are all agreed." (1770)

G. Wallas: "The party is, in fact, the most effective political entity in the modern national state." (1908)

C. Merriam: "The broad basis of the party is the interests, individual and group which struggle to translate themselves into types of social control acting through the political process." (1922)

E. E. Schattschneider: "The parties are not . . . merely appendages of modern government, they are in the center of it and play a determinative and creative role in it." (1948)

M. Duverger: "Present-day parties are distinguished far less by their programme or the class of their members than by the nature of their organization. A party is a community with a particular structure." (1951)

M. Weber: "But 'parties' live in a house of 'power.' . . . their action is oriented toward the acquisition of social 'power,' that is to say, toward influencing a communal action." (1953)

A. Downs: "A political party is a team of men seeking to control the governing apparatus by gaining office in a duly constituted election. . . . parties formulate policies in order to win elections, rather than win elections in order to formulate policies." (1957)

A. Leiserson: "The modern political party is . . . an agency of informal, indirect representation of social groups and classes." (1958)

S. Huntington: "The distinctive institution of the modern polity . . . is the political party. . . . the function of the party is to organize participation, to aggregate interests, to serve as the link between social forces and the government." (1968)

M. P. Fiorina: "The only way collective responsibility has ever existed, and can exist, given our institutions, is through the agency of party." (1980)

is "distinguished from other political organizations by its concentration on the contesting of elections."[8]

This view of parties focuses on elections and the struggle for formal public offices. And it is true that most parties, and certainly the American Republican and Democratic parties, engage in this activity. In this definition, however, there are two types of inadequacies. First, scholars in some countries do not accept this conception as identifying the primary role of parties and hence question the comparative usefulness of the definition. Second, some scholars take exception to this concept by arguing that it does not really explain the *essence* of a party, or it explains only a small part of the essence of a party, because parties engage in other important types of activities and pursue other types of goals than just electoral activities and political power. Those who have studied third-world politics have noted that these party leaders, while not uninterested in power, are often inclined to place primary emphasis on developmental economic and social activities, building schools and roads, teaching illiterate peasants to use modern methods of farming, improving health conditions in villages, and educating people in family planning.[9] While elections are not unimportant, what parties actually emphasize in developing societies may differ greatly from what parties emphasize in our society. In many American communities, also, parties often engage in many activities other than contesting elections. Our local parties can be preoccupied with addressing social problems and human welfare needs in the community and with providing people an opportunity for civic involvement. These activities will be discussed in detail later.

The Second Image of Party: A Group Processing Interest-Group Demands

A second, long-existing image of a party is a group that represents social interests and often is actually a coalition of interests seeking community actions. Some of the earliest scholars saw parties in this light, emphasizing that they were different from other groups because they adjusted interest-group demands in the context of the welfare of the total community. Thus Max Weber, the famous German sociologist, distinguished in an early work between "the communal actions" of parties and those of social classes and status groups, which were concerned, he said, with only one segment of the society.[10] And Charles Merriam, an American scholar, wrote in 1922, "The broad basis of the party is the interests, individual or group, usually group interests, which struggle to translate themselves into types of social control acting through the political process of government."[11] The major emphasis of these writers is on the adjustment of group conflicts, or as G. Almond puts it, "interest aggregation" (the converting of demands of groups into major policy alternatives).[12] V. O. Key also saw the party as having "its foundation in sectional, class, or group interests."[13] The party, then, is unique because it organizes support for governmental leadership from a variety of groups.

Thus this second image sees parties as apparatuses of communal action linked to the groups making up the community, responding to and processing their demands, and thereby translating social reality into political reality. In this conception of a party the acquisition and use of power to contest elections is not inconsequential, but the nature of such a power process is given a broader meaning: Parties are groups which are integrated with the social process.

The Third Image of Party: An Ideological Competitor

A third image of a party is a group communicating an ideology to the public and, at least in democratic societies, competing with other parties on ideological terms. Anthony Downs is a classic exponent of this position; to Downs the most significant aspect of parties is that "they formulate policies in order to win elections." That is, he says, the parties look carefully at the distribution of public opinion among the voters ("the political market") and develop a stance on policy in relation to their calculations of how they can best maximize votes in relation to this political market of public opinion.[14] Figure 1.2 illustrates this conception, both for the United States and for a multiparty system. How parties deal with the political market varies greatly from one country to the next; in the United States the "market" of opinion and of votes differs in its basic nature from that of other countries. Parties have to adapt to these differences. In a society such as the United States, with a "normal" distribution of public opinion, with the majority of the public near the center, parties will crowd that ideological center. In the Netherlands, with a "multimodal" distribution of ideology for the public, parties do not crowd the center but develop clienteles at various points along the ideological continuum.

The essence of this view of a party is ideological competition. Both the parties and the individual voters find ideologies useful. Every party must differentiate its position from other parties in order to attract votes. And voters find party ideologies useful because they help focus attention on the differences between parties. Parties develop ideologies and issue positions in relation to voter preference patterns on issues. The party is a dynamic structure; it adjusts *over time* to changes in mass preferences in order to maintain and maximize its power. If public prefer-

FIGURE 1.2 ■ The Downs Model for Two Contrasting Party Systems

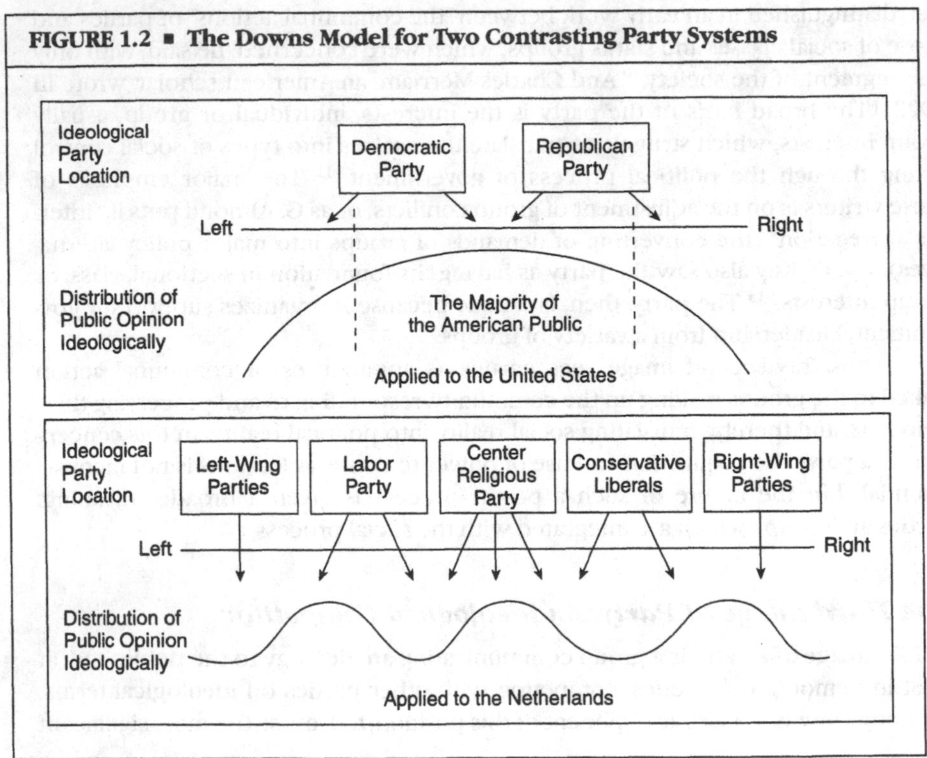

ence swings left, there will be a perceptible shift in party ideological positions; similarly, if public opinion shifts to the right, the parties will shift. Parties are presumably pragmatic, within reason, in their linkage to mass preferences on policy. V. O. Key called this tendency for the two American parties "dualism in a moving consensus."[15]

Certain key assumptions of the Downs concept may not prevail in the real world. Among these is that all voters have policy positions, that they know the policy positions of the parties, and that they vote on the basis of policy preferences, not on the basis of habitual patterns of behavior such as a long-standing party identification. He assumes also that parties know the political market, develop distinctive ideological positions, communicate these effectively to the public, adhere to them consistently, and, above all, are able to change to left or right to adjust to that market without alienating the hard core of their supporters and activists. Obviously voters and parties do not always behave in these ways.

Despite such reservations, it is an ideal-type model of parties and has a real value. In some societies with party systems where ideology is indeed salient, this theory may account for much more than in other societies where ideology is less basic. But parties in any society must compete for power by developing distinctive ideologies and policy positions, whether moderate or extreme, in the minds of the public. Thus weakly in some societies and strongly and effectively in others, parties bring about the intensification and crystallization of ideological conflict or its dilution and deflation. Parties may either clarify and polarize ideological differences or obscure and depolarize them. The relevance of this function for all societies cannot be overstated.

Based on the three images described so far, the authors suggest the following definition of a party: A party in a democracy is a group that competes for political power by contesting elections, mobilizing social interests, and advocating ideological positions, thus linking citizens to the political system.

PARTIES IN COMPARISON WITH OTHER GROUPS

To clearly distinguish parties from nonparty groups is not a simple task. Attempts that have focused on the stability of party organizations over time or on their preoccupation with elections are not convincing enough about the special character of parties.

This problem of definition in relation to other groups becomes most difficult, of course, when the attempt is made to distinguish a political party from another kind of group such as a labor union, an inclusive business association, or a general-purpose civic group (Common Cause). Max Weber had no difficulty in distinguishing a party from a class or a status group; he saw the latter two as representing only a segment of the community, while a party was community wide in its orientation. This same context perhaps can distinguish between a party and other political groups in modern societies. Most nonparty groups have a relatively narrow following and a limited, specific interest. Obviously, the Westside Neighborhood Improvement

Association, the Catholic Young Men's Club, and the Association of University Professors are groups with limited perspectives. They are not community wide in their concerns; they do not compete ideologically for people's votes, and they do not have the special patterns of relationships with other interest groups (which parties have). Common Cause, a "peak" or "bridge" association, however, comes somewhat closer to fulfilling the conception of a party elaborated here. Yet there are significant differences between a bridge association and a party — in style, action strategy, and function.

Emphasis must be placed on the special meaning and purpose of parties in modern societies; the three images outlined here elucidate that meaning. Parties are specialized, community-oriented apparatuses of action distinguished by their special relationships both to other groups in the community and to the public. While seeking formal governmental power, they represent and manage group conflict and at the same time mobilize and compete ideologically for mass support. This in the purest form of a party is its unique goal orientation, its strategy for action, and its community relevance. Parties vary from society to society in the relative importance of their concern with power, their interest-group conflict, and their involvement in ideological competition. But the probabilities are that the further parties diverge from these three differentia, the less likely they are to be considered in empirical terms as party structures. They may be ideological movements or interest groups or flash-in-the-pan ad hoc contenders for governmental office. They are not, then, as likely to be durable, power relevant, and societally functional structures with specialized action strategies and roles in relation to interest groups and the public delimited above. A party is a group that cannot be conceptualized in a power vacuum or a policy vacuum or a social vacuum.

THE RELEVANCE OF PARTIES FOR DEMOCRACY

There is a continuing argument over the need for political parties in democracies. The differing views of two scholars will illustrate this point. Elmer E. Schattschneider says that "political parties created democracy and . . . modern democracy is unthinkable save in terms of the parties."[16] Leon Epstein counters, "There is . . . a serious question whether parties must perform it [the governing function] in every democratic political system" in order for the system to be effective.[17] In trying to resolve this controversy, it is necessary to specify what is meant by democracy.

Democracy has many different meanings, depending on the particular interests and values of the person explaining it. To Aristotle, one major aspect was equality: "the most pure democracy is that which is so called principally from the equality which prevails in it. . . ."[18] Democracy had several meanings, including equality, to de Tocqueville, but at one point he says that "the very essence of democratic government consists in the absolute sovereignty of the majority."[19] The Russian scholar M. Ostrogorski, observing the American system, stated that "the first postulate of democratic government is the active participation of the great mass of the citizens."[20] To Joseph Schumpeter, democracy was a special method, "that institutional arrangement for arriving at political decisions in which individuals acquire the power to decide by means of a competitive struggle for the people's vote."[21] Robert Dahl, advocate of the pluralist thesis, the democratic governmental process that evolved especially in the United States, finds democracy "a political system in which all the active and legitimate groups in the population can make themselves heard at some crucial stage in the process of decision."[22] To Dahl, leaders make the decisions in a democracy, but they must not only "listen to the noise, but expect to suffer in some significant way if they do not placate the group, its leaders or its most vociferous members."[23] Bachrach, however, who rejects the emphases of both elitist and pluralist theses, states the essence of democracy differently:

> I believe that a theory of democracy should be based upon the following assumptions and principles: the majority of individuals stand to gain in self-esteem and growth toward a fuller affirmation of their personalities by participating more actively in meaningful community decisions; people generally, therefore, have a twofold interest in politics — interest in end results and interest in the process of participation. . . .[24]

This collection of views of democracy illustrates the variety of possible emphases. But upon reflection of these democratic theories, one notices a common thread in the ideas of these theorists, with an emphasis on the *citizen public's role in the political process.* Elitist theorists stress the need for political leaders in a democracy to compete for and secure support from the masses in order to stay in office. The pluralistic theorists emphasize the need for elites to be constantly responsive to group demands and to censure. And ultrademocratic theorists place importance on the self-esteem of the ordinary person in a democratic society and the value of participation. Thus elite competition, elite responsiveness, and public participation (all involving the public in some way) are key elements in a full-bodied theory of democracy. One of these elements may be concentrated on to the neglect of others, but it is difficult to advance a decent and acceptable theory of democracy by ignoring all three. Democracy means conflict among ideas, interests, ideologies, and elites; it means a concern by leaders for public problems and demands and needs. Also democracy means a realistic opportunity for non-elites and the public to be involved in the political process.

Given these conceptions of democracy — including even the most elitist (but democratic) conception — it is necessary to ask what political parties have to do with the functioning of the democratic system. Why is it that parties are central to the system? Historically, when democracies were born, parties came into existence to perform two somewhat contradictory roles: (1) to provide the organizational base by which elites could mobilize resources and compete with each other for votes under the new democratic elections and thus maintain themselves in power, and (2) to provide the organizational base by which new claimants to elite status could mobilize support and thus oppose those in power and eventually dislodge them from power. A later discussion of the origins of the American party system before 1800 will show how parties were functional to both elite power maintenance and to elite displacement. Thus parties historically and today are to be seen as being functional to the acquisition and loss of political power.

In a more specific sense, however, parties are relevant to the functioning of democratic systems (see Figure 1.3). First, Huntington finds that the pressure for political participation demands that political institutions cope with and channel such participatory pressure. And parties are one such type of institution, a key type through which the demands for participation can be handled. "Parties organize political participation, party systems affect the rate at which participation expands."[25] And, according to Huntington, the participation of new groups and their integration into the new democratic order is basic for the stability of the system:

> In modernizing society "building the state" means in part the creation of an effective bureaucracy, but, more importantly, the establishment of an effective party system capable of structuring the participation of new groups in politics.[26]

Linked to this role of organizing public participation is a second role, that of providing popular control over elected officials. Many writers have emphasized this function of parties. Indeed, it is hard to visualize how elites would be held accountable in a noncoercive state in which there were no parties or similar types of groups. Parties provide the opportunity for such control in at least four ways:

1. They select, encourage, and support individuals who are seeking leadership positions in society.
2. They "structure the competition" among these candidates.
3. They establish an organizational tie to these candidates on the basis of which their performance in office can be judged.
4. They serve as channels for potential reprisal against, and defeat of, officials whose behavior the public rejects.

As Robert Dahl says, "One of the strongest claims for political parties is that they assist the electorate in gaining some degree of control over elected officials and, thus, over the decisions of government."[27]

FIGURE 1.3 ▪ The Roles of Parties in the Democratic Political System

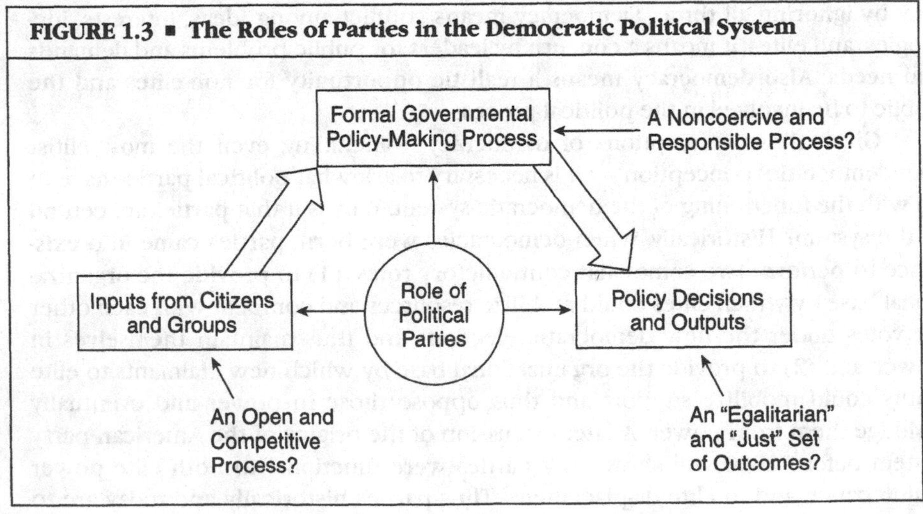

Third, the competition for power in a democracy is closely linked to the party system. Basic cleavages exist in the socioeconomic interests in any society, whether or not one wishes to acknowledge their existence. And these interests collide in seeking governmental power and policy implementations. In authoritarian and totalitarian systems such conflicts are suppressed or ignored, and parties are considered unnecessary or dysfunctional. In democratic societies, however, there are open conflicts and parties are organizational instruments and channels for these conflicting interests. The parties consist of rival cadres of activists and organizational personnel representing policies and programs and making demands to be heard. These conflicts must be managed, compromised, and resolved. As Dahl states, in a democracy there are "multiple centers of power" and between them "constant negotiations . . . are necessary in order to make decisions."[28] Parties provide a major arena for combat among these conflicting interests as well as for bargaining, negotiation, and resolution among the conflicting parties.

It can well be argued that "democracy involves a balance between the forces of conflict and consensus."[29] Parties play a fourth key role in a democracy, that of reflecting and articulating conflict while making agreement and consensus possible. Thus, parties perform both competitive and integrative functions in the democratic society. As Lipset points out, Karl Marx saw the modern political system primarily in conflict (class conflict) terms, while theorists like de Tocqueville were the first to see the system in terms of both conflict and consensus. And the solidarity or consensus of the system is enhanced by conflict. It is in this perspective that political parties must be visualized in the modern democratic system. So Dahl argues:

> Prior to politics, beneath it, enveloping it, restricting it, conditioning it, is the underlying consensus on policy that usually exists in the society among a predominant portion of the politically active members. Without such a consensus no democratic system would long survive the endless irritation and frustrations of elections and party competition.[30]

Huntington, too, sees parties in modernizing societies as sources for stability: "The development of political parties parallels the development of modern government. . . . The Party [in certain developing societies] is not just a supplementary organization, it is instead the source of legitimacy and authority."[31]

From a variety of scholarly perspectives, therefore, political parties are critical for the democratic system. Certainly in the early stages of the formation of the democratic system, the beginnings of a party system were necessary. In the later stages of the system, parties developed in a more complex and differentiated way. There are, to be sure, considerable areas of disagreement among scholars who address themselves to this question, but there is strong support and agreement among them as to what democracy means: the enhancement of citizen welfare through political participation; opportunities for citizens to control elites in power; freedom and channels for the articulation of conflict among diverse interests; effective procedures for the accommodation, aggregation, and translation of these conflicts into public policy; the structuring of meaningful competition among rival cadres of activists seeking governmental power; the continuous responsibility of elites for the development of new policies that reflect the claims and needs of new social forces; and the achievement through conflict of an underlying consensus, a commitment to the democratic "rules of the game."

Accepting that these are the characteristics (or meanings) of democracy, then, most scholars would argue that parties not only are relevant, but probably are central. For parties are structures for political action which have from the beginning in modern democracies provided constitutional channels for citizen participation, control of elites, interest aggregation, conflict management, competition maximization, policy innovation, and system consensus. Whether political parties do effectively constitute institutional channels performing these roles in the American democracy is both the basic and open question in the study of parties and of this book. Lord Bryce said long ago that "party is king."[32] V. O. Key claims, "Political parties constitute a basic element of democratic institutional apparatus."[33] Whether and in what ways parties are central to the functioning of the democratic state are the critical questions that will constantly recur in this analysis of the American party system.

CAN ONLY DEMOCRACIES HAVE PARTY SYSTEMS?

Some people assume that a party system can exist only in a democratic society, because the system consists of two or more parties competing freely with each other for influence and power. This position is based on the belief that a certain type and amount of competition between autonomous groups is necessary before there can be a party system. In an authoritarian society like the USSR in the past, or China today, therefore, it is tempting to conclude that the Communist party is not, or was not, a party system. Care must be taken before jumping to this conclusion. The existence of party systems in these countries may be credible because either they do have a certain amount of competitive interaction among subgroups or factions or such groups perform certain key roles in the system. Consider, for example, the one-party system in the past in some southern states in the United States. The Democratic party was often a highly fragmented structure, consisting of several factions. Although there may not have been any viable opposition from the Republican party, there was competition among factions within the Democratic party. Single-party structures in authoritarian societies functioning as intermediaries between government and the public — communicating, representing, and aggregating the interests of the masses — may indeed be construed as constituting party system, even though competition may be severely limited.

One-party systems differ from plural-party systems in many respects. Aside from the limited competitiveness of one-party systems, their linkage to the social structure is essentially different. They perform functions other than those manifest in plural-party societies, such as being the arm of the bureaucratic or military operation of the state, but functionally they may be in certain respects similar to party systems elsewhere. Scholars of African societies maintain that the one-party system there reflects the intense tribal division of the society, which leads to one party, representing one tribal or geographical interest, gaining control at the expense of all others. Or, they argue, there are no conflicting interests, "the idea of class is something entirely foreign to Africa," and therefore, since "there is no fundamental opposition," there is no need for more than one party to represent a homogeneous society. In either case, the one-party system is linked to the socioeconomic interest-group infrastructure quite differently from that in openly competitive democratic systems. In one-party states, interest cleavages are either unrecognized and under-represented or suppressed. As Huntington states, "a one party system is, in effect, the product of the efforts of a political elite to organize and to legitimate rule by one social force over another in a bifurcated society."[34] Thus, one-party systems perform some of the same basic types of functions as democratic party systems, but perform them in a very distinctive, perhaps authoritarian, style.

THE HISTORICAL RELEVANCE OF PARTIES

Today there are political scientists who would argue that parties are not functionally central to modern government, or that they are ineffectual in dealing with our problems, or that they are rivaled by or being replaced by other groups. Some scholars claim that for anyone interested in social change and social progress, parties are not of primary relevance.

If party decline is indeed occurring, an explanation must be attempted in conjunction with a consideration of the basic argument of the theories of the "party functionalists." What is their basic position? They argue essentially that parties have been, and are, important for the political development and thus for the social and economic development of any society that is modernizing. Historically, these scholars argue, parties came into existence as structures necessary to perform functions of a special nature, particularly as these societies became openly competitive and democratic. The leadership of a new nation then develops a program of policy goals to which it is committed, a set of institutions designed to implement these goals, and a strategy to mobilize public support for these goals. The polity has to be expanded to include sub-elites and citizens, and the role of the political party system in this development process is crucial: cementing a structure of leadership, linking central and local elites, evolving a program of action, and mobilizing public support. It is hard to see how such a development process can occur without a party system. William Chambers, in describing the historical origins of the American system, argues this position effectively:

> As the American founders resolved problem after problem in the shaping of the republic, they not only established the first modern political parties. They were also involved, if most unknowingly, in a general process of political modernization in which parties were at once an element and a catalyst in a broader change from older to newer things.[35]

In the functionalist approach, as the system develops, parties perform critical functions for maintenance as well as change in the system. They provide opportunities for citizens to be involved in political group life and to participate in the political process. They recruit individuals and train them for public office. Attempting to resolve differences between conflicting claims, they represent and respond to the demands and needs of interest groups in the society. They socialize the public to the acceptance of the emerging political system and inform it about governmental programs, then they seek to mobilize public support on behalf of these programs. Above all, parties attempt to inject responsibility into the system, by providing integration both between the branches of government and between governmental leaders and their constituencies. In managing government and in providing an opposition to the government, parties provide a focus of responsibility and accountability. Of the role of parties, David Apter says that "the political party is such a critical force for modernization in all contemporary societies that the particular pattern of modernization adopted by each is often determined by its parties."[36]

As we proceed with the description and analysis of the American party system, we will be concerned not only with the characteristics of that party system but also with the role that system plays in American society.

NOTES

1. Elmer E. Schattschneider, *Party Government* (New York: Rinehart, 1942), 1.
2. Samuel Huntington, *Political Order in Changing Societies* (New Haven: Yale University Press, 1968), 89, 91–92.
3. Avery Leiserson, *Parties and Politics* (New York: Knopf, 1958), 35. For the most recent discussion of linkage see Kay Lawson, ed., *Political Parties and Linkage* (New Haven: Yale University Press, 1980).
4. Leon Epstein, *Political Parties in the American Mold* (Madison: University of Wisconsin Press, 1986), 18–23.
5. Morris P. Fiorina, "The decline of collective responsibility in American Politics," *Daedalus* 109 (Summer 1980): 26.
6. John H. Aldrich, *Why Parties?* (Chicago: University of Chicago Press, 1995), 21–26.
7. Kenneth Janda, *A Conceptual Framework for the Comparative Analysis of Political Parties* (Beverly Hills: Sage, 1970), 83.
8. Frank Sorauf, *Party Politics in America* (Boston: Little, Brown, 1967), 18.
9. See, for example, Thomas Hodgkin, *African Political Parties* (London: Penguin, 1961), 146; and David E. Apter, *The Politics of Modernization* (Chicago: University of Chicago Press, 1965), 183.
10. Max Weber, *From Max Weber: Essays in Sociology*, trans. Hans Gerth and C. W. Mills (New York: Oxford University Press, 1946), 194–95.
11. Charles Merriam and Harold F. Gosnell, *The American Party System* (New York: Macmillan, 1922), 3.
12. Gabriel Almond and G. B. Powell, Jr., *Comparative Politics: System, Process, and Policy*, 2nd ed. (Boston: Little, Brown, 1978), 198–205.
13. V. O. Key, Jr., *Southern Politics* (New York: Knopf, 1949), 15.
14. See Anthony Downs, *An Economic Theory of Democracy* (New York: Harper and Row, 1965), 24–31, for the basic model of parties.
15. V. O. Key, Jr., *Politics, Parties and Pressure Groups*, 5th ed. (New York: Crowell, 1964), 222–25.
16. Schattschneider, *Party Government*, 1.
17. Leon D. Epstein, *Political Parties in Western Democracies* (New York: Praeger, 1967), 315.
18. Aristotle, *Politics*, quoted in Robert Dahl, *A Preface to Democratic Theory* (Chicago: University of Chicago Press), 34.
19. Alexis de Toqueville, *Democracy in America*, quoted in Dahl, *Preface*, 35.
20. See introduction by Seymour M. Lipset to M. Ostrogorski, *Democracy and the Organization of Political Parties* (New York: Doubleday, Anchor, 1964), xxii.
21. J. A. Schumpeter, *Capitalism, Socialism, and Democracy* (New York: Harper, 1950), 269.

22. Dahl, *Preface*, 137.
23. Ibid., 145.
24. Peter Bachrach, *The Theory of Democratic Elitism* (Boston: Little, Brown, 1967), 101.
25. Huntington, *Political Order*, 401.
26. Ibid.
27. Robert A. Dahl, *Pluralist Democracy in the U.S.* (Chicago: Rand McNally, 1967), 244.
28. Ibid., 24.
29. Seymour M. Lipset, *Political Man* (New York: Doubleday, 1960), 24.
30. Dahl, *Preface*, 132.
31. Huntington, *Political Order*, 90–91.
32. James Bryce, *The American Commonwealth* (New York: Macmillan, 1916), 3.
33. Key, *Parties*, 9.
34. Samuel Huntington and Barrington Moore, *Authoritarian Politics in Modern Society* (New York: Basic Books, 1979), 11.
35. William Chambers, *Political Parties in a New Nation* (New York: Oxford University Press, 1963), 14.
36. David E. Apter, *The Politics of Modernization* (Chicago: University of Chicago Press), 179.

CHAPTER 2

The American Party System Viewed Comparatively

The United States has a unique party system. We call it a two-party system. By this we mean a party system in which only two parties share political power; that is, only two parties participate in the actual governing process. Third parties, or minor parties, exist from time to time, and people do vote for them, but it is extremely rare that they get enough votes to secure a seat in Congress, and of course they do not win the presidency. This is quite different from the multiparty systems of countries such as France, Germany, the Netherlands, Israel, and Scandinavia. In these countries, many more than two parties secure seats in their parliaments. For example, in Germany today six parties hold seats in their Bundestag; this number is usually the case also for Sweden and France, while twelve parties sit in the Dutch parliament. Even Great Britain, which many still see as a country with a "two-party system," as in the past, today has eleven parties with seats in the House of Commons. There are practically no systems today that are genuinely "two-party," as the U.S. system is.

It is important to understand clearly what a "two-party system" means. What are the distinguishing marks of a two-party system such as ours? Many scholars have discussed this matter at length and suggested a variety of differentia (for example, Sartori 1976, 185–92; Epstein 1986, 241–50). From such discussions, particularly of the nature of the U.S. system in contrast to other systems, the following observations have emerged concerning the key characteristics of our genuine two-party system.

1. *Majority Control.* One of the two major parties will win the presidency; one of them also will win a majority of seats in the House or the Senate or both. A pluralized legislature is conceivable but rare.
2. *Exclusive Succession.* If one party loses an election, the major opposition party, and only that party, accedes to office.
3. *Power Sharing.* The two major parties (and only these two) divide up the seats in the legislative chambers, and they may also divide majority control over these legislative chambers. Power may also be divided between the party that controls the executive and the party that controls one or both chambers of the legislature. However, there is no sharing of executive authority; only one party controls the cabinet.

4. *Extensive Territorial and Functional Competition.* The two parties compete for political offices throughout the country and at all levels of the system, sharing control of state and local governmental offices. The two-party system is thus pervasive and continuous.
5. *Credible Turnover Probabilities.* Over a long or short time period, it is expected that party A will replace party B in control of government, actually or as a reasonable possibility.
6. *Noncollusive Interrelationships.* Each party wants to govern alone, and the two major parties never join together to run the government. On particular legislative actions, however, members of the two parties may join forces, and occasionally a member of party B may join the cabinet of the president of party A, but this is done without the major parties losing their separate competitive and institutional identities.
7. *Third-Party Futility.* Third parties will appear from time to time to challenge this system and will be able to run candidates for office who may secure considerable votes, but they will fail to penetrate the two-party duopoly.
8. *Termination Conditions.* A third party may displace one of the two major parties if a major party ceases to be representative of the voters, performs ineffectively, fails to adapt to changing socioeconomic conditions, and provides unacceptable leadership for the country. A well-organized and well-led third party, competing extensively for Congress and the presidency on the relevant issues at a time of a major party default in performance can theoretically win the status of "major party." The obstacles are, however, formidable. It has not happened in the United States since 1860.

Party systems differ in many ways, but an initial method of distinguishing among them is by *number* of parties in the system and their relationships to each other. It is important to understand the distinctions between our system and others based on the number of parties. Such an understanding clarifies the context within which politics is conducted in the United States. It helps us realize what confronts citizens when they vote or otherwise participate in politics and what confronts leaders when they seek political office. After discussing the numerical differences, we will discuss other important ways in which party systems differ.

Based on the criterion of number of parties (see Table 2.1), there are five possible types of party systems in democracies. These are (with examples of each):

A. *Two Types of Majoritarian Systems*
   1. Majoritarian two-party — United States
   2. Majoritarian multiparty — Britain

B. *Two Types of Non-Majoritarian Systems*
   1. Moderately pluralist multiparty — Germany
   2. Highly pluralist multiparty — Netherlands

C. *One-Party or Predominant-Party Systems* — Mexico.

TABLE 2.1 ▪ Numerical Indicators of Differences in Party Systems

Indicators	*United States* *1996*	*Great Britain* *1997*	*Germany* *1994*	*Netherlands* *1998*
1. Percent of vote for the largest party	49.0	43.0	42.0	29.0
2. Percent of vote for the two largest parties	90.0	73.0	79.0	54.0
3. Percent of national legislative (lower house) seats held by the largest party	52.0	63.0	44.0	30.0
4. Percent of national legislative seats (lower house) held by the two largest parties	99.8	88.0	80.0	56.0
5. Number of parties on the ballot which won more than 2 percent of the vote	3.0	3.0	6.0	7.0

Note: These data are based on the last national election in each country: 1996 (United States), 1998 (Netherlands), 1994 (Germany), 1997 (Britain).

The British system is distinguished from the American system by the presence in the national legislature (House of Commons) of several parties (eleven in 1997) that have won votes. Thus, even though in Britain one of the two major parties does usually secure a majority of seats in Parliament, legislative representation is more pluralized. And it is possible that no party wins an absolute majority of seats, necessitating a minority government.

For non-majoritarian types, there is a wide variation in the number of parties that secure votes and seats. The one-party or predominant-party system is a rarity in democracies, but it is found today (Mexico) and occurred in the past (India until 1967). In such a system, one party dominates for long periods even though small parties formally exist. The indicators in the accompanying table (Table 2.1) illustrate the differences in party systems if we use some variation of the numerical criterion. It is clear from these indicators that the United States is the least fragmented system on all measures, followed by Britain; the German system is more pluralized, and the Netherlands clearly has the most diversified and fragmented party system of these countries. In the Netherlands, representatives of twelve parties sit in a parliament of 150 members. In contrast, in the United States all 435 seats in the House of Representatives are held by Republicans or Democrats, except for one socialist-independent from Vermont.

A Comparative Perspective:
An Example of a Pluralized Multiparty System

The Netherlands party system is considerably pluralized. In the 1998 parliamentary election, twenty-two parties received votes. There were five relatively major parties ("The Big Five"). Their votes were:

Party	*Percent of the Vote*
Labour Party	29.0
Christian Democrats (CDA)	18.4
Liberal Party	24.7
Democrats 66 (D66)	9.0
The Green Left	7.3

In addition, there were five other small religious parties (with a total of 5.3 percent of the vote), three parties representing "seniors," or elders, two other environmental parties, a small socialist party, a middle-class party, and even a party of "idealists," plus some parties with very few supporters. Over a million votes were cast by the Dutch voters for these seventeen small parties (out of a total of 8.6 million votes). Four of them got seats in the parliament, in addition to the "Big Five," for a total of nine parties with seats in a parliament with only 150 seats in all.

Note: The CDA, which emerged in 1974, is a composite of three earlier parties, two of which were Protestant and the other Catholic. It has been a "center" party. The Liberal party is ideologically right of center. The D66 is a party formed originally by intellectuals in 1966 and is on the moderate "left." The Green Left is a recent party, formed in 1989 and surprisingly successful in elections since that time.

One-party politics, it should be noted, can exist in certain geographical areas of systems where a two-party or multiparty system exists nationally. In the United States we have had the one-party Democratic South in the past, which has now become to some extent very Republican. The North also has districts dominated by one party. In the 1996 elections there were fourteen uncontested House districts, overwhelmingly one-party areas.

The 1996 election in the United States reveals the significant features of our two-party system. First, the two major parties, Democrats and Republicans, received 90 percent of the vote and all but one seat in the Congress (out of 435 House and 100 Senate members). Second, the largest minor party, Perot's Reform party, received 8 percent of the vote, a decline from 19 percent in 1992, and no seats in Congress. In the 18 percent of the 435 congressional districts in which third-party candidates presented themselves to the voters, only one candidate was successful and the combined vote of the others in those districts averaged 7 percent of the votes cast. This result illustrates that minor parties can get access to the ballot, but only in a rare case can this access lead to victory. There were actually nineteen candidates "running" for president in some states, on such party labels as the Green party, Libertarian party, and U.S. Tax party. Perot received 8 percent of the vote, the others a combined total of 2 percent. Third, the two large parties also dominated state politics. They ran candidates for governor and for representatives to state houses and senates. They were the only parties that won seats; forty-four state chambers were won by Republicans, forty-nine by Democrats. Fourth, in the states and in Washington, DC, the parties shared governmental office and power. The Republicans won the House and the Senate, the Democrats the presidency. In thirty-one states at least one of the state legislative chambers was won by a party (Republican or Democratic) different from the party controlling the governorship. Thus, power sharing is a nationwide party system phenomenon. It is also interesting to see evidence in the 1996 election results of how the pendulum swings to party B when party A loses. The Republicans had won back many seats in 1994, but the Democrats recouped somewhat in 1996, winning back a net of thirteen seats in Congress and a net of ninety-six seats in the state legislatures.

So far we have discussed only the *number* of parties that can and do play an effective part in our politics and how that criterion sets our party system apart from other systems. The numerical criterion, however, only scratches the surface of what our party system is really like. A variety of other insights can help us understand our system in contrast to others. We can organize these into six key dimensions.

Party Volatility but System Equilibrium

Party systems differ in the frequency and extremeness of the shifts in electoral strength of the parties. Some systems change often and much, while others are more stable. The U.S. party system ranks fairly high in the extent of fluctuations in presidential elections. In the past twenty years the Democrats won with 51 percent in 1976 (Carter), lost in three successive elections (in 1980 with only 40 percent of the vote), lost in the two succeeding elections (to Reagan and Bush), and finally recouped with Clinton (43 percent in 1992 and 49 percent in 1996). The Republican party's strength fluctuated even more, because that party was affected more by the Perot vote. The Republicans dropped to 48 percent in 1976 (Ford), then increased to 51 percent and 59 percent with Reagan (1980, 1984), dropped a bit to 53 percent with Bush (1988), but fell to 38 percent in 1992 (Bush), recouping only slightly with Dole in 1996 (41 percent). These average inter-election shifts, while considerable, were lower than previously, as the following summary reveals.

Average Percentage Shifts between Presidential Elections

Years	*Democrats*	*Republicans*	*Total*
1956–1976	11.2	11.0	22.2
1976–1996	4.8	7.0	11.8

While lower today, the party shifts in the United States are still higher than for the two largest parties in many European systems. The comparisons for the 1976–1996 period are United States, 11.8 percent; Britain, including 1997, 9.2 percent; Germany, 5.3 percent; and Netherlands, 8.3 percent.

The party shifts in the vote for Congress have also been considerable in recent years. The Democrats held 59 percent of the seats in the House of Representatives from 1986 to 1992, but lost to the Republicans, who won 54 percent in 1994 (continuing in 1996 with over 52 percent).

Clearly, considerable change in party fortunes can occur even in the short term in the United States and in these other countries. Turnover in party control of governments, while less frequent, can also occur. The defeat of the British Conservatives by the Labour party in 1997, the Clinton Democratic victory in 1992, the Republican reversal of control in the House of Representatives in 1994 (from 41 percent of the seats to 54 percent), and the major change in the cabinet of the Netherlands after the 1994 election (from Christian-Liberal to Socialist-Liberal) are all examples of the opposition potential in these systems and the genuine possibility of party turnover. The U.S. system is one of the most dynamic of these systems in the sense of the very real possibility of major party system change. Why this occurs is not an easy question to answer. And we shall spend considerable time analyzing campaigns, leadership, issues, and voting behavior to attempt to answer it. Each system has a large body of independent (or "floating") voters who can swing to and from parties. They are impelled to do so depending on the comparative records and popularity of candidates, their positions on issues, the quality of the campaigns, the social and economic conditions of the country and the way in which voters are affected by these conditions, the work of the parties in communicating with voters and mobilizing their votes, the presence of new alternative parties, and a host of other factors. Change is constantly occurring in the society, and how well the parties and their leadership respond to these changes and how well they persuade the public of their responsiveness to social and economic change are crucial aspects of the party shift phenomenon.

Despite electoral variations, the two major parties in the United States, like two giant corporations competing against each other, have maintained themselves as the preferred parties of a majority of Americans in spite of all threats. They have divided the popular vote ("controlled the voter market") as well as virtually all partisan political offices. Since the Civil War the Republicans and Democrats have constituted a party system in a state of dynamic equilibrium.[1] It is a system in which change takes place but the basic units in the system maintain themselves, and the basic character of the system does not change essentially over time. Great change takes place in voting support for the parties, but despite these changes the system does not disintegrate. Donald E. Stokes and G. R. Iversen, analyzing and reporting on this feature of American party history, note a negative correlation of -.55 between the division of the vote for one presidential election and the change of the vote from that election to the next. "In other words," they conclude, "the greater a party's share of the vote at one election, the greater is its share likely to be reduced at the next." Further, their analysis leads them to the conclusion that the probability "that the party division could have stayed within the historic boundaries of the vote for President *without the influence of equilibrium forces* [emphasis added] is less than four in a hundred."[2]

What is the nature of this equilibrium and what are these "equilibrium forces"? Figure 2.1 presents dramatically the nature of the equilibrium since the

FIGURE 2.1 ■ The Equilibrium of the American Party System

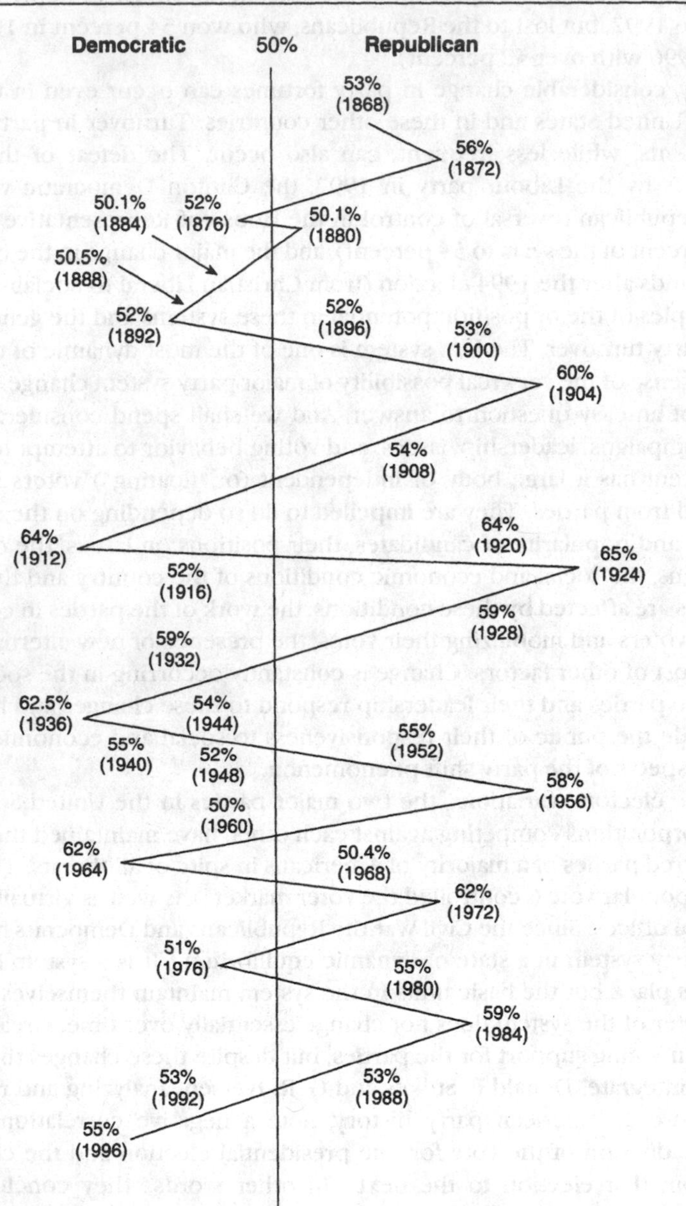

Note: Based on the percentage distribution of the vote for the two major parties.

Civil War, showing a "pendulum" swing in the party vote in presidential elections. The extreme point in major party strength is 65 percent, usually somewhat less than that, after which the strength of the winning major party recedes. There is a continuous movement toward the 50-50 distribution point. When a recession in party fortunes develops, it usually continues so that the opposing party gradually reduces the overwhelming support of the winning party. Then the opposition wins with, usually, a victory at the 50.1 percent to 52 percent level. This victory then becomes greater, until its strength reaches an outer limit and then begins to recede.

Competition for votes is fairly constant. The actual shift in absolute percentages of the two-party presidential vote, as Stokes and Iversen point out, is 5.7 percentage points (up to 1960; it is larger, 6.8 percent, if elections through 1996 are included). Yet significant fluctuations in the vote can indeed occur. Elections such as those of 1912, 1920, 1932, and 1972 might even be called "voting upheavals" or "revolutionary landslides." What is interesting, however, is the rapid comeback of a major party that has lost heavily. These shifts in the strength of the major parties over short eight-year spans are remarkable and attest to the volatility of the system, even though the system remains in a state of equilibrium. The party system since the Civil War has demonstrated a remarkable recuperative capacity, particularly after a staggering defeat of a major party that has often led to dire predictions about the demise of the system. What seems to happen is that the winning party in a landslide election does not continue to cumulate more support, but rather gradually loses it. Further, the landslide election results are not as manifest in the elections to the House of Representatives (the winning party securing a smaller proportion of the vote), and this introduces a balancing factor that leads the system to "right" itself and the pendulum to begin a swing back to center.

The key question, What are the restoring forces? in a sense contains three questions: (1) Why doesn't the dominant party grow cumulatively and continually stronger until it reaches 70 percent or more of the vote, driving its opposition into oblivion (as happened to the Federalists after 1812)? (2) Why and how does the losing major party maintain its status as the focus of the opposition? and (3) Why has not a third force replaced one of the two major parties in the equilibrium system (as happened in Britain after World War I when the Labour party replaced the Liberal party as the second major party)?

A variety of theories has been advanced to explain this equilibrium. One of them, based on observations about the nature of the coalition of supporters of the winning party, argues that when a party wins overwhelmingly it finds that there are too many interests, many of them contradictory, that must be served and satisfied. Because a party in office has to make hard policy decisions, favoring some interests over others, discontent sets in among certain sets of supporters, who eventually leave the coalition and go over to the opposition. Another theory argues that there is an inevitable "surge and decline" in party support, that a winning party in a presidential election attracts "peripheral voters" (previous nonvoters) who deviate for one election. These supporters drop out or desert the party in the next off-year elections for Congress. Then they either do not turn out to vote, or, more importantly, when they do vote in ensuing elections, they revert to their basic party loyalties.[3] Still other scholars have suggested that there are cycles in the public's attitude toward politics: liberal-conservative reversals in ideology over time, periodic reactions against those in power, and a belief that rotation in office is good for the society. The extent of solid support for such theories is certainly arguable, however. More important is the shift in interest-group support for parties over time, leading to a gradual alteration in the group coalitional character of the parties linked to the specific issue appeals or to the personalities of the candidates in a particular campaign. And then, of course, specific circumstances of party competition in particular campaigns should not be ignored. The 1912 election, with the split in the Republican party and the appearance of Wilson as the Democratic party's leader, was very special. So, too, was 1932 because of the depression, the personality of Roosevelt, and the economic program of the Democratic party. Again, in 1972, the perception of George McGovern's candidacy as too far to the left and the success of Richard Nixon in Peking and Moscow were special circumstances influencing the vote. The Iran hostage crisis hurt Carter and the Democrats in 1980. And in 1992 President Bush's repudiation of his "no new taxes" pledge, as well as Perot's third-party campaign, probably swung the election to the Democrats.

The equilibrium character of the American system suggests that two major parties have been perceived by the public as sufficient for channeling and representing their divergent demands and interests. This is a proposition that many people will resist. Nevertheless, while third parties have been important, they simply have not been able to mobilize committed workers for long periods of time or to secure adequate funds or to attract able candidates for public office at different levels of the system. See Chapter 4 for a thorough discussion of third parties.

There seem to be at least three reasons for this acceptance of two-partyism. Both parties have demonstrated a remarkable capacity for reacting positively to threats, whether external or internal, and accommodating themselves to these threats. This might be called the *capacity for absorption of protest.* The new Republican party in 1854 accommodated Whigs, Democrats, Free Soilers, and Abolitionists. In 1896, by including within their platform the "free and unlimited coinage of both silver and gold," the Democrats adopted the major issue and most of the followers of the Populists in 1892. In 1916 the Roosevelt Progressives were already wooed back into the Republican fold. The La Follette family had gone back to the Republicans by 1928, and their followers had opted for one of the major parties. At the 1952 Democratic convention the Dixiecrats were asking to be allowed back into the Democratic fold. Henry Wallace had already resigned in 1950 from his own party of Progressives. Recently, in the 1996 election over half of Perot's supporters returned to the major parties, particularly the Republicans. Thus, the major parties did not reject the extremists, either within their own ranks or from the outside; they absorbed them. William Jennings Bryan was taken into Wilson's cabinet, and the La Follettes were again considered Republican leaders.

The ideological eclecticism of the major parties has also been a contributory factor. The Democrats moved from left-wing populism in 1896 to reactionary conservatism in 1904, and the Republicans moved from Roosevelt's dynamic progressivism of 1904 to the conservatism of normalcy in 1920. A recent example is the movement of the Clinton Democrats to "the center," appropriating, it is argued, many of the Republican issue positions. Parties also have been able to modify their positions on particular issues over time — the Democrats on the racial equality question since the Civil War, the Republicans on welfare-state issues since the Depression. This is not to say that there are not distinguishing central and majority ideological tendencies in each party, but rather that there is ideological compatibility within the two major parties. While in the national image the Democrats are more left of center, liberal, and welfare-state oriented, and the Republicans more right of center, conservative, and oriented to a freely competitive marketplace, both parties embrace a broad spectrum of ideological positions. The emphasis is not ideological rigidity, but nonconformity. This means that those who crusade for ideological causes have difficulty in claiming that the system rejects their causes.

The coalitional flexibility of the major parties has also been noticeable. That is, neither party has consisted of irrevocable, subordinate social- and economic-interest subgroups whose loyalty to the party is constant and to whom the party in turn is exclusively beholden. Since the Civil War the phenomenon of "the reversibility of coalitional support" has occurred on several occasions. Thus, business interests were not always on the side of the Republican party nor the lower classes and the labor unions always in the Democratic coalition. The history of American parties is one of the breakup, modification, and reconstruction of socioeconomic coalitions. This limited fidelity of such subgroups to the party, as well as the coalitional expansionism of the parties, means that the major parties are penetrable by new social forces and amenable to their overtures. Again, this does not mean that both parties are but transient aggregations of subcoalitions that desert the party readily; rather, it means that there is no coalitional rigidity, and continual dynamic renovation of the party takes place as a result of coalitional alliances.

Whenever parties have ceased to be absorptive, eclectic, and flexible, major party system crises have occurred, for example in 1824 and in 1854–60. Individual parties have had many other particular crises reflecting that party's incapacity to adapt to threats: the Democrats in 1904, 1924, and 1948; the Republicans in 1912 and 1932. When a party refuses to be ideologically tolerant, ignores what is happening within its coalitional substructure, and refuses to react to significant voices of protest outside the two parties, apparently it can lose an important proportion of popular support. Ordinarily, however, it loses this support only to the other major party, unless that party also, as in 1854–60, is oblivious to the threats to the system. Usually, one of the two parties is sufficiently adaptive and absorptive to respond to new demands, as in the periods 1892–1912 and 1928–36. Our two major parties have generally alternated in their capacity to maintain the two-party system. Seldom have both major parties been insensitive to new social forces and developing protest. No sizable body of political "have nots" has been mobilized for a new party and persuaded that the party was a permanent necessity. By alternatively playing the roles of adapter to social change, catalytic broker for new social protest, and equilibrator of political conflict, the two parties have preserved each other's place in the system by preserving the two-party system itself.

Cohesiveness of the Opposition Vote

One of the most important distinctions among countries is in the extent of pluralization of the party system. How many parties compete for power, or to put it another way, how cohesive is the opposition? How many party channels are available to the public through which to express protest, dissent, or policy preferences? In one-party states, of course, there is usually very limited or no opportunity to protest through opposition parties — still true in China, Cuba, North Korea, and other authoritarian systems. But where the idea of competition and opposition is accepted as a legitimate part of the democratic process, a variety of systems emerges. These systems differ in the number of opposition parties, their durability, their effectiveness in securing governmental positions, and their electoral strength.

In the United States the opposition is essentially concentrated in one large, major party that has existed for a long time (the Democratic or Republican party, as the case may be), and this opposition party is the only credible threat to the party holding the presidency, the House, or the Senate. There are other parties on the ballot and occasionally one of them breaks through to win a large portion of the vote (as did Perot in 1992 with 19 percent of the vote). But these parties are essentially periodic aberrations with no long-term staying power. Above all, they usually command less than 10 percent of the vote. In the House races of 1996, there were eighteen districts in which more than two candidates contested. In seven of these, a three-way split of the vote produced a plurality result for the winner (less than 50 percent of the total vote). But only one third-party candidate won: a socialist in Vermont who normally supports the Democrats in Congress.

In contrast to the U.S. system of cohesive, nonpluralized opposition are the European systems. In the last election (1997) in Britain, which also has a district system of electing members to Parliament, many more parties won seats in the House of Commons. The opposition in the British system, which remains a relatively strong two-party-oriented system, is thus more fragmented than in the U.S. system. As indicated earlier, multiparty systems like those in the Netherlands and Germany are even more fragmented. Scholars have calculated "fractionalization index scores" for many party systems and found the United States is low, followed by Britain and Germany, and then the Netherlands.[4]

The significance of these differences in party systems has not been adequately studied. But it is highly probable that the pluralization of opposition parties in a system, unless these parties coalesce at election time, makes it difficult to dislodge the incumbent party from its control of the system. The convergence of protest forces into a major opposition party or movement maximizes the chances of dislodging the establishment. On the other hand, the comprehending of all protest into too few opposition parties may mean the frustration of dissidents and their withdrawal from political involvement. In societies with multiparty opposition systems, as in Israel and the Netherlands, the opportunities for small protest sections of the population to organize politically and elect representatives to parliament are important cathartic aspects of the system. In the United States such pluralized opposition is normally difficult to organize, but when it does occur, as in 1992, it reflects disaffection with the major parties and is a vehicle for many citizens to express their frustrations.

The Social Group Coalitional Character of American Parties

Parties have linkages to social groups in all systems. In fact, the party is in one sense a collection of organized interests seeking to influence government. The citizens and leaders in these social groups express their needs and preferences through parties, while the candidates and party leaders communicate with, and seek to mobilize the support of, these groups in order to win elections. As one scholar put it, the political party is "the single most important instrument for the translation of social power into political power."[5] So the party must be seen as an "alliance" of groups.

In all countries parties differ in the social groups they appeal to. The appeal depends on the affinity of groups for parties. And this in turn depends on the way the party originally came into existence, the ideology of the party, its leadership, and its performance. Parties usually receive disproportionately more support from one group than from another. Indeed, party leaders and candidates for office usually develop careful strategies as to how to put together a winning coalition of social groups. This coalition may change over time, as the party adapts to socioeconomic and population changes. This process of developing coalitions and maintaining coalitions is crucial for party success.

The patterns of group relationships to parties vary considerably by party systems. Systems differ in four major ways: (1) the extent to which there is exclusiveness of attachment of social groups to parties; (2) the success of party leaders in mobilizing votes "across the board" from many groups; (3) the durability of this relationship of social groups to parties over time, revealed especially in consistent voting support; and (4) the degree of intraparty factional conflict that is generated by competing social groups within the same party. Two basic types of party systems, distinctive in these terms, can be then identified:

A. *The Aggregative (or "Catchall") Party System*
Few or no groups with exclusive attachments for only one party; extensive efforts by parties to mobilize votes from many groups; unreliability over time in social group support for parties; and considerable intraparty conflict and factionalism.

B. *The Clientelistic Party System*
Strong ties of social groups to parties; more limited efforts by parties to mobilize votes widely in the electorate; durable support for parties by certain social groups over time; less extreme intraparty factionalism based on social group ties.

The U.S. party system looks like the first of these systems. Multiparty systems often look like the second type. However, usually a party system reveals a mixed pattern of social group relationships to parties. While the United States is essentially an "aggregative" system, a few social groups have distinctive and almost exclusive ties to a particular party. A close look at the voting behavior of social groups reveals the social group basis of American parties more precisely.

First, one should note the breadth of the appeal of the Democratic and Republican parties to major social categories of the population (Table 2.2). The Republicans in 1996, despite their defeat, won a rather high proportion of votes in

TABLE 2.2 ▪ The Breadth of Support for U.S. Parties by Age, Education Level, and Income (1996 vote for president in percent)

Candidate Voted for	Age *Young (18-29)*	*Middle*	*Old (60 +)*	Education *Grade School*	*High School*	*College*	Income *Low*	*Middle*	*High*
Clinton	53	48	48	59	51	44	55	48	40
Dole	34	41	44	28	35	46	34	40	52
Perot	10	9	7	11	13	8	10	10	7

SOURCE: Pomper et al., *The Election of 1996* (Chatham, NJ: Chatham House, 1997), 180–81.

many social sections of the population, even among the young, less educated, and poorer voters. Similarly, the Democrats did relatively well, even among the more affluent, better educated, and older groups in the population. Both parties thus reach out to citizens at all levels of the electorate.

Second, however, there are certain evidences of "clientelism" in social group ties to parties in the United States. Examples are found in Table 2.3. African Americans are an extreme example of high *and* durable behavioral attachments to the Democratic party. Those of the Jewish faith are almost as high and consistent. On the Republican side, the "born again" religious "right" is clearly strongly attached to the Republicans, only a few in 1992 and 1996 supporting Perot. White Protestants and the wealthy are also skewed in their voting behavior toward the Republicans, as are the low-income families toward the Democrats. But the presence of Perot on the ballot did reduce somewhat the control by both parties of these voters. Nevertheless, it is clear that we do have certain groups strongly committed over time to one or the other of the parties — at the 60 percent to 80 percent level of support.

TABLE 2.3 ▪ Illustrations of Close Group Ties to Either Democratic or Republican Party (by presidential election voting, in percent)

Social Group	1988 *D*	*R*	1992 *D*	*R*	*I*	1996 *D*	*R*	*I*
African Americans	86	12	83	10	7	84	12	4
The "religious right"	18	81	23	61	15	26	65	8
Jews	64	35	80	11	9	78	16	3
White protestants	33	66	33	47	21	36	53	10
Those with high incomes	37	62	38	45	17	40	52	7
Those with low incomes	64	33	58	23	19	59	28	11

SOURCES: Pomper, et al. ed., *The Election of 1988*, 133–34; Pomper, et al., *The Election of 1996*, 179–81.

A third point that must be made is that even in these "clientelistic" party-group relationships, the opposition party still is often able to win the vote of a sizable minority of voters. Examples of this phenomenon follow.

Percent of the Vote Cast for Democrats in these "Republican" Groups	*1992*	*1996*	*Percent of the Vote Cast for Republicans in these "Democratic" Groups*	*1992*	*1996*
Religious right	23	26	Low-income families	23	28
Protestants	33	36	Grade school education only	28	28
High-income families	38	40	Jews	11	16

These examples suggest the extensiveness of the mobilization effort and the appeal of both of our major parties, even though certain groups are more closely tied to one than the other.

In a multiparty system, as in the Netherlands, social groups can be more closely and strongly affiliated with the parties, although again this is not an all-or-none phenomenon, and it can change over time (Table 2.4). The critical social-group cleavages in Dutch politics over the years have been based on religious and class differences (working class versus the middle class). In the early days after World War II, as the data reveal, the ties of these religious and class groups to parties were very strong and persistent. Since the late sixties, the Dutch system has undergone considerable social change, with a decline in religion and softening of class differences. Nevertheless, there is considerable continuity in the clientelistic attachments of Dutch religious and class groups to parties, as manifested in their more recent voting behavior.

TABLE 2.4 ▪ Religion and Class Ties to Parties in the Netherlands (percentage voting for the National Parliament)

Religion or Class Tie	*1956*	*1968*	*1977*	*1986*
Percentage of Catholics supporting the Catholic party	95	72	67	66
Percentage of conservative Protestants supporting the Protestant party	93	78	75	59
Percentage of the secular working class supporting the left-wing Labor party	68	65	68	60

SOURCE: B. A. Irwin and J. J. M. Van Holsteyn, in Hans Daalder and Galen Irwin, eds., "Politics in the Netherlands: How Much Change?" *Western European Politics* 12 (1989), 39.

Ideological Conflicts and the Parties

The way in which the parties differ in their policy agendas and in their basic ideological positions as they compete for power is of major significance in understanding how party systems function. In democratic societies parties are supposed to present alternative programs for dealing with the problems in their societies. That is, they are programmatic and ideological competitors. In some systems the parties are consensual on issues; they are close to agreement. In other systems there is moderate or even extreme dissensus. When we find the latter to be true, we say the party system is "polarized." How consensual or polarized is the American party system on basic ideology and on specific current issues?

There is general agreement that today there are more issue differences, and probably more intense issue conflicts, between the parties than previously. In the early postwar period it was claimed that the positions of voters on issues were not closely linked to voting decisions, but that such decisions were more the result of party identification and candidate evaluations. Since then, scholars see increased issue voting and "growing electoral impact of issues."[6] Further, new issues have appeared in recent years. It is argued that as the public becomes better educated and also more sophisticated about politics, it evaluates parties and campaigns more in issue terms. The proportion of the American public which sees differences between the Republican and Democratic parties has increased from 50 percent in the 1950s to over 60 percent today.[7] And the correlation of the vote with attitudes on key issues has been documented for some time (see Table 2.5).

Ideologically, Americans in the aggregate have changed somewhat over the years. When asked whether they are "liberals," "conservatives," or "moderates," respondents evidenced a small decline in liberalism and an increase in conservatism. When one looks at ideology *by party,* there *has* been considerable change (Table 2.6). The party differences on liberalism now are 34 percent; they were only 20 percent. On conservatism, the party differences now are 40 percent; they were 24 percent. And voters perceive the Republicans as more conservative, and the Democrats as more liberal, today than in earlier years.

Of even more importance, probably, is the way in which the link between the ideological positions of voters and their voting behavior has changed over the years (Table 2.7). Generally, of course, liberals support Democratic candidates and

TABLE 2.5 ▪ Correlations between the Vote and the Issues, 1956–1972

Issue	*1956*	*1964*	*1968*	*1972*
Welfare, including redistribution policies (job guarantees, minimum wage, collective bargaining)	.29	.61	.51	.54
Foreign policy	-.03	.66	.36	.64
Race	.08	.44	.41	.47

SOURCE: Norman H. Nie, Sidney Verba, and John R. Petrocik, *The Changing American Voter* (Cambridge: Harvard University Press, 1976), 188.

TABLE 2.6 ▪ Ideological Self-Placement of Party Identifiers (in percent)

	Democrats			Republicans		
Ideology	*1972*	*1980*	*1992*	*1972*	*1980*	*1992*
Liberal	33	38	44	13	10	10
Moderate	41	34	33	37	23	27
Conservative	26	28	24	50	67	64

Note: Democrats and Republicans include "leaners." The "don't knows" (27 percent in 1992) are excluded here.

SOURCE: National Election Studies.

conservatives support Republican candidates. The moderates supported Reagan slightly more than Democratic candidates, but this was reversed in the 1988–1996 period. Perhaps the best summary of the trend during the past twenty years is the change in the ideological voting "overlap," the percentage of liberals supporting Republican candidates and vice versa. We find as follows:

	1976	*1996*
Percentage of liberals voting Republican	26	11
Percentage of conservatives voting Democratic	30	20

In voting trends the system seems to have become more "ideological."

On particular issues the parties can differ considerably. The question is, To what extent are these issue differences extreme? Are there issues on which the parties are polarized? The answer is that on certain issues the parties agree closely, in 1996 as in previous years, while on other issues there is much disagreement. And there is considerable continuity for some of these issue differences over the years.

TABLE 2.7 ▪ Presidential Vote of Voters by Ideological Orientation (in percent)

	1976		1980			1988	
Ideology	*Carter*	*Ford*	*Carter*	*Reagan*	*Anderson*	*Dukakis*	*Bush*
Liberal	74	26	60	28	12	82	18
Moderate	53	47	43	49	8	51	49
Conservative	30	70	23	73	4	19	81

	1992			1996		
Ideology	*Clinton*	*Bush*	*Perot*	*Clinton*	*Dole*	*Perot*
Liberal	68	14	18	78	11	7
Moderate	47	31	21	57	33	9
Conservative	18	64	18	20	71	8

SOURCE: National Election Studies, University of Michigan, Center for Political Studies for 1976–1992; *New York Times*/CBS 1996 election exit polls, *New York Times*, November 10, 1996.

TABLE 2.8 ▪ Issue Positions of Democrats and Republicans for Five Selected Issues (in percent)

	1980		1992		Party Differences	
Issue Position	*D*	*R*	*D*	*R*	*1980*	*1992*
Favor aid to minorities	31	11	30	17	20	13
Believe government should guarantee jobs/standard of living	41	16	40	17	25	23
Believe government services should be increased	60	27	48	23	33	25
Believe defense spending should be increased	65	81	16	24	16	8
Believe in equal rights for women	64	60	78	69	4	9

SOURCE: National Election Studies.

In Table 2.8 we compare five key issues, which were tests of party agreement in 1980, and look at party support patterns in 1992. One notices that on two issues there is close agreement: women's rights and defense expenditures. There is much more distance between the parties on the other three issues in Table 2.8, but it is not extreme.

However, this is not the whole story. The American public is sharply divided today on certain "new politics" issues, and the parties are likewise divided. Table 2.9 provides examples of both "consensus" and "dissensus" issues which confront us today. While there is strong support for balancing the budget, reducing the size of the federal government bureaucracy, and even guaranteeing some sort of medical care for everyone, there are other issues on which public opinion is clearly split. Indeed, if one reflects on the dissensus issues, it appears that there are certain key issue cleavages in American society.

There are cleavages over minority rights, environmental protection, social welfare, and foreign aid. One might also possibly argue that there is a cleavage over morality issues. The division of opinion in the party system expresses or represents these cleavages to some extent. For example, on the specific question of the role of the government in providing health insurance, one sees today this division of opinion among the party supporters:

Opinion	*Democrats*	*Republicans*
Favors a government health insurance plan	64%	34%
Neutral — uncertain	19%	22%
Favors a private health insurance plan	17%	44%

TABLE 2.9 ▪ Issues on which the American Public Agreed and Disagreed in 1996 (in percent)

Issue	*Favors*	*Takes Neutral Position*	*Opposes*
"Consensus" Issues			
Balancing the federal budget	73	16	8
Guaranteeing adequate medical care for everyone	81	8	11
Allowing voluntary prayer in schools	65	18	14
Reducing the size of the federal government	64	21	11
"Dissensus" Issues			
Allowing gays to serve in the military	38	22	35
Ending affirmative action programs for women and minorities	35	23	39
Reducing federal regulations aimed at protecting the environment	34	19	45
Using American soldiers to keep peace in Bosnia	32	21	46
Reducing welfare payments to person living in poverty	31	20	48
Increasing foreign aid to poor nations	29	27	44

SOURCE: Roper Center for Public Opinion Research, *The Public Perspective* 8 (February/March 1997): 17.

One of the most important issues dividing the parties is race. This is an issue on which the parties have reversed their positions. Before 1964, 60 percent of the Republicans in Congress were racial liberals, compared to only 40 percent of the Democrats. After the election of 1964, these positions were reversed. The Republicans sought the support of conservative whites, particularly in the South, and the Democrats became much more liberal. So congressional roll call votes on racial issues reveal that only 30 percent of Republicans are racial liberals, while 55 percent of Democrats support racial legislation. The opinions of party supporters in the public have followed these changes.[8] When people today are asked directly whether they favor increasing programs to assist blacks, this is the division by party:

Opinion	*Democrats*	*Republicans*
Favors increase of aid to blacks	35%	14%
Neutral	49%	53%
Wants to decrease aid to blacks	16%	33%

This example illustrates the important differences by party on a critical American issue. They are not extremely polarized response patterns, but they suggest that there are important issue conflicts, linked to basic ideological differences, which separate our parties, the loyal supporters of each party, and the public.

A Majority-Oriented Party System with Frequent Divided Control

It has long been accepted that the Founding Fathers of our system could not make up their minds about political parties. They were both "pro-party" and "anti-party." They wrote a Constitution which provided freedom for parties to exist but they also included provisions, such as federalism and separation of powers, which were defended as the means to frustrate majority party control. Yet Jefferson later argued strongly that parties, particularly a majority party system, were needed.[9]

In reality, there have been periods of one-party control of the national government as well as periods of divided government, with the president in one party and the House or Senate or both controlled by the other party. Indeed, in recent years the latter pattern of divided control has been much more likely. Since Carter's presidency (1977-1980) we have had only one two-year period of unified party government, the 1992-1994 period of Clinton's presidency when both houses of Congress were Democratic also. Thus, since 1980 we have had divided government almost 90 percent of the time.

This was not always the case. If we go back to the early years of the republic, we note that the Jeffersonian Republicans, beginning with the election of 1800, had twenty-four years of majority control. Following a brief lull, after the election of 1828 the Democrats had twelve years of majority control. Since the Civil War the United States has fluctuated between one-party majority control and divided-party control. The Republicans had majority control of the national government from Lincoln's victory in 1860 to 1874, and again for fourteen years after the 1896 election. Otherwise, there was divided party government in the latter part of the nineteenth century. On the other hand, the Democrats were briefly in control from 1912 to 1918 and dominated the presidency and Congress from 1932 to 1946. Again, after John Kennedy's election in 1960, the Democrats controlled both branches up to 1968. With the 1976 election of Jimmy Carter, the Democrats again attained strong majority control in the House and the Senate, but it did not last long — 1980 again produced divided government.

It is interesting that for the 138 years of our present two-party system, since 1860, the two parties have had periods of dominance of almost equal length (forty-two years of Republican control and forty years of Democratic control). There have been fifty-six years of divided-party control. The periods of one-party control can be divided into three types: periods of *marginal* one-party control (fewer than 55 percent of the seats in Congress held by the same party as the presidency); *moderate* party control (55 percent to 60 percent control of the seats in Congress); and *strong* party control (over 60 percent of the seats held by the dominant party). Table 2.10 shows the proportion of years since 1860 that each type of party control occurred. For the entire period we have had strong party control about one-fifth of the time, divided government 40 percent of the time and marginal or moderate party control a little over one-third of the time. These data also reveal how, in recent years, we have had a very high incidence of divided government.

One factor not entering into this calculation is the margin of victory of the presidential candidate. Carter's marginal victory in 1976 (51 percent of the two-

TABLE 2.10 ▪ The Marginal Majoritarian Nature of the U.S. Party System, 1860–1998

Strength	*1860–1920 (percentage of 60 years)*	*1920–1980 (percentage of 60 years)*	*1980–1998 (percentage of 18 years)*	*Since the Civil War (percentage of 138 years)*
Divided government[a]	36.7	30.0	89.0	40.0
Marginal government[b]	23.3	16.7	0.0	17.4
Majoritarian government (moderately strong)[c]	20.0	23.3	11.0	20.0
Majoritarian government[d] (strong party control)	20.0	30.0	0.0	21.7

[a] No party controls presidency and both houses of Congress.
[b] The party controlling the presidency controls Congress with less than 55 percent of the seats in one or both houses.
[c] One party controls the presidency and up to 60 percent of seats in both houses of Congress.
[d] One party controls the presidency and has a 60 percent or greater majority in both houses of Congress.

party vote) would raise further questions about whether this is a period of strong majority-party control, even though Democrats won over 60 percent of the seats in the House and the Senate. The same could be said for Kennedy's slim victory (51 percent) in 1960, even though Democrats won 60 percent of the House seats and 64 percent of those in the Senate. Similarly, Clinton won the presidency in 1992 with only 43 percent of the vote even though the Democrats controlled both houses of Congress.

How important is it for there to be unified party control in Washington? Can we accomplish more effective policy making when there is such control than when the presidency and the houses of Congress are controlled by different parties? This was a question discussed long ago by early scholars of the American system, with considerable difference of opinion. One of these scholars, after an examination of the volume of legislative business in the Congresses of the twenties and thirties, concluded that "a great mass of business can be transacted across party lines when the necessity arises." He added: "The mere fact that one party is in control of the Senate while the other is in control of the House does not bring the government to a standstill."[10] A recent study by Mayhew demonstrates this conclusion. He argues convincingly and in detail that the most significant laws adopted by the U.S. government from 1946 to 1990 (267 major pieces of legislation) were passed either in periods of unified party control or divided control. It makes "very little difference," he says, what the party control pattern in Congress is. But he adds quickly: "It is not as if parties do not amount to anything. . . . they amount to a great deal."[11] In explaining why legislation can be passed despite divided party control, the roles of parties, elections, and presidential leadership are very important. As a result, major laws such as the Civil Rights Act of 1964, the Clean Air Act of 1970, and the Natural Environment Policy Act of 1969 were passed by large majorities despite divided government.

The myth that divided government necessarily leads to deadlock and less effective legislation has been exploded by this recent research. Divided government may have other results that we may dislike, such as more intensive party conflict and an adversarial context for policy making. Nevertheless, divided government is on the increase in the United States: thirty-one states in 1997 also have it (compared to eighteen states in 1960). More people today seem to be splitting their ballots (voting for one party for president and the other for House or Senate or both). It is interesting that even many Clinton supporters in 1996 preferred divided government: 65 percent of all voters said they were happy that "the Republican Party maintained control of the U.S. Congress," and this was true of 39 percent of the Clinton voters also![12] Americans are a mixed breed: two-thirds identify with one party or the other, but they are also very candidate oriented and enchanted by the doctrine of "checks and balances." The U.S. party system, with the possibility of divided control resulting from deadlock between the chief executive and the legislature, is virtually unique. (France is the only other western country in which that occurs.) Political deadlock in Western European democracies usually occurs because of their multiparty systems and the difficulty of achieving and maintaining agreement among several parties in a politically diverse cabinet. Since the cabinet commands a majority in parliament, normally cabinet-parliament deadlocks are much less common. If they do occur, new elections can be called immediately to resolve the conflict. In the United States such an executive-legislative deadlock must continue to the next, fixed, election.

Variation in Party Strength by State and Region

American party politics is geographically very diverse. We are inclined to think of fifty different party systems rather than one national system. While that is a rather extreme conception, the reality is that state differences in voting patterns are substantial. Consider the presidential vote in 1996 as an example: the vote for Clinton ranged from 62 percent in Massachusetts and 61 percent in New York and Rhode Island, to as low as 34 percent in Idaho and Utah and 35 percent in Alaska and Nebraska. As for the congressional election results, some states were overwhelmingly Republican and some overwhelmingly Democratic. In nine states a Democratic voter had no party representation in Congress at all (Oklahoma, Nebraska, Nevada, Montana, Utah, Idaho, Wyoming, Kansas, and South Dakota). All twenty-seven seats were won by Republicans. In five states, however, a Republican voter had no party representative elected to Congress (Massachusetts, West Virginia, Rhode Island, Hawaii, and Maine). All these nineteen seats were won by Democrats.

Certain sections of the country have been dominated by one party or the other. The best examples are the South and the border states, a total of seventeen states that were controlled by the Democrats for a long time. In fact, in the South the Republicans won only an average of 2 percent of the House seats from 1932 to 1962. Today that trend has been reversed.

Percentage of House Seats Won by Republicans

	Average Percentage 1932–1962	*Percentage of Seats Won in 1976*	*Percentage of Seats Won in 1996*
10 southern states	2.1	23.0	56.9
7 border states	17.5	26.8	58.5

There are areas of Democratic strength still in the South: in Texas 17 percent of the thirty seats were won by Democrats, and in North Carolina the vote was split, as in Virginia, Arkansas, and West Virginia. But in most of these states the Republicans won a majority of the seats. Also Clinton did win the electoral vote in Arkansas, Florida, Kentucky, Louisiana, Tennessee, and West Virginia. But the trend in the South has been strongly pro-Republican. The contrast in the extent of party control of House and Senate seats from the South over the years is dramatic. The Democrats held 100 percent of Senate seats from the South in 1955, today only 31 percent; Democrats held 92 percent of House seats from the South in 1955, today only 40 percent.[13] This reversal is evidence of the potential in the United States for major electoral swings in party fortunes by area.

Scholars have devised classification schemes for state party control patterns. One of these, the Ranney Index, used three measures: the party vote for governor, the party distribution of state legislative seats, and the length of time of unified party control of state government.[14] On the basis of these measures he could differentiate party control patterns by states. This index has several uses. It illustrates the variance among states in party control and the patterns of such control, as well as how these patterns change. Table 2.11 presents the early Ranney classification and how it has changed over the years. The general trends in the past twenty years are interesting to note. There has been a reversal in party strength in the so-called "one-party" states — except for Texas they are now split, or Republican! The ten "modified one-party states" also are Republican (except for West Virginia and splits in North Carolina, Virginia, and Maryland). The middle category, "two-party states inclined to be Democratic," holds up pretty well (except for Nebraska and Nevada, which had no Democratic representatives in Congress in 1997). On the other hand, the "two-party states inclined to be Republican" in Table 2.11 have altered considerably: New York is more Democratic than Republican, Illinois and Wisconsin are split between the parties. The "one-party Republican" states remain solid Republican territory — only one Democrat was elected to Congress in 1996 from this group of states!

This characteristic of the American system, which reveals contrasting patterns of party conflict at the local level and the strong influence of both local traditions and party culture, should not be forgotten in the analysis of the American system. There is a great deal of organizational decentralization and diversity in the system, special patterns of leadership recruitment and institutional mechanisms and processes through which parties function in the sections, states, and localities of the United States. Maine is different from California; Michigan conducts its politics quite differently from even its next-door neighbor Illinois; the Republican and

TABLE 2.11 ▪ Classification of States According to Degree of Interparty Competition

Type of System	*States Included in Ranney Taxonomy by Type*	Mean Percentage of Total Congressional Seats Won by Democrats *1976*	*1996*
One-party Democratic	Louisiana, Alabama, Mississippi, South Carolina, Texas, Georgia, Arkansas (7)	84	43
Modified one-party Democratic	North Carolina, Virginia, Florida, Tennessee, Maryland, Oklahoma, Missouri, Kentucky, West Virginia, New Mexico (10)	69	42
Two-party states inclined to be more Democratic	Hawaii, Rhode Island, Massachusetts, Alaska, California, Nebraska, Washington, Minnesota, Nevada, Connecticut (10)	71	61
Two-party states most even in party control	Delaware, Arizona, Montana, Oregon, New Jersey, Pennsylvania, Colorado, Michigan (8)	67	49
Two-party states inclined to be more Republican	Utah, Indiana, Illinois, Wisconsin, Idaho, Iowa, Ohio, New York, Maine, Wyoming (10)	60	47
Modified one-party Republican states	North Dakota, Kansas, New Hampshire, South Dakota, Vermont (5)	27	11

Note: The Ranney Taxonomy is based on party votes for governor and state legislature, 1956–70.

SOURCE: For Ranney Taxonomy: Austin Ranney, "Parties in State Politics," in Herbert Jacob and Kenneth N. Vines, eds., *Politics in the American States: A Comparative Analysis,* 2nd ed. (Boston: Little, Brown, 1971), 87. Copyright 1971 by Little, Brown and Company, Inc. Reprinted by permission. Ranney based his classification on the work of Richard Dawson and James Robinson, "Interparty Competition, Economic Variables, and Welfare Policies in the American States," *Journal of Politics* 25 (1963): 265–89.

Democratic parties of Georgia and Virginia are quite different, and they in turn contrast with other states in the South, such as Florida or Louisiana or Mississippi. To study American party politics, therefore, is a complex undertaking; to generalize about them confidently must be done with extreme caution.

TYPES OF PARTY SYSTEMS: DIMENSIONS FOR COMPARISON/PROS AND CONS

Party systems in democracies can be differentiated on a variety of dimensions. Scholars have contributed different approaches in attempting to distinguish party systems.[15] We have selected five of these key dimensions that are critical for comparing the U.S. party system with others, and we have discussed these at length in the preceding pages. These dimensions, which pose key questions about each system, are:

1. Majority oriented or plurality oriented
2. Volatile or stable
3. Cohesive or pluralized opposition
4. Aggregative ("catchall") social group relations or clientelistic
5. Moderate or extreme (polarized) ideological conflict.

These alternative characterizations capture the special character of the American system. It is perhaps the most majority oriented of all democratic competitive systems, with the most cohesive and nonpluralized pattern of party opposition. While there is considerable ideological conflict, it is not as polarized as most of the multiparty systems can be. Our two parties do appeal to special groups but they are basically aggregative, trying to put together a heterogeneous coalition. It is a system with considerable potential for short-term and long-term change, in the patterns of competition territorially and in individual voting decisions. Thus, we are distinctive but with cross-national similarities to other systems.

The type of party system may be a key to understanding the character of political life in a society. A majoritarian, aggregative, moderately conflicted, yet potentially volatile system such as ours stands in stark contrast to highly polarized (ideologically) and pluralized, clientelistic party systems one finds elsewhere. Which system is preferable has been the subject of a long-standing debate among scholars, columnists, and citizens. We are inclined to extol the virtues of the two-party system for being nonextremist, giving voters clear choices between two parties, providing governmental stability, and crosscutting group affiliations. We used to claim that effective democracy was more attainable through our two party system than through a multiparty system. No longer! Multiparty systems can be very successful democracies, as Scandinavia, the Netherlands, and other countries have demonstrated. In fact, there are those who extol the virtues of the multiparty system as providing more opportunity for political choice and group representation in the legislative body.[16] And there is evidence also that citizen participation in elections is lowest under our type of two-party system.[17] There are obviously "trade-offs" for each type of party system, advantages and disadvantages. Our party system seems to fit with our political culture and over the years has been durable because it has performed reasonably responsibly. On that we will elaborate throughout this book, and that should be our key focus: the democratic performance of the party system.

NOTES

1. Elmer E. Schattschneider, *Party Government* (New York: Rinehart, 1942), 93–96. He speaks of the equilibrium of the two-party system and the "recuperative power of the second major party."

2. Donald E. Stokes and G. R. Iversen, "On the Existence of Forces Restoring Party Competition," *Public Opinion Quarterly* 26 (Summer 1962).

3. Angus Campbell, "Surge and Decline: A Study of Electoral Change," *Public Opinion Quarterly* 24 (1960): 397-418.

4. Giovanni Sartori, *Parties and Party Systems* (London: Cambridge University Press, 1976), 313.

5. Franz Neumann, *The Democratic and the Authoritarian State* (Glencoe, IL: Free Press, 1976), 12.

6. Russell J. Dalton, *Citizen Politics in Western Democracies* (Chatham, NJ: Chatham House, 1988), 194-95.

7. John H. Aldrich, *Why Parties?* (Chicago: University of Chicago Press, 1995), 174.

8. Edward G. Carmines and James A. Stimson, *Issue Evolution: Race and the Transformation of American Politics* (Princeton, NJ: Princeton University Press, 1989), 63-64, 168.

9. William Chambers, *Political Parties in a New Nation* (New York: Oxford University Press, 1963), 149.

10. Schattschneider, *Party Government,* 90.

11. David R. Mayhew, *Divided We Govern* (New Haven: Yale University Press, 1991), 100.

12. Roper Center for Public Opinion Research, "America at the Polls 1996," University of Connecticut, 1997, 192: Based on *Los Angeles Times* Poll of November 5, 1996.

13. Gerald M. Pomper et al. *The Election of 1996* (Chatham, NJ: Chatham House, 1997), 220.

14. For Ranney Taxonomy: Austin Ranney, "Parties in State Politics," in Herbert Jacob and Kenneth N. Vines, eds., *Politics in the American States: A Comparative Analysis,* 2nd ed. (Boston: Little, Brown, 1971), 87.

15. Sartori, *Party Systems*; Jan-Erik Lane and Svante O. Ersson, *Politics and Society in Western Europe* (London: Sage, 1987); G. Bingham Powell, Jr., *Contemporary Democracies* (Cambridge: Harvard University Press, 1982); Arend Lijphart, *Democracies* (New Haven: Yale University Press, 1984).

16. Lijphart, *Democracies,* 106-14.

17. G. B. Powell, "Party Systems and Political System Performance," *American Political Science Review* 75 (December 1981): 861-79.

CHAPTER 3

The American Party System: Origins and Development

The history of the American party system is fascinating. It has been a complex process of trial and error; even today it is adapting. It was a process characterized by both contradictory, alternating patterns of support and rejection and continuous efforts at reform and revision. In America we have struggled to develop a system acceptable to both the public and the political elites. We have repeatedly questioned the need and function of parties, and we have periodically questioned our type of party system. At the beginning, before 1800, leaders moved only hesitantly toward a competitive party system, and many people warned against it. In the nineteenth century scholars and political leaders attacked parties, developed new ones when the old seemed ineffective, regulated them rigorously, and seriously questioned their value to our society. In the twentieth century, even while the two-party system was maturing and capturing the public's support, we have attempted to rob parties of crucial functions, such as leadership recruitment, and questioned their superiority compared to interest groups. Throughout our history, therefore, we have wavered between two sentiments and two cultures: pro-party and anti-party. In the context of persisting ambivalence about parties, this chapter will review the origins of the early parties and then sketch their later developments.

Party systems emerge from the special historical circumstances and patterns of political, social, and economic conflict attending the development of modern societies. During this gestation period, political elites greatly influence the initial character of the system as they struggle with each other for status in the system and seek to organize supporters on behalf of their goals and beliefs. A system then goes through several periods of historical experimentation with one or more party systems. A variety of party groups may come and go, and different types of party systems may be tested before one system crystallizes with a particular form and character. The American party system went through such a development process and has evolved its own special form.

It is useful to remember the events and circumstances at the time of the origin of our system, which culminated eventually in the party system we have today. In review of this early period, one observation stands out: There was not immediate and overwhelming, enthusiastic support for a competitive party system; such support developed only very gradually.

THE EARLY PARTY ERA: 1790–1800

Political parties first appeared sometime between 1790 and 1800. To some scholars this was a pre-party period; to others the evidence of the definite development of parties is stronger.[1] The events of this fascinating period of our history should be recalled in order to understand the way in which political leaders and social groups met the problems of their time and in the process created a type of party system. Before 1790, before and during the battle over the ratification of the Constitution, parties and factions were deplored. James Madison argued in the Federalist Papers for the adoption of the Constitution on the grounds, among others, that the principles of that document would effectively prevent a political group like a party, no matter what its size, from acquiring the totality of political power. The doctrines of federalism, bicameralism, and separation of powers, he argued, would blunt the power drive of a party.[2] Also critical of parties, Benjamin Franklin wrote at one time of "the infinite mutual abuse of parties, tearing to pieces the best of characters."[3] Yet, despite these arguments and the early antagonism of many of the other Founding Fathers, parties did come into existence early in our country. As Richard Hofstadter says, it was indeed paradoxical that Thomas Jefferson, who played a leading role in "creating the first truly popular party in the history of the Western world," was also the leader who initially "had no use for political parties."[4] It was almost grudgingly, then, that leaders like Jefferson took actions which led to the establishment of an opposition party as a counterpoise to the policies and organizational genius of Hamilton. Political controversy became so profound and conflict among the elites so intense that, despite anti-party sentiment, a party system of sorts was born — perhaps by 1796, probably by 1800.

In 1788 Washington was the virtually unanimous choice for president and, among eleven other candidates, John Adams was chosen as vice president by the Electoral College, consisting of electors elected by the states. People were being called Federalists and Anti-Federalists, as well as Constitutionalists (ten Anti-Federalists were elected to the first House of Representatives). Parties as groups with any semblance of organization, however, did not yet exist. The maneuvering of leaders such as Alexander Hamilton, Aaron Burr, De Witt Clinton, James Madison, Patrick Henry, and James Monroe occupied center stage. Their efforts to put together political coalitions and, at the persuasion of others, to support particular nominees for the House, Senate, and vice presidency constituted the extent of our politics. Estimates of popular participation in this early election ranged from 5 to 8 percent of the white males.[5] Parties as mass mobilization structures had not yet materialized.

For the first year of the new government there was limited controversy, since the cabinet and the Congress were preoccupied with implementing the new Constitution and setting up the judiciary and executive departments. Then in 1790, Hamilton, as the secretary of the treasury and the dominant figure in the cabinet, presented a series of proposed bills dealing with some critical economic problems confronting the new government: the funding of state debts, a protective tariff, and the creation of a national bank. When Hamilton pressed for the adoption of these economic legislative measures, differences of opinion became apparent

Early History: Jefferson's Role as Party Founder

"When Thomas Jefferson thought of setting down the lasting achievements he wanted inscribed on his tombstone, he mentioned the writing of the Declaration of Independence and of the Virginia Statute of Religious Liberty (sic) and the founding of the University of Virginia — thus omitting almost flamboyantly all the accomplishments of his long career in national politics. Yet surely this democrat and libertarian might have taken justifiable pride in his part in creating the first truly popular party in the history of the Western world, and in his leading role in the first popular election of modern times in which the reins of government were peacefully surrendered by a governing party to an opposition. Jefferson did more than assert the claims of democracy: he was also a central figure in developing responsible constitutional opposition, an accomplishment which alone would grace any man's tombstone.

But here we are brought face to face with the primary paradox of this inquiry: Jefferson, the founder, or more accurately, cofounder, of the first modern popular party, had no use for political parties. This seeming inconsistency is but one aspect of a larger problem: the creators of the first American party system on both sides, Federalists and Republicans, were men who looked upon parties as sores on the body politic."

SOURCE: Richard Hofstadter, "The Idea of a Party System," in William N. Chambers, ed., *The First Party System* (New York: John Wiley and Sons, 1972), 67–68.

and a debate ensued. In 1790 Hamilton called the first legislative caucus of those identified as Federalists. Dissent from and opposition to his policies became sharper, but it was a highly individualized opposition; no opposition legislative caucus was even held until 1795. In the meantime, considerable tensions developed in the cabinet between Jefferson and Hamilton, and Jefferson began to ally himself with other leaders in the Congress and outside, notably with James Madison. In the election of 1792 the vice presidency was contested, with the incumbent John Adams pitted against De Witt Clinton, whom the Jeffersonian-Republican leaders had informally designated as their candidate by means of conferences and correspondence. The resulting vote was concentrated, with Adams receiving seventy-seven and Clinton all but five of the sixty opposition votes. Yet Jefferson remained in the cabinet until late 1793.

A variety of party-type activities had been going on since 1790 — informal nominations, informal canvassing of electoral votes, the development by oppositionist leaders in Congress of similar policy positions, correspondence between these leaders and other notables in the states, the appearance of an opposition press led by Philip Freneau's establishment of the *National Gazette* in Philadelphia in October 1791, and the founding of Democratic and Republican local societies (which Washington disapprovingly called "self-created societies").[6] But in the opinion of most scholars, these efforts did not congeal into a clearly articulated national party structure. In the congressional election of 1794, the opposition Republicans won the House for the first time; this, however, was primarily the result of both individual efforts and those of particular local societies. Nevertheless, the movement toward the establishment of two political structures at the national level was continued and intensified. The controversial Jay Treaty evoked violent partisan passions, and the appropriations in support of it were only passed by a small margin — forty-eight votes to fifty-one — in April 1796. (This, in a sense, was an index of the development of party structure: loose, individualized, and informal.) And it was this debate that moved Jefferson to reluctantly agree to be a candidate for the presidency that year. He lost to Adams but again only by the narrowest of margins — seventy-one votes to sixty-eight — and thus became vice president under the Electoral College system that then prevailed.

Clearly parties as loose leadership associations were developing. Thus John Taylor, a planter in Virginia, after consulting with men like Madison and Monroe, declared in a pamphlet he wrote in 1793:

> the existence of two parties in Congress, is apparent. The fact is disclosed almost upon every important question . . . whether the subject be foreign or domestic — relative to war or peace — navigation or commerce — the magnetism of opposite views draws them wide as the poles asunder.[7]

And on the last day of 1795, Jefferson, who had two years previously stepped down from Washington's cabinet in opposition to the policies of Hamilton, wrote:

> were parties here divided merely by a greediness for office as in England, to take a part with either would be unworthy of a reasonable or moral man, but where the principle of difference is as substantial and as strongly pronounced as between the republicans and the Monocrats of our country, I hold it as honorable to take a firm and decided part. . . .[8]

These men, as well as others, not only perceived the existence of parties but also were engaged in a pattern of activities and relationships which made parties a reality.

In the latter part of the 1790s, then, it was clear that two leadership structures had emerged differing in basic respects as to the direction of national policy and beginning to articulate the structures from the national level to the local level. Cadres of active followers were beginning to be mobilized and strategies for organizational action were devised, which became more evident in the period leading to the election of 1800. In that presidential election the Federalists and Republicans competed for national and state power as two loose-knit political structures utilizing some of the techniques that today are assumed to be the genius of modern parties. This time Jefferson defeated Adams — seventy-three votes to sixty-five — and the Republicans controlled the House of Representatives — sixty-six votes to forty.

What were the factors most responsible for the emergence of our early parties? First, there was the deepening of controversy over national policy, particularly over the role of the federal government and its program in dealing with the problems of American society before 1800. The issue of critical conflicts over economic and foreign policy raised serious questions concerning what interests and sectors of society should be regulated, protected, and benefited. Too, the issue cleavages then were tied to interest-group conflicts. Jefferson saw these alignments clearly in May 1793:

> The line is now drawn so clearly as to show on one side:
> 1. The fashionable circle of Philadelphia, New York, Boston, and Charleston; natural aristocrats
> 2. Merchants trading on British capital
> 3. Papermen. All the old Tories are found in some one of these three descriptions.
>
> On the other side are:
> 1. Merchants trading on their own capital
> 2. Irish merchants
> 3. Tradesmen, mechanics, farmers, and every other possible description of our citizens.[9]

The original impetus to party conflict provided by Hamilton's economic proposals was followed by the bitter debates over the French Revolution and what American loyalties should be in that war. As Hofstadter has argued:

> The French Revolutionary War quickly moved the American party contest . . . into a new phase. From the spring of 1793, when the provocative French Minister Citizen Genet arrived at Charleston, there came a series of events which rapidly polarized the leaders and their followers and finally inflamed them to the point at which the entire political system was threatened. . . . The war had brought into one common focus three sets of issues: the domestic issues, which mobilized interest against interest, the foreign policy issues and finally, a set of ideological passions of surprisingly intense kind. . . . Jefferson himself [wrote that] the war "kindled and brought forward the two parties."[10]

Thus the issue cleavages and their linkages to conflicted socioeconomic interests were the seedbed for the origin of these parties. Two persons with great magnetism, Hamilton and Jefferson, assumed leadership of the two contesting social and political forces. Somewhat reluctantly at first but more self-consciously later, they built their parties, drawing to them a variety of other statesmen, sub-elites, and mass supporters. Gradually a substructure made up of sub-elites and the public became more involved in electoral politics. Estimates are that the proportion of turnout for 1792 to 1798 was 24 percent of the white adult male population, which increased to 39 percent by 1800, when Jefferson was elected to the presidency.[11] And while parties were developing local structures that were mobilizing votes at the citizen level, the parties in the government were becoming somewhat more coherent, if not more disciplined. Thus there was a clearer division in party voting in the House of Representatives after 1790. Table 3.1 shows the increase in party voting in the House, according to the calculations of Joseph Charles.

At the elite level the concept of party was indeed taking hold. Generating diverse ideological positions on government, the events of these turbulent years had divided elites and the public. In embryonic form two major parties had emerged as the organizational response to controversy and conflict, parties that were ideological competitors for governmental power and for influence in the society.

TABLE 3.1 ▪ Percentage of Members Consistently Supporting Either the Federalists or Republican Party Positions in the House

Year	66⅔ Percent of the Time	75 Percent of the Time
1790	58	46
1795	93	86

SOURCE: Joseph Charles, *The Origins of the American Party System* (New York: Harper and Row, 1956), 93.

Samuel Huntington has suggested that parties in modern societies pass through four stages:[12]

1. Factionalism
2. Polarization of social forces leading to the emergence of initial party structures
3. Expansion of the electorate and perfection of party organization
4. Institutionalization, the establishment of parties that are coherent and complex structures that are adaptive and viable

By 1800, perhaps, the country was passing through the second of these stages; between 1800 and the Civil War it had moved into stages three and four.

EARLY PARTY CULTURE

In this early period of party history, certain norms or beliefs about how the system of partisan politics should function had begun to emerge. Such norms or beliefs, which each country develops on its own and are called its political culture, seem gradually to become accepted by citizens and political leaders, are transmitted to others, are reinforced by experience, and finally greatly influence how politics is conducted in a particular society.[13]

Over the years in the United States there have been two contrasting attitudes toward parties — in a sense, two party cultures. One of these supported the idea of a party, the other was critical of it. Here the concern is with those norms and beliefs that emerged and were connected with the emergence of a pro-party culture. The first such cultural norm in this early period was *acceptance of the idea of, and indeed the necessity of, political opposition.* Chambers refers to this as "the legitimacy of opposition and of opposition parties."[14] The republic was born as a result of political opposition, and with minor deviations from this principle (and expectations) it was an orientation that was persistently maintained and exercised in the period up to 1800. Although it was true that Washington was elected unanimously, the principle of uncontested elections never was accepted in the United States. The Alien and Sedition Acts came close to suppressing the opposition, but this effort failed. Few Federalists actually rejected this concept of the legitimacy of opposition; rather, they were deeply concerned by the virulence and bitterness of the opposition to their policies. Certainly we never had as large a body of leadership and public opinion adhering to the concept of unanimity and skeptical of the value of political conflict as is found today in many nations. Robert Dahl has pointed that out in the past, and it is still true today.[15] In early America there was a premium on insurgency and protest, and any attempt by government to proscribe organized political action was not to be tolerated. The monopolization of political communication media by the government was rejected, as were serious and sustained efforts to harass the opposition in such a way as to limit its capacity to oppose. This basic orientation was deeply embedded in the American spirit and in its constitutions. It was a political value and a way of thinking about politics which had tremendous implications for action within our political system. In a way, Watergate tested our commitment to this value. Although today we take it for granted, we must remember that it was the basic cultural norm undergirding the development of a competitive party system.

A second critical element in our early cultural patterns, that *the majority had the prerogative of power,* was regarded more ambivalently by people than was the first element. James Madison — at the time of the fight over the ratification of the new national constitution — had taken a contrary position, claiming that no group, even if it had a majority of the population, should be permitted to govern. This view found later exponents in the views of other statesmen, notably John C. Calhoun and his doctrine of the concurrent majority. But, from a practical standpoint, Madison in the 1790s changed his views as he became a leader of the Republican party and later president. Jefferson saw the need for opposite parties, one of which "must for the most part prevail over the other for a longer or shorter time."[16] Washington had sought to govern with a bi-party cabinet, but with Jefferson's resignation in 1793, the cabinet was dominated by the Federalists until 1801. The idea that the party that won elections did have the right to use this power in any reasonable, legal and *legitimate* fashion was a concept that did secure wide currency. And in fact it was practiced. There was no question of this to Hamilton, to Adams, or to Jefferson. When in power, they used their power to control the Congress, seeing the Speaker and standing committees as partisan organs. They felt it was their right and responsibility to adopt legislation in accord with their ideologies and campaign programs. And they used the power of patronage to reward their faithful followers and to guarantee support, even if it meant the appointment of their partisans to the Supreme Court, as Adams did just before midnight of his last day in office. The emphasis here should be on legitimacy, for at the time of Watergate when Nixon and his advisors sought to distort this norm by engaging in illegal activities to spy on and harass the opposition, the majority's prerogatives were stretched too far. Unreasonable excesses in the use of this norm were punished.

An important corollary to this norm was the idea that the party that lost an election should withdraw from power. This in a sense was the meaning of the presidential contest in the House after the 1800 election when Burr and Jefferson were tied in electoral votes. In the last analysis, the Federalists abstained so that the

Republicans in the House could give the presidency to Jefferson, thus ratifying the electorate's choice. This peaceful transfer of power, supplemented by the expectation that the majority party would in fact govern, had significant meaning for American political behavior.

A third general expectation, which developed early, was that *political competition would be bipolar,* but not necessarily extremely polarized. Many people began to expect a duopoly as proper. From 1790 to 1800 people became accustomed to two ideological, leadership, and political organizational tendencies in politics, constellations of social and political forces that differed in their philosophies and appeals but were not extremely distant from each other. And the parties — at that time Federalists and Republicans — acted in such a way as to monopolize the political terrain. Often a multiplicity of candidates ran for office (at least nine candidates secured electoral votes in the 1796 election for the presidency and vice presidency), but there were only two parties. Both of these were open structures, appealing widely for support among diverse sections and social interests. In pivotal states like New York, Pennsylvania, and New Jersey, the contest was close. But the competition was not shared with any third force. Despite maneuverings for advantage by individual leaders, the party system was developing into a competitive system in which two distinctive parties sought power by their exclusive efforts, and the winning party acceded to power without enduring a bargaining process. No system developed in which parties negotiated over the accession to power.

A fourth expectation, or attitude pattern, in early American political culture was *the tolerance of almost heretical factionalism* within political groups, particularly political parties. In colonial days and in early state politics, as Chambers has pointed out, there were cliques, juntas, social elites, and ad hoc caucuses. And these tendencies continued during the period of Federalist-Republican party development. In 1796 three factions were discernible, perhaps the most notable being the anti-Adams "High Federalists," some of whom had actually connived at displacing Adams as the presidential candidate. During Jefferson's period the most notable faction was the Quids who opposed the party policy in the Yazoo lands controversy and felt that the party was too subservient to northern and business interests. And in 1811 the Young Republicans led by men like Clay and Calhoun became a real threat to the viability of the Republican party. Upon analysis, these factional developments reflected ideological differences, personal power ambitions, and social-interest connections. They were not merely differences between the president and his party in Congress. They apparently had grassroots popular support and were often extremist and clearly rebellious against party policy and leadership. Significantly, each party represented a very broad spectrum of issue positions and socioeconomic interests. These party spectra overlapped considerably, so that the "haves" and the "have-nots" in American society could find a place within one or both of the political parties. Those people clamoring for social status, political rights, or economic betterment in early America did not have to go outside the two major party structures and ideologies.

The openness and breadth of party interests served the cause of the two-party system and had implications for the goals and tactics of each of the parties.

But the factionalism, while extremist and virtually heretical, was contained within the parties. Politicians, concerned about groupism, sought to contain these deviationists and to manipulate them but not to expel them. This tolerance became an important cultural political orientation. Leaders resisted attempts to place them under party discipline. Thus in the first Republican congressional caucus of 1795, Albert Gallatin observed that the members were left "to vote as they pleased." There was no early expectation that America would have a parliamentary party system, with the president and his cabinet leading a parliamentary group bound to support presidential actions or the majority sentiments of the legislative party group. It was indeed a highly pluralized elite structure, with leaders interacting informally, at times almost surprised to discover like-minded souls, and retaining their independence of judgment. As scholars demonstrate, there was a high degree of party voting. But there were defections — critical ones for both parties. And, more important, this cohesion in voting was not the product of a discipline exercised by the party leadership and accompanied by sanctions. Leadership independence was the accepted pattern of behavior for party elites in early America.

A fifth type of cultural orientation had a localist-populist content and perspective: the expectation that *political leaders would be accessible to the public,* would consult the public, and to a certain extent would defer to the public. In reality this is a combination of several attitudes about how politics should be practiced. It emphasized the prerogatives of the local community, the state, and the section. Too, it stressed the responsibility of the representative to the constituency and the potentiality of reprisal by "the home folks." Finally, the feeling that control over groups such as political parties should be local and popular, not in the hands of a centrist cabal of leaders whose actions were clandestine, was markedly important. Early American history is replete with examples of local protest movements, not all as violent as the Whiskey Rebellion in Pennsylvania in 1794 and the Fries Uprising in 1799, or as comprehensive as the Virginia and Kentucky resolutions of 1798. Perhaps more significant than these dramatic populist crises were the developments of indigenous political action movements and party structures at the local level, such as the Democratic and Republican Societies in 1793 and 1794, the Myrmidons of Burr in New York in 1800, and the development in the Middle Atlantic states in the early 1800s of county conventions of delegates to conduct party business and make party nominations. The idea of secret and unrepresentative caucuses of state leaders or of congressmen making decisions for the party was considered undemocratic. Above all, a vote-conscious egalitarianism was the emerging orientation of American politics. Politicians learned to "treat" the voters, as well as to entreat them; those who were not willing to treat the voter as king soon learned the consequences. The story is told of James Madison who, after having lost one election because of his refusal to "treat," campaigned in the subsequent election with great appreciation for individual votes. He traveled in midwinter over frozen roads to state his views at a meeting at a country church. "I then had to ride in the night twelve miles to quarters and got my nose frostbitten, of which I bear the marks now."[17] This vote-conscious and localist-oriented populism became an essential ingredient of American political practice, despite the early contemptuous attitudes of many of the aristocrats for the masses. This cultural orientation required considerable adaptive political ingenuity, and it affected the organization of parties and their strategies of appeal.

These incipient orientations constituted a set of expectations that might be called the developing party subculture. These images of party politics, representing values that were incorporated into practice, gradually became fixed in the minds of American elites and citizens. Though ambiguous and obscure and not unanimously accepted, taken together they represented a cultural context for party behavior, which if compared to contexts in other societies, reveals both equivalent characteristics and striking differences. Thus, the English code was much less localist and populist; the French more pluralist and multipolar; and the Dutch much less vote deferential, less factionalized, and more ideologically purist. From country to country there are both overlapping cultural patterns and great dissimilarities. These values, orientations, and expectations in each country, however, fundamentally conditioned how the game of party politics was — and is — to be played in that society.

EARLY PARTY ORGANIZATION

Besides cultivating the concept of partisanship, the early party leaders sought to develop a comprehensive organization from the national capital in Philadelphia to the grass roots. Both parties were fully aware of this need, yet developing a neatly pyramided structure was well-nigh impossible. The twin problems were coordination and cohesion. A variety of expedients was employed to weld a national party together: circular letters, committees of correspondence, use of patronage, legislative and local caucuses, and even purges of the disloyal. In 1796 a remarkable purge of those Republicans who had dissented in the vote on the Jay Treaty appropriation was conducted. Four of the seven were replaced (three by Federalists!), and the behavior of two was subsequently much more Republican.[18] Much earlier the device of "inter-visitation" had come into vogue, with travelers going from one local society to another. Finally, in 1808 the Republican congressional caucus set up the first national committee of correspondence consisting of one member from each state. Thus, through informal and formal means the party struggled to articulate a coherent and loyal organization. But geographical diversities of the time, local jealousies, and the sheer problems of communication meant that the early parties were loose aggregations of local groups of leaders and their followings. These local groups showed great independence of spirit, performed with varying degrees of efficiency, and pressured their governmental representatives to conform to local demands.

Whatever their structural defects, in relative terms both early parties opposed each other vigorously and vehemently. Historians of this period have uncovered a great deal of evidence indicating the intensity of party competition. The parties "treated" the voters, helped immigrants to be naturalized quickly (if it was to their advantage), marched their tenants to the polls, held rallies and ox roasts, and hanged their opponents in effigy. They even engaged in the questionable practice of repeating (voting more than once) on election day, and "the Lame, Crippled, diseased and blind were either led, lifted or brought in carriages to the Poll" [in Charleston].[19] In New York in 1800, Aaron Burr worked out an ingenious method for evading the state's property tax requirement for voters by seeing to it that city workingmen were granted joint land tenancies and thus were qualified to vote.[20] Further, a developing body of volunteer professionals — county committeemen — began to appear at the local level. In addition to mobilizing the vote, their task was to "fight that most formidable enemy of civilized men, political ignorance."[21]

These were no "mass" parties (those with large followings that could be relied on each election); they were still embryonic party structures. Only white males could vote, depending on their property tax qualifications, and fewer than 40 percent did so by 1800.[22] But the patterns of parties' activities strongly suggest that they were assuming certain key fundamental responsibilities in American society. Chief among their functions were determining public policy, recruiting leadership for the top policy-making positions, mobilizing support for these leaders, and influencing public attitude and preferences. In addition, they saw the administrative and adjudicative processes as proper provinces for party control and activity. Finally, their activities can be construed as having an important socializing function. The party leaders were identified with and believed in the developing American system of democracy. Though intensely partisan, they began to communicate their understandings of that system to the public.

MAJOR ERAS IN PARTY HISTORY

By 1800 not only did two major political parties exist, they were competing as political structures in fairly systematic fashion for national and local power. But the enduring outlines and characteristics of the party system were by no means irrevocably set. The ebb and flow of American party politics from 1800 to the present is challenging to study in great detail. The objective here, however, is not to deal with the details of each period or each election campaign, but rather to identify the major patterns of change and stability in the party system in form, process, and function. The aim is to describe the alternations in types of party competition and in party system form which occurred, a type of historical experimentation that led eventually to our present party system.

The history of the American party can be divided into five basic periods, called five "party systems" by some scholars.[23] These periods are roughly as follows.

From 1788 to 1824

This was the period of the building of the party system. There were three subperiods within it: that of the competition between Federalists and Republicans, that of one-party dominance by the Republicans after the Federalists began to disappear from effective competition after 1808, and that of transitional pluralism and factionalism within the Republican party from 1820 on, between the Jackson, Adams, Crawford, and Clay subunits within the party.

Under Jefferson, Madison, and Monroe — a period of one-party dominance — the Republicans won from 53 percent to 92 percent of the electoral college votes and from 61 percent to 85 percent of the House seats. By 1824 the Republicans were split, however, and four different presidential candidates received the following Electoral College votes: Jackson, 37.9 percent; Adams, 32.2 percent; Crawford, 15.7 percent; and Clay, 14.2 percent. The House of Representatives picked Adams.

From 1824 to 1854

The masses were mobilized and the parties were democratized during this period due largely to the efforts of Jackson, who won in 1828 and 1832. The delegate convention replaced the legislative caucus (or the mixed caucus) as the basic form of party organization. Sectional and ideological tensions increased in the party system and led to the Kansas-Nebraska Act of 1854 and the breakdown of the two-party competitive system between the Democrats and Whigs (or National Republicans).

From 1854 to 1860, the parties again went through a transitional pluralistic crisis before the two-party system was reestablished. This was the period during which the present Republican party was formed, but no one party could secure a majority. In the 1856 election the vote was distributed as follows: Democrats (Buchanan) received 45.3 percent of the popular vote, Republicans (Fremont) received 33.1 percent, and Whigs (Fillmore) received 21.6 percent. Even in 1860 Lincoln received only 39.9 percent of the popular votes, although he won 59.4 percent of the Electoral College vote. This was clearly a period of realignment.

From 1860 to 1892

After a period of Republican control to 1874, the two parties alternated control of the presidency and Congress. The major threat to the two-party system was posed by the Populist party in 1892, which secured over a million popular votes and twenty-two Electoral College votes. This threat was short-lived, however, and by 1896 the two-party system was clearly the dominant pattern.

The salient events during this period were (1) the development of urban political organizations or machines; (2) the adoption of legislation to protect the integrity of elections (the Australian ballot system, for example), as well as the regulation of party organization; and (3) the public involvement in the party process both in the strengthening of party loyalties and the participation of people in voting.

From 1896 to 1932

There was Republican dominance until 1912, when Roosevelt's Bull Moose Progressive Movement split the party and led to two national presidential victories for the Democrats with Woodrow Wilson. This was a minor pluralization crisis, but at the time it did split the presidential vote three ways, as follows: Wilson received

435 Electoral College votes and 41.8 percent of the popular vote; Taft received 88 Electoral College votes and 23.2 percent of the popular vote; and Roosevelt received 8 Electoral College votes and 27.5 percent of the popular vote. Republican dominance was restored in 1920 and lasted until Franklin D. Roosevelt's victory in 1932.

Crises or threats to the two-party system occurred twice, in 1912 (Roosevelt Progressives) and in 1924 (La Follette Progressives); the former secured 27 percent of the vote, the latter 17 percent. But they, again, were short-lived threats to the system. A major development during this period was the reform movement, designed to take the leadership-nomination function out of the hands of party organization leaders and to put it in the hands of the rank-and-file party members (the direct primary). In addition, there was the movement toward nonpartisan elections and the regulation of campaign finance activities of parties. The emphasis during this period was to treat parties as public-interest organizations that should be subjected to legislative control.

Since 1932

The New Deal coalition put together by Franklin Roosevelt presumably produced a basic party realignment and ushered in a whole new period of party life, a new party system. The Democrats were well in control until 1952, despite a minor pluralization crisis in 1948 when two factions within the Democratic party nominated their own candidates for the presidency (Henry Wallace and Strom Thurmond). Together, however, they received less than 5 percent of the popular vote. The Eisenhower victories of 1952 and 1956 interrupted Democratic control, and from 1960 on the parties have alternated, sometimes winning in landslides (1964 and 1972), sometimes in very close elections (1960, 1968, and 1976). In 1980 Reagan received 51 percent, Carter 41 percent, and Anderson 6.5 percent of the votes.

After 1980 Reagan was decisively reelected in 1984 (59 percent), but then a Republican decline in presidential vote set in as Bush won with 53 percent in 1988 but lost to Clinton in 1992, as did Senator Bob Dole in 1996. The tradition of the third-party threat reemerged with Ross Perot in both of these last elections. His 19 percent of the vote in 1992 was a major shock to the system, larger than any minor party vote since the Progressives in 1912. In a sense American politics was more pluralized and volatile in this period. And the Republicans shocked the world with their capture of both houses of Congress in the election of 1994.

In this era the most important changes have been the technological developments in campaigning (particularly the use of television), the new awareness of the role of interest groups in party politics and the attempts to regulate their role, and the emergence of ideological conflict in party campaigns. George Wallace's threat to the two-party system in 1968 when he secured over 13 percent of the popular vote was a significant event, but again a short-lived phenomenon. Some scholars argue that 1968 was the year that the fifth party era ended and a major realignment in parties began, launching a sixth party era. Others refuse to accept this interpretation. This question of "realignment" will be dealt with later; now, these basic, formative periods of the party system will be evaluated to clarify what historical circumstances molded America's party system.

THREE MAJOR PATTERNS OF PARTY COMPETITION

This glimpse at the historical eras of the American party system has revealed that in the early days, as well as more recently, *three basic types of party politics* have been experienced: one-party dominance, two-party competition, and transitional pluralism. Since 1800 there have been five periods of one-party control (totaling seventy-six years) when the same party controlled the presidency, the House, and the Senate. There have been four periods of sustained and evenly competitive conflict between two major parties, periods of balanced party competition (ninety-eight years in total). And there have been three pluralization periods (totaling twenty-two years) when the popular vote was dispersed among more than two parties or candidates, internal party factionalism and breakup existed, and/or "minority" presidents held office (see Table 3.2). Thus, the American experience has not been a homogeneous or single-directional experience. In a real sense there has been consistent movement in the direction of the basic two-party pattern. But all three patterns have occurred in the twentieth century, just as they did in the

TABLE 3.2 ■ Three Basic Patterns of American Party Politics since 1800

Years	Number of Years for Each Pattern: *One-Party Dominance*	*Pluralized Party Politics*	*Balanced Two-Party Competition*
1800–24	24 (Jeffersonian Republican)		
1824–36		12	
1836–54			18
1854–62		8	
1862–74	12 (Republican)		
1874–96			22
1896–1910	14 (Republican)		
1910–12		2	
1912–20			8
1920–32	12 (Republican)		
1932–46	14 (Democratic)		
1946–96			50
Total number of years	76	22	98
Number of "marginal" presidential elections[a]	0	5	16

[a] A marginal election is one in which the winning candidate for president won with less than 52 percent of the vote.

nineteenth century. And these historical patterns, reflecting as they must the desires, tolerances, and preferences of the people, constitute the mold and context from which American party politics have developed and which will shape their future.

Yet as the grand sweep of parties from 1800 to 1996 is viewed, these patterns have a set of discernible modifications. The periods of one-party control are not as long nor as completely dominant today as they were from 1800 to 1822. Also the pluralization crises are not as deep, as protracted or as recurrent.

The 1912 crisis (and certainly the 1924, 1948, and 1980 episodes) in the fragmentation of the party system was mild compared to those before. The Perot third-party candidacy was certainly not mild in 1992, but by 1996 had subsided. Perhaps the party system has developed a capacity for absorbing and containing threats to the system whether from the outside or as a result of internal dissension. Thus, while one-party control, duopolistic competition, and pluralization crises are still genuine possibilities today, the two-party system itself appears very viable.

DETERMINANTS OF PARTY SYSTEMS

Mysteries attend the origins and development of the party system in any society, and they are not easy to unravel. Why did parties come into existence when they did? Why did one type of party system appear? Why did parties with particular characteristics appear, representing particular group interests (for example, working-class or religious parties) and having particular competitive or organizational features? Particular aspects or circumstances of democratic societies made parties both necessary and inevitable, and also molded the emerging nature of the party system. Thomas Jefferson, writing in 1798, very early came to this position: "In every free and deliberating society there must, from the nature of man, be opposite parties and violent dissensions and discords."[24] As discussed in Chapter 1, this is a basic postulate; conflict in the early history of democracy has meant the emergence of parties. Open competition for power, tolerance of opposition, and the expansion of participation in the political system to larger proportions of the public, all make parties necessary and natural. The opportunities for power, as well as the necessities for the performance of certain indispensable system functions (recruiting leaders, formulating policies, and mobilizing support), led to the appearance of groups called parties. Given this basic position, however, little has been explained in answer to the question, Why did particular types of party systems appear? The analysis and research of scholars concerned with this question have led them to develop a set of positions that together may constitute an explanation.

Position 1: The Early Patterns of Interest-Group Conflicts

The basic cleavage structures in the formative years of a new democratic society establish the setting within which party conflict develops, and this original cleavage pattern leaves its imprint on a party system as it matures. Thus, Stein Rokkan and S. M. Lipset see for European countries three early cleavage patterns over the role of the Catholic church (a result of the Reformation), the question of secularism (a result of the revolution of 1789), and the primacy of agricultural or industrial interests (a result of the industrial revolution of the nineteenth century), to which subsequently was added the conflict between owner and worker in the capitalist system.[25] Parties there presumably were organized in terms of these basic cleavages, each particular form of which varied by society. Representing the interests of the landowners or urban commercial and industrial entrepreneurs, a dominant party emerged in the hands of established elites. And then an opposition party or set of parties emerged that reflected interests in opposition to these established elites, an opposition that could consist of religious or secular dissidents, landowners or urban entrepreneurs, or interests on the periphery opposing central national control.

The early political cleavages in the United States before 1800 conform somewhat to this conception. The differences between Hamilton and the Federalists on the one hand, and Jefferson and the Republicans on the other, were both interest-group oriented and ideological. Led by two magnetic personalities, the dominant elites split into two camps, disagreeing basically over how the power of the new federal government was to be exercised, and mobilized support from different socioeconomic interests. They disagreed over national economic policy, over sympathy for the French Revolution, and finally over the political rights of citizens threatened by the Alien and Sedition Acts. These were ideological concerns of great moment. Above all, the Jefferson Republicans opposed favoritism for the mercantile class — they opposed government for the privileged few. As Jefferson saw the party conflict, it was a dualistic confrontation between different sets of economic interests with distinctive ideologies — "republicans" against the "monocrats" — seeking to impose their will on the people.

Position 2: The Conditions under Which the Suffrage Was Extended

A major factor in the development of democratic party systems was the timing and circumstances under which suffrage was extended to all white male adults, at first, and eventually to all adults. As Leon Epstein observes, "There is every reason to believe that modern political parties emerged with the extension of the vote to a fairly large proportion of the populace."[26] He goes on to assert that the United States was "the first to enfranchise fairly large numbers" and "had the first modern parties, . . . in response to the nearly universal white manhood suffrage" by the second quarter of the nineteenth century.[27] It is important to remember that by the mid-nineteenth century America had eliminated religious, taxpaying, and property-holding qualifications for voting and had extended the suffrage to all white males. There was a delay in extension of the vote to blacks; the Fifteenth Amendment adopted in 1879 finally forbade abridgment of the right to vote on the basis of "race, color, or previous conditions of servitude."

Although the United States extended the suffrage early, in other democracies in Europe there was often considerable delay in the expansion of the right to vote and, therefore, of greater participation in the political process. Following are the dates when universal male suffrage was achieved in European systems and in Canada:[28]

Country	*Year Universal Male Suffrage Achieved*
Germany	1871
France	1875
Belgium	1893
Canada	1896
Norway	1898
Switzerland	1902
Finland	1906
Denmark	1915
Netherlands	1917
Britain	1918
Sweden	1921

The size of the actual eligible electorate is a major consideration for a party, since it is a vote-mobilization structure. As that electorate enlarges, the organizational character and competitive nature of the parties in the system are transformed. From the 1830s, parties in the United States had to appeal to rapidly expanding numbers of eligible voters. This made them behave quite differently than they had before in appealing only to the small aristocratic groups of property-holding and taxpaying notables who alone had the vote before 1800.

The most important point here is that the fight over the extension of the suffrage in a sense reinforced the early basic cleavages. In the United States the battle was between an aristocratic group with special status and a "frontier" group demanding inclusion in the electorate. This, too, became an ideological conflict, just as it did in Europe, where throughout the nineteenth century (in France earlier, at the time of the Revolution) usually there were two sets of elites — liberals and conservatives — divided on this very question: Who should have the voting franchise? The way in which that question was resolved had a considerable impact on the type of party system that emerged.

Position 3: The Treatment of New Claimants for Power by Established Elites

During the nineteenth century in all Western democratic societies, new groups made demands for inclusion in the electorate, in a sense claiming admission as full members of the polity. These groups included the urban working class, farm workers, religious nonconformists, Catholics (treated as a minority in certain countries after the Reformation), and particular ethnic, national, and language groups. Several scholars argue that the treatment of these new claimants for power varied greatly from one society to another, based on two factors: the rigidity of the status system and the extent to which the old dominant elites were tolerant of these new claimants. One critical test of the attitude of the established elites, of course, was their willingness to give these new claimants the right to vote. When the old elites were intolerant and the status system rigid, new parties formed to present the demands of the new claimants — religious parties, farmers' parties, workers parties, as well as ideological splinter parties — that either broke away from the parent liberal and conservative coalitions or mobilized new sectors of the society. New, small parties emerged in all European countries. Britain was spared this development (primarily because in 1867 and 1885 it expanded suffrage) until World War I when the Liberal party split and the new Labour party organized effectively enough to replace the Liberal party as the second major party.

The United States after its independence never had a rigid social status system of any consequence; it was relatively egalitarian. Very early, the established elites accepted the idea of a liberalized suffrage. Because of a variety of pressures, calculations of competitive advantage, and arguments based on principle, new claimants did not have to fight their way into the system but were given the right to vote early.[29] As a consequence, urban workers, farmers, Catholics, and other minority groups (with the exception of Blacks) did not develop the organizational solidarity, the political consciousness, or the leadership that was necessary for the establishment of the type of minor parties found in Europe.

Position 4: The Political Constitution, Particularly the Election System, as Determinant

Much has been made of the role of the arrangement of political institutions in determining party systems. In the United States the federal system, the presidential office, the bicameral division of legislative power, and the election system itself, it is said, are responsible for our type of party system.[30] Some scholars have suggested that there is virtually an immutable "law" concerning the relationship between election systems and party systems. The simple plurality, single-member district system (United States and Britain) encourages a two-party system; the proportional election system (and the two-ballot system) encourages a multiparty system, it is claimed.

On the surface the arguments that certain basic features of the governmental system determine the nature of the party system seem logical. Indeed, at the time of the campaign for the adoption of the Constitution, the Founding Fathers reasoned similarly. James Madison argued that federalism, bicameralism, and separation of powers would influence the way parties or factions operated. And the election-system argument does appear to have a certain rationality. When you can elect only one person from a geographic area, a multiplicity of parties is discouraged. The trouble with the argument (whether for election systems or other institutional arrangements) as an explanation for the *emergence* of types of party systems is threefold:

1. *There are many exceptions to the rule.* Nations with the same institutional features as the United States may have quite a different type of party system. For example, Canada is a federal system but has four durable parties (two majors and two somewhat smaller). France has an elected president, but a very fragmented party system. Britain has a single-member district election and a two-and-one-half or three-party system. India has an election system like the United States but a much different party system.

2. ***The causal relationship may be reversed.*** The parties may be visualized as determining the election system, rather than vice versa. The United States really had established its party system first and then adopted its election system. A good example for a multiparty system is the Netherlands, which already had seven parties in Parliament before adopting a proportional representation system.
3. ***Other more basic social and political forces determine the character of a party system.*** There may be forces deeply imbedded in the nature of the society, and both the party system and the election system are a reflection of those basic forces. Just as Americans normally would be appalled at the idea of a five- or six- or seven-party system, so the French, Dutch, Norwegians, and Swiss generally would dislike the idea of a two-party system. Why? Because that type of party system would not be perceived as functional to the needs of their societies.

Once a party system is established, however, the type of institutional framework within which it has to function will affect its operation considerably. And that framework may well work to maintain the basic character of that party system over time. The United States is a good example of this.

Federalism, for instance, may have certain effects relevant to the equilibrium of our two-party system. As Madison predicted, federalism makes the assumption of total political power through the country well-nigh inconceivable, since no party wins total power at the national and state levels simultaneously. This represents a check on the dominant party's coming to national power, and this check is manifest in the actions of state governors and state legislatures in actual opposition to the party in power. This means that federalism provides a sheltering place for the losing party and enables it to use state office as a springboard for restoring the balance in the system. A Woodrow Wilson in New Jersey, a Franklin Roosevelt in New York, or a Nelson Rockefeller in New York can keep attention focused on the activities of the major parties' top leadership potential and keep the losing party a visible political reality throughout the nation. Further, federalism invites minor and third parties to test their wings and to achieve some measure of satisfaction attempting to capture power in state capitals. In fact, any party having designs at the national level realizes that it must operate at the state level first. Potential national third parties — the Populists, the Farmer Labor party, the American Labor and Liberal parties of New York, and even the Dixiecrats — found the federal system with its decentralization of power an outlet for protest. But these third parties expend much of their activities at the state and regional levels, and their impact is rarely felt at the national level. Thus, while the federal system may discourage too much one-party dominance, it also severely limits the development of minor parties that are serious national threats. Mounting an effective minor-party organization and campaign in Minnesota or South Carolina or New York is one thing. Doing it nationally is something far more complex.

The contention that the presidential system is a deterrent to the development of third parties no doubt has some truth. This is claimed to be so in the American system because the presidency is not divisible, cannot be easily won by a third party, and is so powerful. It can be argued otherwise that the presidency is after all not the entire system of government, that Congress is a very powerful counterbalance to the presidency, and for a minor party to achieve some status in the Congress would in itself be of considerable consequence. In order to constitute threats to the system, minor parties need not have the presidency but could serve obstructive or balancing roles in Congress. The British Liberal party certainly does not expect in the near future to capture the government and provide England with a prime minister; it is but a faint hope. Nevertheless, the British Liberal party contests each parliamentary election, elects some members to Parliament, and sometimes can be successful enough to be a force with which the major parties must reckon. Indeed, at one time, in order to stay in power, the Labour party had to work out an agreement with the Liberals. The presidency, nevertheless, is obviously a chief prize for any U.S. party; any genuine forces of protest in our society eventually must be interested in capturing the presidency.

Theoretically, the plurality feature of the American system should provide some encouragement and opportunity to minor parties, but it has not, except in certain areas where state third parties have had a limited success, such as New York with its Liberal, Labor, and Conservative parties. The important point is that this election system has not necessarily inhibited the sporadic minor-party movement, but the vote outside the two-party system has not been able to organize into a durable third party. If a significant social need and determined leadership arose, third parties could be organized despite the system of electing legislators from single-member districts. It has been the case in Canada, England, and India. But the election system is certainly not facilitative. Third parties, without established clienteles, can easily be discouraged by the election system, unless the conditions of party life and the problems in the society change radically.

Position 5: The Responsiveness of Established Parties to Social Problems

As parties develop organizationally, they become complex structures with both organs of leadership and decision making at a variety of levels (from the national to the local), and also vertical and horizontal relations articulated in order to make them progressively more coherent action mechanisms. One characteristic such structures must have is adaptability; that is, parties must have the capacity for responding to the needs of society, particularly when crises (social, economic, or moral) occur which threaten their stability. The extent to which the established parties develop as continuously responsive institutions will determine whether new parties appear and secure the support of the voting public. Thus when working class Catholic or farmers' parties emerged in Europe, it must have been the case that the existing parties were not adequately sympathetic to the needs and demands of these groups. The Labour party in Britain became a major party at the time of World War I because neither the Liberal party, which was badly split, nor the Conservative party was considered adequately interested in the problems of the working class. Too, the same can be said for virtually every country in Western Europe.

The Republican and Democratic parties since the Civil War have been responsive enough to the new problems of American society to quash the threats of minor parties. However, in the period preceding the Civil War the Democrats and National Republicans could not solve the crucial, moral issue of slavery, and as a result, from 1854 to 1860 America went through a pluralization crisis that led eventually to a new two-party system. Since that time threats have recurred periodically, but the Republicans and Democrats have survived them, so that no labor, Catholic, or farmers' party has lasted for any length of time. The character of our party system cannot be comprehended without an awareness of these twin phenomena: the inability of special-interest parties to take root and the responsive capability of our major parties. A major concern today is whether our parties are responsive enough to the problems of the poor, the disadvantaged, and the minorities, particularly in the big cities, to continue to receive the electoral support of such groups.

Position 6: Psycho-Cultural Socialization of Citizens to the System

Accepted by many scholars, the proposition has been advanced that long exposure to the same party system without any interruptions (wars or nondemocratic regimes) will tend to habituate the public to the acceptance of a party system and the cultural expectations connected to it. Philip Converse argues that exposure for two and one-half generations is probably necessary for any party system to be firmly imbedded in society — that is, a system with which people identify for a long enough time and that is transmitted intergenerationally.[31] This may very well have happened in the United States. The period before the Civil War was considerably volatile and too short for firm party loyalties to have developed. After the war, the system seems to have stabilized so that even though there still was considerable volatility in party successes, a psychological commitment to the parties that had great durability appeared to emerge. This commitment has declined in recent years; estimates suggest that as much as 90 percent of the public identified with the two major parties at the time of World War I, while only two-thirds or less are party identifiers in recent years. Nevertheless, there is strong indication that the two major parties have retained a hold on the loyalties of the majority of American citizens, a hold that may make a transformation or realignment of significant modification of the party system difficult.

Perhaps this argument should not be taken too far, because studies have shown that high proportions of the pubic identify with parties in other countries that have not had a party system for as long as the United States has. The evidence is certainly not completely clear on this point, and it certainly cannot be assumed that with the passage of time the public becomes so irreversibly committed to the party system that significant change is impossible. Major party-system change is certainly possible, but such change is more difficult and will not be revolutionary in societies where party systems have survived long enough to evoke sentiments of identification and loyalty from the majority of the public.

CONCLUSION

Since 1789 American democracy has gone through a great developmental and maturation process. Despite constant challenges and much anti-party sentiment, the party system was a part of this modification and transformation. Early parties were loose coalitions of social interests, with poor organization and very little discipline, led by elites with very pragmatic styles. As they became mass-oriented, complex, and durable structures, ever more diverse in composition and socially adaptable, parties also assumed responsibility for the performance of a variety of tasks and functions. And over time, particularly from the end of the nineteenth century to the present, the public developed important attachments to the parties.

In the meantime, however, interest groups appeared that also became involved with the political process. Even as the pro-party culture matured, anti-party developments reappeared. Parties began to share with other sectors certain functions, such as the articulation of public opinion, representation of group demands, political communication with the public, and socialization. Parties lost control over the recruitment of leadership for the top policy-making positions in society. Yet the parties continue to play a major role in the mobilization of support for political leaders; they were actively involved in the intergroup bargaining activities that became a prominent part of the American political scene. Over the years the parties were constantly harassed by anti-party sentiment. People demanded the regulation and reform of parties, and some atrophying in party functional performance occurred. Yet, despite all these transformations, today the two-party system remains viable and responsive.

Perhaps a basic dilemma developed in the role of parties in the American society as our system matured during the nineteenth and twentieth centuries, a dilemma of organizational cohesion and effective action or lack of discipline and delayed action. Pro-party sentiment pressed for the former; anti-party sentiment resulted in the latter. A two-party equilibrium had developed after the Civil War, a party system that was relatively integrated with the social order. But the pragmatism and coalition bargaining that attended the development of this stable party system made it difficult for our parties always to act effectively and expeditiously. And the demand for reform that weakened party organizations may have contributed to a decline in party system performance. This may indeed be the irony of American parties.[32]

NOTES

1. A valuable review of the arguments pro and con is presented in Ronald P. Formisano, "Deferential-Participants Politics: The Early Republic's Political Culture, 1789-1840," *American Political Science Review* 68 (June 1974): 473-87. Formisano argues that early events and activities "did not constitute purposive party building" (p. 476) and the evidence of activity in these early days was "incipient party-like behavior" which "Burnham correctly characterized . . . as 'pre-party' (p. 486)." Other scholars have been somewhat more inclined to see 1790-1800 as the formative party era.

2. *The Federalist*, No. 10.

3. Richard Hofstadter, *The Idea of a Party System* (Berkeley: University of California Press, 1969). Franklin is quoted on p. 2.

4. Ibid., 1-2.
5. William Chambers, *Political Parties in the New Nation: The American Experience 1776-1809* (New York: Oxford University Press, 1963), 32.
6. See particularly Ibid., chs. 1-3, for a description of these developments. See also Joseph Charles, *The Origins of the American Party System* (New York: Harper and Row, 1956), and Hofstadter, *Party System*, chaps. 1-3.
7. Chambers, *Political Parties*, 64-65.
8. Ibid., 93.
9. Ibid., 75.
10. Hofstadter, *Party System*, 88-89.
11. Chambers, *Political Parties*, 32.
12. Samuel Huntington, *Political Order in Changing Societies* (New Haven: Yale University Press, 1965), 412-20.
13. Gabriel Almond, "Comparative Political Systems," *Journal of Politics* 18 (1956): 391-409; Sidney Verba, "Comparative Political Culture," in Lucian W. Pye and Sidney Verba, eds., *Political Culture and Political Development* (Princeton: Princeton University Press, 1965), 518.
14. Chambers, *Political Parties*, 145. I have relied heavily on the analysis of Chambers in his lucid discussion of this early period.
15. Robert A. Dahl, ed., *Political Opposition in Western Democracies* (New Haven: Yale University Press, 1966), 333, quoted in Hofstadter, *Party System*, 7-8.
16. Chambers, *Political Parties*, 149.
17. Quoted in Ibid., 4.
18. Ibid., 89.
19. Ibid., 157.
20. Ibid., 155.
21. Ibid., 165.
22. Ibid., 32-33.
23. William Chambers and Walter D. Burnham, *The American Party Systems* (New York: Oxford University Press, 1967). They also suggest there may be a "cyclic" characteristic to these party eras — that each lasts thirty years or so, each ends in a crisis of realignment, each period accomplishes certain goals, and then a new period seems to be necessary to meet the needs of American society. See pp. 29-30, 288-89 for a discussion of these patterns.
24. Quoted in Chambers, *Political Parties*, 149.
25. Seymore M. Lipset and Stein Rokkan, eds., "Cleavage Structures, Party Systems, and Voter Alignments: An Introduction," in Seymour M. Lipset and Stein Rokkan, eds., *Party Systems and Voter Alignments: Cross National Perspectives* (New York: Free Press, 1967), 26-50.
26. Leon D. Epstein, *Political Parties in Western Democracies* (New York: Praeger, 1967), 19.
27. Ibid., 20-21.
28. Scholars differ somewhat in their dating of suffrage extensions. Perhaps the most careful analysis is in Stein Rokkan and Seymour M. Lipset, *Citizens, Elections, Parties* (New York: David McKay, and Oslo: Universitetsforlager, 1970), 84-85.
29. For an extremely lucid discussion on this argument, see Seymour M. Lipset, "Political Cleavages in 'Developed' and 'Emerging' Politics," in Erick Allardt and Stein Rokkan, eds., *Mass Politics* (New York: Free Press, 1970), 23-44.
30. See, for example, Maurice Duverger, *Political Parties: Their Organization and Activity in the Modern State* (New York: Methuen, 1951), 204-5.
31. See Converse, "Partisan Stability," *Comparative Political Studies* 2 (July 1969): 1142-45.
32. See Chambers and Burnham, *Party Systems*, for a discussion of this same problem, particularly p. 280; Morton Grodzins, to whom they refer, also saw this paradox, suggesting our parties are "anti-parties," in "American Political Parties and the American System," *Western Political Quarterly* 13 (1960): 974-98.

CHAPTER 4

Third Parties in American Politics

Third parties have been, and continue to be, part of our political culture. To understand the two-party system, we need to know the roles which these third parties play. They are important because they have become fairly frequent phenomena, because they reveal the vulnerabilities of the two major parties, and because they can influence the results of elections and the direction of public policy. Rather than criticize them, we should understand their contribution to the dynamism of our politics. And finally, studying them can help us understand how and why the two party system has survived as long as it has.

THE HISTORY OF THIRD PARTIES

Third parties are nearly as old as the nation itself. In the early state legislatures and Congress there was considerable factional conflict. Perhaps the first serious threat of a third party occurred in 1806 when John Randolph of Virginia, a member of the Republican party of President Jefferson, opposed the president over his land policy and other questions. He and his supporters said they sought a "tertium quid"— a "third position"— and came to be known as the Quids.[1] Randolph and his followers failed as a third force. But other factional associations appeared. In 1828 the first organized third party emerged; known as the Anti-Masonic party, it was a native American party opposed to foreigners and Catholics. Dissenters soon developed a tradition of organizing in opposition to the major parties.

There were many third-party efforts in the nineteenth century.[2] (See Table 4.1.) After the Anti-Masonic party (which secured 8 percent of the votes in 1832), the antislavery movement spawned third parties in five elections from 1840 to 1860. The Liberty party was the first and in 1840 helped President Polk win the state of New York. This led in turn to the Free-Soilers in 1848, who opposed the extension of slavery into new western territories. They won 10 percent of the popular vote. The influx of immigrants from 1840, whose votes allegedly helped Democratic President Pierce in 1850, produced a reaction and a new third party — the American (or Know-Nothing) party, which was particularly anti-Catholic. It secured one of the largest third-party votes ever, 21 percent in 1856. But the slavery issue split this party. And two new third parties emerged, both in the South: the Constitutional Union party, which refused to even discuss the issue of slavery,

TABLE 4.1 ▪ Third Parties Winning the Most Votes in Presidential Elections: 1832–1996

Year	*Party*	*Percentage of Total Votes Cast*
1832	Anti-Masonic	8.0
1848	Free-Soil	10.1
1856	American	21.4
1860	John Breckinridge — Southern Democratic	18.2
	Constitutional Union	12.6
1892	Populist	8.5
1912	Theodore Roosevelt — Progressive	27.4
	Socialist	6.0
1924	Robert La Follette — Progressive	16.6
1968	George Wallace — American Independent	13.5
1980	John Anderson — Independent	6.6
1992	H. Ross Perot — United We Stand	18.9
1996	H. Ross Perot — Reform Party	8.0

SOURCE: Adapted from Daniel Mazmanian, *Third Parties in Presidential Elections* (Washington, DC: Brookings Institution, 1974), 5, for the years 1832–1968; Steven Rosenstone, Roy Behr, Edward Lazarus, *Third Parties in America,* 2nd ed. (Princeton: Princeton University Press, 1996), 278–79, for the years 1980–1992; and "Portrait of the Electorate," *New York Times* (November 10, 1996), 16.

and the Breckinridge Southern Democratic party, which openly defended slavery. These two parties contested the 1860 election against the Republicans (Lincoln) and Democrats (Douglas) and did remarkably well — 12.6 percent for the Constitutional Union and 18.2 percent for the Breckinridge Democrats. Thus Lincoln won with only 39.9 percent of the popular vote.

In the latter part of the nineteenth century, a wide variety of third parties appeared on the political scene, ran candidates for president and also for Congress, then died after securing very small percentages of the vote: Greenbackers (3.3 percent in 1880), Anti-Monopoly party (1.2 percent, 1884), Union Labor party (1.3 percent, 1888). But a more successful third party did contest in 1892 — the Populists, or People's party. It arose in the West and South because of the economic plight of farmers due to a decline in prices, railroad and grain elevator monopolies, oppressive farm-lien systems, and high farmer debt. Farmers saw bankers as manipulating the supply of money, and they saw as the remedy an elastic currency. A variety of farm organizations came into existence and pressed for reform. They banded together into a national union, leading to a national party convention in 1892, when the People's party was formed. Although they secured only 8.5 percent of the vote, the new party won in four states, elected many state legislators, and continued over the next two years. In 1894 its vote increased to 11.2 percent. But by 1896 the Populists' platform of social and economic reform appealed to the major parties, particularly the Democrats, who took over many Populist ideas and nominated as their candidate William Jennings Bryan, who many Populists accepted as their own spokesman. As the major parties adapted to this threat by co-opting minor party ideas, the Populist challenge to the system disappeared.

In the twentieth century, we have continued to have many third parties, but four of them have been particularly significant in that they have taken a relatively large percentage of votes away from the two major parties.[3] These four are:

Major Third Parties	*Percentage of Vote*
1912 — Roosevelt Progressives	27.4
1924 — La Follette Progressives	16.6
1968 — Wallace, American Independent Party	13.5
1992 — H. Ross Perot, United We Stand	18.9

There were other interesting third parties that did not do as well, including the Socialists (6 percent in 1912), the States Rights party of Strom Thurmond in 1948 (2.4 percent), Henry Wallace Progressives in 1948 (2.4 percent), and John Anderson's Independent party in 1980 (6.6 percent). But the four major third-party efforts of this century had certain special characteristics. Roosevelt's (Bull Moose) party of 1912 was a result of a split in the Republican party at their national convention over policies concerning business regulation. It relied primarily on Roosevelt's ideas, personality, energy, and reputation. Robert La Follette was a prestigious U.S. Senator who launched his party in 1924 on a platform of agricultural reform, antimonopoly regulation, and political change (direct primaries and popular referenda). A strong coalition of 140 progressive members of Congress, as well as nonpartisan and liberal groups, supported his party. However, problems of finance, ballot access, and party organization in the end doomed this effort. George Wallace, Governor of Alabama, in 1968 led a very different movement. His emphases were on law and order and doing something about "urban unrest," appealing particularly to the economically disadvantaged while also exploiting racial conflict and the opposition to the civil rights movement. He ran well in several primaries, but his appeal in the last analysis was primarily in the South where he won in five states (forty-six Electoral College votes).

These significant third-party performances were notable for their different origins, their different ideologies and issues positions, and their common failure to draw enough leaders and citizens away from the major parties to sustain their campaigns, and their parties, beyond one election. They were led by prominent leaders disaffected with their own parties (whether Republican or Democratic) during times of real crisis in the society. But the obstacles to weaning voters away from their traditional party loyalties and voting behavior patterns, in all cases, were too high for third parties to overcome. Up to one-fifth of the vote could be won by third parties, at the outside, but the average vote percentage was actually much lower. Yet they enlivened our politics from the early 1900s until today and forced the major parties to change.

H. ROSS PEROT, THE MOST RECENT THIRD-PARTY PHENOMENON

On February 20, 1992, in an appearance on the television talk show *Larry King Live,* H. Ross Perot, a billionaire industrialist, announced that he would run for the presidency "if voters in all fifty states put me on the ballot." He said he was fed up with "the mess in Washington" and irked at the budget deficit. He wanted the government to cut its discretionary spending, increase certain taxes, eliminate tax breaks, and adopt policies to stimulate the economy, thus opening opportunities to all Americans. He also wanted political reform, "to overhaul the political system, particularly by enacting campaign finance reform."[4] He encouraged people to join his crusade, and thousands called in to his headquarters offering support and joined state and local groups that became part of his organization, United We Stand. In August of 1992 a group of scholars accessed a data base of almost 500,000 volunteers at the Dallas headquarters alone.[5] Perot set up an organization, provided $73 million of his own funds to finance his campaign, hired top personnel to direct the campaign and decide on the media messages and programs, and continued in long talk shows to present his positions to the American people. In July he suddenly pulled out of the race, but then reentered in early October. In the meantime he sent letters to state election officials stating his desire to remain on the ballot. He published a book, *United We Stand: How We Can Take Back Our Country,* which became a best-seller. Perot conducted an expensive advertising campaign, committing $45 million to television. He also was invited to participate in the three presidential television debates in October, which gave him tremendous recognition, exposure to sixty million Americans, and a chance to get his message across, thus distinguishing himself from Clinton and Bush. Viewers rated his performance high, as well as or better than that of the other two candidates. During this period, his rating in the polls went up (in early October), but then dropped by election day. During the period from April to November, Perot's Gallup rating rose to 39 percent by early June (he was then the front-runner), dropped precipitously by July to 15 percent, where it stood in November despite a 5 percent temporary increase after the debates in October. His vote on election day was 18.9 percent, the highest for a third party since 1912. After the 1992 election, Perot remained active, converting his organization into United We Stand America, Inc., a nonprofit group presumably pushing Perot's agenda. The group did not enter the 1994 elections, but encouraged people to vote for a Republican Congress — 63 percent of Perot's supporters did so. By late 1995 he created his new party, called first the Independence party and then the Reform party. While other potential presidential candidates (Colin Powell, Bill Bradley, Lowell Weicker) decided not to run, Perot pushed his party's efforts, again gaining access to the ballots in all states. Despite his efforts, fewer people thought his campaign was credible in 1996. Polls in July that asked the public, "Do you think Ross Perot can win the presidential election in November, or don't you think so?" found only 7 percent responding, "Yes, he can win"; in 1992 the percent in July had been 49 percent. Perot received only 8 percent of the vote in the election and lost support among all major groups in the population (Table 4.2).[6]

TABLE 4.2 ▪ Demographic Characteristics of H. Ross Perot's Third-Party Votes: 1992 and 1996 (in percent)

Characteristic	*Perot 1992*	*Perot 1996*	*Difference*
Total	19	8	-11
Race			
White	20	9	-11
African American	7	4	-3
Latino	14	6	-8
Asian	15	8	-7
Ages			
18-29	22	10	-12
30-44	21	9	-12
45-59	19	9	-10
60+	12	7	-5
Gender			
Men	21	10	-11
Women	17	7	-10
Party Identification			
Republican	17	6	-11
Independent	30	17	-13
Democratic	13	5	-8
Ideology			
Liberal	18	7	-11
Moderate	21	9	-12
Conservative	18	8	-10
Religion			
White Protestant	21	10	-11
Catholic	20	9	-11
Jewish	9	3	-6
Union Household	21	9	-12
Region			
East	18	9	-9
Midwest	21	10	-11
South	16	7	-9
West	23	8	-15
Income			
Under 15,000	19	11	-8
15-29,999	20	9	-11
30-49,000	21	10	-11
50-69,999	17	7	-10
70-99,999	16	6	-9
100,000+	—	6	—

SOURCE: Adapted from "Portrait of Electorate," *New York Times* (November 10, 1996), 16. Based on Interviews with those leaving the polls: 16,627 voters in 1996, 15,490 in 1992.

However, he probably kept President Clinton from getting 50 percent of the vote, as he had in 1992. Perot suffered the fate of predecessors after a significant mobilization effort four years previously.

THEORIES TO EXPLAIN THIRD PARTIES

Many scholars have described our third parties and have advanced theories to explain these phenomena.[7] These theories have addressed a series of questions: Why do third parties appear? Why do people vote for them, or, what explains their success or failure? Why do they not endure? How do the two major parties cope with them? What is the impact on, or meaning of, third parties for the nature of our politics?

In attempting to explain the origins of third parties, the general point should be made first that we should not be surprised that third parties appear frequently. In a two-party system such as ours, some citizens feel constricted by having only two political choices at election time. It is inevitable that there will be some who feel unrepresented, because the major parties often focus on "the median voter." Periodically voters feel the inadequacy of the Democratic and Republican parties to reflect the wide spectrum of interests and sentiments of both the public and also some political leaders. This is true not only of our system but also of others. Multiparty systems encourage this more than our system does and often reward new parties with votes and legislative seats.

It is generally recognized that certain special circumstances trigger the emergence of the most demonstrative and effective third parties. An economic crisis (such as the agricultural depression of the 1890s) or a very divisive new issue (such as the slavery question before the Civil War) or a feeling of failure of the major parties to deal with a basic problem of governance (such as the budget issue in 1992) or law and order (and the Vietnam War in 1968) may set the stage for a third-party movement. Usually this is combined with two other developments: (1) the perception by a fairly large estranged sector of the public that the major parties cannot or will not cope with the crisis, and (2) the emergence of able leadership, often rejected by the major parties, as charismatic spokespersons for the new third-party movement. If this is also accompanied by the development of a "grassroots" organization in support of the movement, the third party may flourish early in its existence. Another, somewhat different, scenario is that of a major leadership fight within one of the major parties and the secession of one leadership group for the purpose of running as a third party. This happened with the breakup of both the Whigs and Democrats in the 1850s. It happened again in 1912, when Theodore Roosevelt and his Bull Moose Progressives seceded from the Republican party. On a minor scale this also happened to the Democrats in 1948, when the States Rights faction, opposed to the civil rights proposals, walked out of the Democratic convention and nominated Strom Thurmond as its presidential candidate. In each case, issue conflict was important, and an unresolvable internal party battle led to a breakup that resulted in the formation of a third-party candidate for the presidency.

Thus, the conditions for the emergence of third parties are clear: an issue or ideological crisis, the alleged incapacity of major parties to contain or resolve the conflict, and the appearance of capable, often prestigious, national leadership to assume control of a new movement, appealing for the votes of a presumably alienated and discontented public.

How can we explain the vote for third parties, or rather the variation in the success of third parties? Why did some third parties do well (such as the La Follette Progressives in 1924, with 16.6 percent, and Perot in 1992 with 18.9 percent) while some did rather poorly (such as Anderson in 1980 with 6.6 percent, and Perot in 1996 with 8.0 percent)? The most direct and convincing theoretical argument is that of Rosenstone and his colleagues.[8] The model they present consists of *motivations* of voters to support a third-party alternative and, on the other hand, the system *constraints* to a successful third-party candidacy that must be overcome. The argument is that three key motivations are involved in voting for a third party: citizen conviction that the major parties have deteriorated functionally, attraction and commitment to the person leading the third party as a credible candidate, and a strong sense of allegiance with this third party.[9] But there are severe constraints which may influence a voter not to support a third party: strong party identification with a major party, loyalty to the two-party system, and doubts about the capability of a third-party candidate to overcome the barriers (legal, constitutional, financial) facing third-party candidates. If the voter feels the barriers to success are too high and thus the third-party candidate is not a possible winner, despite the voter's pro-third-party motivation, the chances of a vote for a third party will dwindle.

One must include in any theory of third-party voting the ways in which our system handicaps third parties. The single-member district system and presidential Electoral College system, both of which are winner-take-all electoral systems, discriminate against third parties. If you cannot get a vote higher than any other party, you lose the district or the state. And of course the major parties not only protect that electoral system, they also make success difficult for third parties in other ways. They pass state laws making access to the ballot difficult for third-party or independent candidates. They demean third parties by saying they can't win, so "don't waste your vote." They may adopt some of the proposals on which third parties campaign, a practice called "co-opting issues." They also usually have more money, spending on the average ten times more for media coverage than third parties can afford. And through their organizations, they usually have many more resources for the campaign, such as personnel, than third parties can possibly muster. The efforts to "delegitimize" third parties means that the playing field is by no means level. If voters are aware of this, they may seriously doubt the credibility of the third-party candidate, no matter how capable, attractive, and reputable that candidate may be.

To illustrate this theory, we can use the Perot candidacies in 1992 and 1996 as examples. Excellent research has been done on these campaigns, and the findings are very revealing. Many people were disillusioned with the major party candidates and parties in 1992, as in 1980 and previous years. Perot did exceptionally well in mobilizing this disaffected vote in 1992; one in five voters voted for him. There was wide support for him, except among blacks, Jews, and those over age sixty (see Table 4.2).

In addition, there was strong support for Perot among voters disappointed by the candidates of the two major parties: 27 percent of those who evaluated Clinton and Bush negatively voted for Perot.[10] The basic reason that Perot did so much better in 1992 than his predecessors (particularly Anderson in 1980) was traced to his ability to "break through the constraints that third-party and independent candidates before him had faced."[11] While declining party identification with the major parties and the public's political alienation certainly provided the opportunity for Perot, as it did for other third-party candidates before him, the single most important factor in 1992 was that Perot had the resources to get national visibility, access to the public, a place on the ballot in all fifty states, participation at the presidential debates, and the extensive media coverage and attention which other candidates usually could not get. He spent $73 million of his own fortune, and, as the Rosenstone study put it bluntly, "money buys votes."[12]

In addition, Perot had a fairly extensive group of activists in his United We Stand organization. A sample survey of those activists revealed the motivational and belief patterns of Perot workers (Table 4.3). The distinctive position of the Perot activists, who presumably were contacting voters and communicating the Perot message to them, is clear: frustration over the economy, distrust of the major parties, and very weak party loyalty, if any, to the Democrats and Republicans. Seventy percent of these activists worked for Perot before the conventions and 57 percent later in the campaign, after Perot returned to active candidacy. Despite this "drop-off," Perot apparently had a large activist cadre working on his behalf.[13]

TABLE 4.3 ▪ A Comparison of the Views of Perot Activists and Democratic and Republican Caucus Members in 1992 (in percent)

Attitude	*Perot Activists*	*Democratic Activists*	*Republican Activists*	*National Electorate*
1. Believes the economy has become worse in the past twelve months	61.5	42.5	13.3	72.3
2. Almost never trusts the government	40.6	7.2	24.1	1.9
3. Believes the major parties confuse the issues	76.3	36.4	36.0	No data
4. Has strong party identification	17.0	63.0	64.0	29.0
Number	1,353	764	385	2,487

SOURCE: A sample of the almost 500,000 people who called to support Perot in spring and summer; data base developed by Perot's Dallas headquarters. The Democrats and Republicans were from a sample of individuals who attended presidential nomination caucuses in 1988 and 1992. The National Election Studies 1992 survey provided the national voter data. Ronald B. Rapoport et al., "Activists and Voters in the 1992 Perot Campaign: A Preliminary Report."

By 1996, however, as Table 4.2 shows, Perot's support dropped in almost all groups, dramatically. He was not the viable candidate he had been four years previously. Only independents still gave him a rather sizable vote, 17 percent. Young voters deserted him, as did those with a college education, white Protestants and Catholics, and voters in all regions of the country (for example, 23 percent of the Western voters backed him in 1992, only 8 percent in 1996). His credibility had disappeared, his program did not have a distinctive appeal, the major parties were pushing "centrist" policy options successfully, and thus only 7 percent of the voting public thought he could win (49 percent in 1992 had said they thought he could win!). In 1996, Perot's impact was negligible, while in 1992 he had important consequences, cutting Clinton's victory to 43 percent and significantly affecting Clinton's presidency.

THIRD-PARTY ATTEMPTS BY BLACKS AND OTHER GROUPS

There have been many interesting efforts over the years by special groups to launch new political parties. But these have all failed to control many votes and did not survive for long. In Western European systems, on the other hand, there are, or have been, parties of labor, farmers, fishermen, refugees, regional interests (Scotland, Ireland, Wales, for example), Catholics, Protestants (fundamentalist or liberal), even elderly people's parties! And they usually win seats in parliaments and on city councils. In the 1997 British election, for example, eight smaller parties won seats in Parliament, in addition to the Conservative and Labour parties.

In the United States we have had efforts by such special-group parties, but they never last. We had the Farmer-Labor party in Minnesota (1920–1932) and the American Labor party (1936) in New York. But no nationwide labor party movement ever materialized. In fact, in the nineteenth century Samuel Gompers, head of the American Federation of Labor, "firmly established the doctrine that it was inexpedient for labor to attempt to form an independent labor party." The CIO (Congress of Industrial Organizations) adopted the same doctrine, adding the "nonpartisan doctrine" of rejecting affiliation with either the Democrats or Republicans.[14]

Other groups have created parties that played important roles in generating political interest and raising political consciousness, but they rarely last very long. The farmers had their Greenback party in the 1870s and did elect fourteen to Congress in 1878. The southern Dixiecrats seceded from the Democratic party in 1948 and won thirty-nine electoral college votes. Women were instrumental in organizing the Equal Rights party in 1872. There have been, as a matter of fact, quite a few female presidential candidates for minor parties (see box on Gender and Third Parties).[15] And Latinos created La Raza Unida party, active in recent years in five Southwest states.[16]

The efforts over the years by blacks to launch their own party is particularly important to keep in mind. This struggle has a longer history than most of us realize.[17] It began after the Revolutionary War when many blacks who had won their

Gender and Third Parties

The role of women, as voters and candidates, has become increasingly important in elections. But they have not headed the tickets of the two major parties, although Geraldine Ferraro was Mondale's running mate in 1984. The election in 2000 may change that.

Third parties have been ahead of the Republicans and Democrats in selecting women as presidential candidates. Since 1872 there have been seventeen female presidential candidates, running on minor party tickets. Six of these have been women of color. One of them got over 200,000 votes in 1988 (the New Alliance party).

As for the origins of these third parties, both gender-based and non-gender-based issues have been responsible. The abortion issue led to the formation of the Right-to-Life third party in 1980. The party did not capture many votes in that contest and it ceased to exist. It was revived eight years later, in 1988. But this time the party was headed by a man and a female served as his vice-presidential running mate. In this election, the party captured 20,497 votes.

Women of color have essentially led left-of-center third parties. Of these third parties, the recent (1988) New Alliance party that combined gender issues with matters of economic protest became the first of such parties to get on the ballot in all fifty states. George Wallace's American Independent party had done this in 1968. Only H. Ross Perot in 1992 and 1996 could equal that feat. And because the New Alliance party reached this rare plateau, to date it has outpolled all other third parties led by women. The central issues for parties led by women of color have been economic improvement of the poor, civil rights, peace, and the role of workers. And part of the reason for this emphasis on economic issues is that most black female-headed households fall below the poverty line.

Long before the Democrats and Republicans placed females on their presidential tickets, third parties innovated in this manner. And this occurred long before women were enfranchised.

SOURCE: Hanes Walton, Jr., "Black Female Presidential Candidates: Bass, Mitchell, Chisholm, Wright, Reid, Davis, and Fulani," in Hanes Walton, Jr., ed., *Black Politics and Black Political Behavior* (Connecticut: Praeger Publishers, 1994), 251–73.

freedom were given the right to vote in quite a few states, particularly in New England. The early parties, Federalists and Anti-Federalists, contested for black votes, with the latter usually being more successful. In some southern states (North Carolina and Tennessee, for example) free Negroes voted also. Third parties such as the Liberty party in 1840 and the Free Soil party in 1848 received black support. But by the 1850s this involvement by blacks became minimal. The major parties ignored blacks, in effect disfranchising them. And blacks went through a dismal period of American history of political isolation, lasting into the twentieth century. Despite the Fifteenth Amendment and "black reconstruction" activities, we suffered politically through the "era of white supremacy," both parties ignoring black civil and political rights. The Populist party in the 1880s and 1890s was no help. Even in the early twentieth century, progressives like Theodore Roosevelt were not supportive of black interests and Woodrow Wilson was only slightly interested. It was really not until FDR in 1936 that blacks became involved in party politics because of the social and economic policies of his administration. Yet it took another thirty years before a meaningful civil rights act was adopted, in 1964, primarily by Democratic votes. This was a comprehensive antidiscrimination law. It dramatically transformed the racial policies of the two parties, the Republicans generally critical, the Democrats basically supportive.[18] In a later chapter we will discuss racial politics in greater detail.

The limited interest of the major parties in blacks after the Civil War, indeed their efforts to exclude them from a role in party politics, led black leaders to think of a third-party alternative, or other nonparty alternatives. The creation of black interest groups like the National Association for the Advancement of Colored People (NAACP) in 1910 and later the Congress of Racial Equality and Southern Christian Leadership Conference was part of the answer. But the possibility of a third party was increasingly entertained. Indeed, as Walton puts it, "Fed by frustration and desperation, the belief soon grew that only all-black organizations could make a meaningful change in Negro life."[19] Walton describes at great length the variety of efforts initiated which sought to bring black leaders together, to reach consensus. These were not too successful. In the meantime, two types of third-party efforts were launched: "satellite parties," which were offshoots of Republican or Democratic state party organizations, and "separate parties" unattached to the major parties. An example of the latter was the National Liberty party, created in 1904, which nominated a candidate for the presidency. The United Citizens party organized in 1969 in South Carolina to contest state elections there. An example of "satellite parties" was the Mississippi Freedom Democratic party, a dissident group fighting the "lily white" politics of the majority of Mississippi Democrats. It won recognition in 1964, after an historical confrontation at the national convention, as the legitimate Democratic party of Mississippi. Other "satellite parties" were the "Black and Tan Republicans" in Alabama and the "South Carolina Progressive Democratic party," which were engaged in the same types of conflicts and organizational efforts that occurred in Mississippi. These examples of black third-party efforts suggest that blacks through them have demonstrated "the potential for independence" and "the unique roles which black parties give blacks."[20] They have been more successful at the local and state levels than at the national level, where they still seek today to work effectively, if at all, through the Democratic party.

CONCLUDING OBSERVATIONS

These third parties have been, since the Civil War at least, transitory phenomena. That does not mean they are not important. Most scholars attest to their values as channels for protest and as opportunities for certain types of citizens to be involved who otherwise would not be. Third parties develop new policy agendas, which the major parties often adopt. And new leadership or disaffected old leadership can emerge with the inauguration of a new third party. One senses that there is always the outside possibility that a third party will endure and become a major party, as the Labour party in Britain did after World War I. We have seen this happen at the state level, most recently with the election of Jesse Ventura, a professional wrestler, head of a so-called "Reform party," to the governorship of Minnesota in 1998 — a truly amazing and stunning rejection of both major parties' top leadership in that state!

We sense, however, that our national political culture predisposes us to retain the two party system. Our election system works to that end, although it certainly does not close out the possibility of a third party breaking through that system (as they have in Britain). Special groups such as farmers, labor, blacks, and Catholics have found it difficult to mobilize their members to penetrate the two-party system through a third party. In the nineteenth century one reason this was true was that suffrage was given to all such groups (except women and blacks, who since then have secured voting rights). The factors which might *facilitate* third-party success are usually not present, such as special-group unity and political consciousness, well-developed organizations, and resources. On the other hand, there are many factors which *handicap* third-party development: strong major-party identification, commitment to our system as is, apathy, and creative adaptive strategies by the major parties. There is more to handicap third-party efforts than to facilitate them!

Finally, however, as one reflects on the great amount of third-party effort over the past two centuries, and all the people who have been involved in these efforts, one is more inclined to say that these types of political action are more a creative, dynamic part of our system than a waste and loss. The message should be clear: many citizens have been, and are, and will continue to be, very dissatisfied, and many leaders are also from time to time chafing to change things. There is a real need for third parties from time to time. And change does occur, as the result of third parties. To get out of our two-party straitjacket periodically is very good for us, and for the two large parties to face a serious challenge is also very good for us. It improves the quality of our public life and, hopefully, makes our leaders more responsive.

NOTES

1. Noble Cunningham, Jr., "Who Were the Quids," *Mississippi Valley Historical Review* 2 (September 1963): 252-63.

2. See the excellent review of those parties in Steven J. Rosenstone, Roy L. Behr, and Edward H. Lazarus, *Third Parties in America* (Princeton, NJ: Princeton University Press, 1984 or 1996).

3. Ibid.

4. Ibid., 231-73, for a thorough review of Perot's 1992 campaign.

5. Ronald Rapoport et al., "Activists and Voters in the 1992 Perot Campaign: A Preliminary Report," unpublished paper (paper prepared for the Institution of Behavioral Science, University of Colorado).

6. Roper Center for Public Opinion Research, *America at the Polls 1996* (Storrs, CT: Roper Center, 1997), 80-83.

7. Fred Haynes, *Third Party Movements Since the Civil War* (Iowa City: State Historical Society of Iowa, 1916); Paul H. Douglas, *The Coming of a New Party* (New York: McGraw Hill, 1932); William B.

Hesseltine, *The Rise and Fall of Third Parties* (Princeton, NJ: VanNostrand, 1962); David A. Mazmanian, *Third Parties in Presidential Elections* (Washington, DC: Brookings Institution, 1974); J. David Gillespie, *Politics at the Periphery* (Columbia: University of South Carolina Press, 1993); Rosenstone, Behr, and Lazarus, *Third Parties in America*, 1984, 1996.

8. Rosenstone, Behr, and Lazarus, *Third Parties in America*, 1984, 126–50.
9. Rosenstone, Behr, and Lazarus, *Third Parties in America*, 1996, 126.
10. Ibid., 245.
11. Ibid., 246.
12. Ibid., 271.
13. Ronald B. Rapoport et al., "Activists and Voters in the 1992 Perot Campaign," paper prepared for the Institution of Behavioral Science, University of Colorado.
14. V. O. Key, Jr., *Politics, Parties, and Pressure Groups* (New York: Crowell, 1964), 61–62.
15. Hanes Walton, Jr., "Black Female Presidential Candidates," in *"Black Politics and Black Political Behavior* (Connecticut: Praeger, 1994), 251–74.
16. Rodney Hero, *Latinos and U.S. Political System* (Philadelphia: Temple University Press, 1992), 37–38.
17. Hanes Walton, Jr., *Black Political Parties: An Historical and Political Analysis*, 1972. His chapter 1 is the basis for much of our narrative.
18. Edward G. Carmines and James A. Stinson, *Issue Evolution: Race and the Transformation of American Politics* (Princeton, NJ: Princeton University Press, 1989), 37–58
19. Walton, *Black Political Parties*, 37.
20. Ibid., 204.

CHAPTER 5

The Public's Support for the Parties

In studying parties in a democratic society, there is a need for a clear understanding of the nature of the public's involvement with, and support of, the party system. This includes attitudinal support as well as behavioral support. Citizen attitudes toward the parties probably condition citizen behavior and thus provide part of the context within which parties must function. Citizens' feelings toward parties also constitute a final test of the effectiveness of the party system. If citizens do not want to affiliate with the parties or have only a weak commitment to them, or if citizens do not feel that parties are effective institutions and do not want to participate in them, the task of parties in providing linkages between citizens and government may be extremely difficult. On the other hand, where citizens join parties readily and identify with them, the parties' tasks will be easier. Systems may differ considerably in these respects. Before describing party organizations and what they do, mention should be made of what appears to be the state of the American public's attitudes toward parties.

Anyone studying American politics these days can be alarmed by the negative tone of the attacks. Reading or listening to the reports by the media and noting the results of political polls and surveys, one can come to the conclusion that the majority of citizens are disillusioned about politics. When asked by competent survey organizations in 1996, "How much of the time do you think you can trust the government to do what is right?" only 27 percent responded "most of the time" or "always"; thirty years ago the response was very positive: 73 percent.[1] When people are asked how much confidence they have in American institutions, the pattern in recent years has been increasingly negative. Here are a few examples.[2]

Level of Confidence (in percent)	Congress *1986*	*1996*	Executive Branch of Government *1986*	*1996*	Television *1986*	*1996*
A great deal	16	8	21	10	15	10
Some	61	47	53	56	56	46
Hardly any	20	43	24	42	28	42

In the last decade our national political institutions (and media) have suffered considerably in public confidence. This is partially due to a general increase in cynicism, but it is clear that Congress and the presidency as institutions are particular targets of public concern. Further, the public seems pessimistic about the nation's political future. Over one-half feel that we are in a decline, and less than 10 percent feel that we are improving.[3] On the other hand, it is *not* true that Americans have lost faith in our country's ideals, principles, or the political system generally. The 1997 Roper report claims that only 4 percent "are seriously disaffected."[4] Rather, the performance of politicians and the institutions in which they work are the primary target of the public's negative evaluations.

The important question is whether there is associated with this public discontent for American politics a decline in the public's positive evaluations of parties and the party system. And if there is such a linkage, how great is the decline in confidence about parties, and what groups in the public are most disillusioned. Many scholars previously have spoken of the deterioration of the party system (see Preface). Recently again this type of negative assessment has been made. Everett Ladd declared in 1997: "Political parties . . . just aren't important institutions for most Americans today. . . . [They] are perhaps less relevant now than ever before."[5] This is the type of sweeping statement that is rather common today. It is not new. If one looks back in American history over the past two hundred years, one finds this type of attack made from time to time with even more polemic virulence. George Washington warned of "the baneful effects of the spirit of party." And our second president, John Adams, wrote: "a division of the republic into two great parties . . . is to be dreaded as the greatest political evil under our Constitution."[6] The attack on parties by scholars and the media has been persistent in our history. It is part of the periodic ambivalence, frustration, and disenchantment of the American people with their politics.[7] Politicians have always been fair game. When George Washington stepped down from the presidency in 1797, here is the commentary of the Philadelphia *Aurora*: "This day ought to be a jubilee in the United States . . . for the man who is the source of all the misfortune of our country, is this day reduced to a level with his fellow citizens."[8]

It has become fashionable among students of American politics to belittle political parties. And we must ask, therefore, is it true that "for most Americans" parties are "irrelevant" today? If this sweeping attack on parties truly reflects the attitudes of the American public, there should be hard evidence to support this charge; otherwise it is an irresponsible attack. Is there such evidence? Four specific arguments need to be examined and the evidence pertaining to each evaluated. These arguments are:

1. Have most Americans lost their "affect" for parties; in other words, do they "like" parties less, care less for party politics, feel more negative about parties?
2. Is it true that a majority of Americans no longer want to identify with, belong to, or feel any loyalty toward a political party?
3. Is it true that a majority of Americans no longer find parties useful in providing cues or guidance in taking positions on the issues of politics? Has the mass media completely taken over this function?

4. Is it true that the voting decisions of Americans are now made without reference to party identification, that the parties do not "structure the vote" today as in the past?

These are propositions which define the argument over the "relevance" of parties. Are these propositions empirically true? Attitudinally, motivationally, and behaviorally, are members of the public "turned off" by parties today, more than previously? To this one might add one final question: Insofar as there is a decline in public support for parties in the United States, is that uniquely our experience, or is it part of a "dealignment" that is happening generally in Western democratic systems as a postmodern phenomenon? We shall in the ensuing discussion seek to deal with each of these questions, presenting the available evidence for evaluation.

THE PUBLIC'S GENERAL BELIEF IN PARTIES IS POSITIVE

Before we get into the specifics of these four arguments, the point should be made that Americans are strong supporters of our democratic system. A highly regarded study in 1984 concluded that the "norms relating to democratic values . . . are now more firmly entrenched . . . than they were in the nineteenth century."[9] Another survey of our political culture, in 1996, concluded similarly that there is no real evidence that "the ideals of the American political system" are rejected.[10] And the same *general* support has been found for the party system. When asked, "Do you think having parties is good or bad for the country?" 70 percent replied "good" or "very good."[11] When pressed, people will say we "need parties."[12] Recent attempts to probe what people think of the Democratic and Republican parties reported that on a thermometer rating scale 71 percent rated the parties positively.[13] So a variety of evidence indicates that ***in general terms*** the majority of the public continues to support the idea of party and our two parties.

However, what is disturbing to some is the counterevidence that suggests people want to be independent politically. This emerges from responses to short agree-disagree statements which people are asked to answer, such as:[14]

Statement	*Percent Agree*
"The best rule in voting is to pick the man regardless of his party label."	73
"The parties do more to confuse the issues than to provide a clear choice on them."	56

What do these data, and others like it, suggest? Above all, that there is no blind, automatic, habitual support for a party by many citizens. People insist they are politically sophisticated and will make voting decisions carefully by evaluating candidates' credentials and performance. There is no perceived need to vote a straight party ticket. In 1996, in fact, one study reported that 65 percent of Americans said,

"Typically I split my ticket."[15] In the past, fewer than 30 percent would make such a claim.[16] So Americans have adopted a very independent orientation toward voting, even though their behavior is not all that nonconformist. In 1996 15 percent of the Democrats and 23 percent of the Republicans split their ballots.[17]

While it is clear that in the past the bonds between voters and their parties appear to have been considerably loosened, the evidence of decline is not all that consistent nor monotonic. The bonds of the public to parties have not been shattered, as some would have us believe. There is limited evidence over time to rely on, but what we do have does not suggest the precipitous deterioration of public support for parties. The Rosenstone evidence is very suggestive of the contrary. Here are responses to agree-disagree questions put to samples of the public in 1980 and 1992.[18]

Agree-Disagree Question	*1980*	*1992*
1. Are there important differences in what the Republicans and Democrats stand for? (percent agree)	63%	63%
2. Are you concerned about which party wins the presidential election? (percent agree)	56%	75%
3. Ranking on thermometer scale (0 to 100) of the parties	73%	71%

Thus, while the public is more "independent" in attitude, the evidence of deterioration in the citizen's relationship to the party system is not strong.

THE PUBLIC'S "AFFECT" TOWARD PARTIES

In national surveys for a long time, researchers have asked people what they "like" or "dislike" about the Republican and Democratic parties. This type of question permits the respondent to express the content of his or her *feelings* toward the parties. From these responses we get more than a general expression of support or opposition to the parties. We learn whether people are on balance positive about both parties, negative about both, split positive and negative, indifferent and neutral, or uninformed. Some people are very adversarial about their politics — they hate Democrats and are happy with Republicans, or vice versa. Some are antagonistic — they dislike both parties (a "plague on both parties"). Some people are euphoric — they praise both parties and seem to enjoy the competitive party struggle. Some people are blasé, neutral, or indifferent to both parties and can't say much good or bad about either. Where does the American public stand? How are these four types of "affect" distributed over time in the adult population?

The basic trend since 1952, as Wattenberg pointed out (Table 5.1), is an increase in neutrality toward the parties. There has also been a periodic tendency to negative attitudes toward both parties in 1965–72 and again up to 25 percent in 1996. Adversarial attitudes have declined regularly since 1952, and yet the percentage of voters with positive views of both parties has remained steady.

TABLE 5.1 ▪ Variation since 1952 in the Public's Affect toward the Parties (in percent)

Types of "Affect"	*1952*	*1956*	*1960*	*1964*	*1968*	*1972*	*1976*	*1980*	*1984*	*1988*	*1996*
1. *Euphoric* Only positive, no negative, replies	24	32	33	26	22	19	18	23	19	22	21
2. *Adversarial* Positive comments for one party, negative for the other	50	40	41	38	38	30	31	27	31	34	22
3. *Nihilistic* Only negative, no positive, comments for either party	13	12	9	16	24	21	19	14	14	14	25
4. *Neutral* Neither positive nor negative comments about either party	13	16	17	20	17	30	31	37	36	30	32

SOURCE: Martin P. Wattenberg, *The Decline of American Political Parties 1952–1980* (Cambridge: Harvard University Press, 1984), 61; also, Wattenberg, "From a Partisan to a Candidate-centered Electorate," in Anthony King, ed., *The New American Political System* (American Enterprise Institute, 1990), 150. Based on National Election Studies data. One should emphasize that these calculations of "affect" are based only on those who responded to the likes-dislikes questions *and* who gave responses which are consistent and fit the above categories. In 1996, for example, over one-half of the sample did not provide meaningful and useful responses for these purposes.

Here are data that one can interpret in different ways, depending on one's predispositions. One could become concerned because the public is less adversarial (a decline from 50 to 22 percent), or one could feel fairly sanguine about that finding. One could be satisfied that the number of those with positive comments about the parties has remained consistently between one-fourth and one-fifth since 1952, or one can be concerned that the percentage is so low. Again, we can note the low proportion with only negative comments and say that is a good sign. Or one can view with concern the high proportion that is neutral (or uninformed), an increase from 13 to more than 32 percent. But whatever one does with these data, it is doubtful that one can conclude that they definitively prove that parties have "declined", or are "irrelevant" or that party images have "dissipated" over time. What does seem to emerge are these conclusions: (1) the public is somewhat more negative toward parties today than in 1952; (2) the public is more neutral, increasing in recent years to one-third of the electorate; (3) on balance the partisan images of the voters today as previously have been more positive or neutral than negative. Above

The Differential Electoral Effects of Two Presidential Scandals

Presidential scandals can have a significant impact on election results, but the impact depends a great deal on the public's perception of the scandal and the president's performance during the period of the scandal. During the Watergate scandal crisis resulting in President Nixon's resignation, the Republican party lost heavily in congressional seats. The scandal involving President Clinton appears to have had a strikingly different effect, as follows:

Number of House Seats Won by the Parties

	Nixon Period			*Clinton Period*		
Party	*1972*	*1974*	*1976*	*1994*	*1996*	*1998*
Republicans	192	144	143	230	227	222
Democrats	239	291	292	204	207	212

The public apparently reacted quite differently in the nineties compared to the seventies. We will have to wait until the election year 2000 to see what the eventual "fallout" of the 1999 Clinton scandal will be.

all, these data suggest that the public's "affect" toward parties fluctuates by election. For example, negative comments decline in the 1960 Kennedy-Nixon campaign, but increase again in the 1964 Goldwater-Johnson campaign, and really increase in 1968, 1972, and 1976 (Vietnam, McGovern, Watergate). The trend toward more neutrality began in 1972 and has seemed to hold since that time.

There has been a great deal of research also on the public's view as to which party can better deal with the problems confronting us. And this varies, of course, by issue. Thus the Democrats in 1996 were perceived to have the advantage on issues such as the environment, health care, medicine, education, and abortion. The Republicans in 1996 were given the advantage on issues such as spending control, promotion of moral values, foreign policy, crime, and taxes. In studying these responses, it is interesting to find that usually from 70 percent to 85 percent of the respondents *do* select one or the other of the two parties as most likely to deal with the problem. And when asked directly, "Tell me whether you have a favorable or unfavorable opinion of each of the following parties," in the latest survey reported before the 1996 election, 55 percent had a favorable opinion of the Republican party and 55 percent had a favorable opinion of the Democratic party.[19]

On balance, then, on party affect we find no increased negativism, more increased indifference or neutrality, and less of a confrontational image than previously. While, as our previous analysis suggests, many citizens are not bonded to parties as strongly as before, there is still much positive affect, much willingness to acknowledge party difference, and much favorable opinion of parties. Citizens have a self-image of political independence and sophistication, but they still can think more positively than negatively about our parties.

THE STRENGTH OF THE PUBLIC'S IDENTIFICATION WITH PARTIES

In our surveys for many years now, we have asked the famous "party identification question": "Generally speaking, do you think of yourself as a Republican, a Democrat, an Independent, or what?" This was followed with the question, for those who classified themselves as Republicans or Democrats, "Would you call yourself a strong (Republican, Democrat) or a not very strong (Republican, Democrat)?" Those who called themselves Independents were asked, "Do you think of yourself as closer to the Republican or Democratic party?" Note the words: "think of yourself," "call yourself," "closer to." One could, of course, substitute other questions to discover a person's party attachment, but this wording has been accepted and has been used a long time, thus permitting an analysis in depth and over time. This approach to determining a person's partisanship, conceived in 1952, has resulted in forty-five years of voluminous data. It has been argued that responses to this series of questions provide an insight into a person's psychological attachment to a political party. From these questions, a sevenfold set of categories was derived, revealing the distributions found in Table 5.2. Over the span of years since 1952, one can see the degree to which Americans "identify" with the parties or insist on being "independents."

From the table, one can see how amazingly similar the party distributions are today compared to forty-four years ago.

There is a net decline of Democrats (for all three categories) of 5 percent and a net increase of Republicans of 4 percent. In one sense, the Independents have increased (using all three categories) 11 percent, but for "pure" Independents (who do not admit to "leaning") it is only 4 percent. Whereas there were 35 percent with

TABLE 5.2 ▪ An Overview of Party Identification, 1952–1996 (in percent)

Ideology	*1952*		*1972*		*1996*		*Change 1952–1996*
Strong Democrat	22		15		19		-3
Weak Democrat	25		25		20		-5
Independent-leaning Democrat	10		11		13		+3
Independent	5	22	13	35	9	33	+4
Independent-leaning Republican	7		11		11		+4
Weak Republican	14		13		15		+1
Strong Republican	13		10		12		-1
Apolitical/Don't Know	4		2		1		-3

SOURCE: National Election Studies, University of Michigan.

strong party identification in 1952, there are 31 percent today. There is no reason to claim that such data reveal a staggering decline in party identification, as some have claimed. This is a system in fairly stable party identification equilibrium. (See Figures 5.1 and 5.2.)

FIGURE 5.1 ▪ Fluctuations in Strong Party Identification, 1952–1996 (in percent)

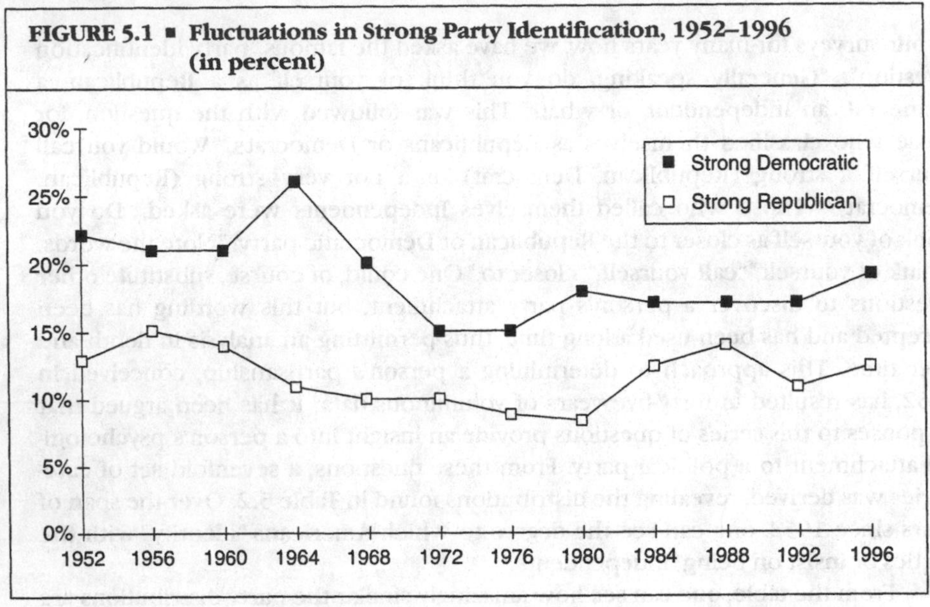

FIGURE 5.2 ▪ Fluctuations in Party Independence, 1952–1996 (in percent)

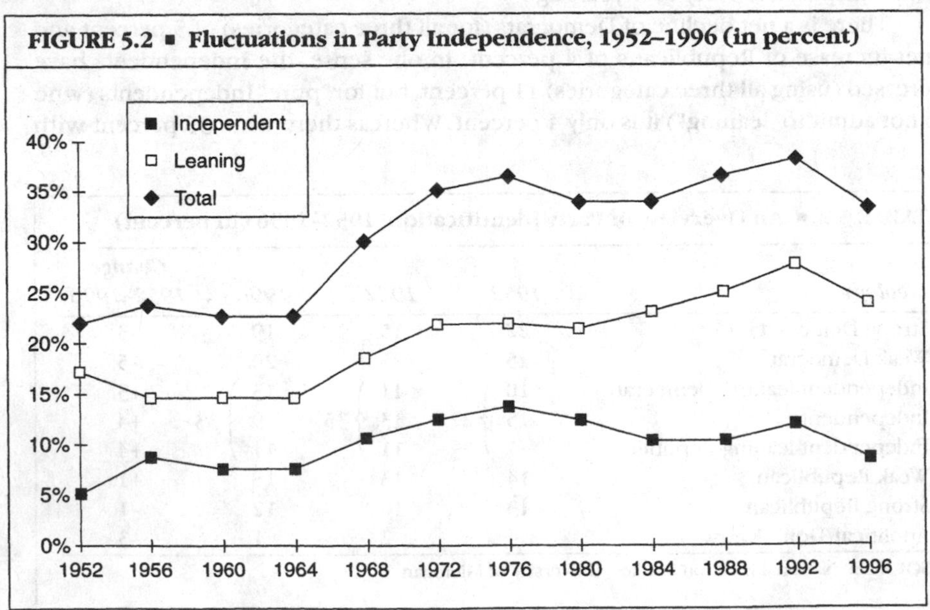

TABLE 5.3 ▪ Party Identification for Each Presidential Election Year, 1952–1996 (in percent)

Party Identification	*1952*	*1956*	*1960*	*1964*	*1968*	*1972*	*1976*	*1980*	*1984*	*1988*	*1992*	*1996*
Strong Democrat	22	21	21	26	20	15	15	18	17	17	18	19
Weak Democrat	25	23	25	25	25	25	25	23	20	18	17	20
Independent Democrat	10	7	8	9	10	11	12	11	11	12	14	13
Independent	5	9	8	8	11	13	14	13	11	11	12	9
Independent Republican	7	8	7	6	9	11	10	10	12	13	12	11
Weak Republican	14	14	13	13	14	13	14	14	15	14	14	15
Strong Republican	13	15	14	11	10	10	9	9	12	14	11	12
Apolitical/ Don't Know	4	3	4	2	1	2	1	2	3	2	1	1

SOURCE: National Election Studies.

Admittedly, if we look at the full range of data by year, we do see interesting fluctuations (Table 5.3). A decline occurred in the 1968–1976 period when an average of only 26 percent were "strong" (compared to an average of 36 percent in the preceding four elections). The George Wallace third-party challenge, Watergate, and the divisiveness in the Democratic party over the race issue and over reform seem associated with this decline. The increase of "independents" (pure or leaning) to 30 percent in 1968 has been maintained until today. But in the eighties, a resurgence of identification began, resulting in the strengthening of party affiliation. Contrary to some reports, there has *not* been a steady decline, but instead fluctuations due to the vicissitudes of politics. It is true that more people want to be known as "independents," although the majority of these "lean" toward, or admit that they prefer, one or the other major party. Some scholars, after careful study of this phenomenon, refer to the "myth of the independent voter."[20] Yet the number of "pure" independents has increased from 5 percent in 1952 to as high as 14 percent in 1976. However, this trend is also on the downward slope and in 1996 was only 9 percent. It is interesting that the periodic decline in party identification in the United States in the late 1960s and 1970s was paralleled to some extent by the same development in Europe, linked as in the United States to "political events and crises" as well as to "broader patterns of social and economic change."[21]

Party identification change is thus a cross-national phenomenon. But change appears episodic and not necessarily continuous. In fact, in the United States, affiliation with parties is today somewhat restabilizing. Yet there is no question that if one looks at *strong* party identification and *pure* independents, party ties continue to be weaker than fifty years ago, but not as weak as twenty-five years ago.

VARIATIONS BY GROUPS IN POLITICAL INDEPENDENCE OVER TIME

The assumption that party loyalty has declined uniformly in the public and also continuously over the years, and that the increase in self-declared independence from partisanship is also uniform, is not borne out by the data. If we look at the party identification patterns of different groups, we can see how the patterns differ.[22] Jews, for example, were early independents, for already in the 1950s and 1960s many were not willing to commit to a party (47 percent independent by 1960). Yet in recent years there has been a decline in Jewish self-declared independence. The Catholics were much later in their withdrawal of commitment to either major party (only 20 percent independents in 1960) and today are relatively quite independent (32 percent in 1996). Also the variations in party identification by age groups is striking (Figure 5.3). The youngest age group already in 1956 had increased in independence to 37 percent and continued in this trend of with-

FIGURE 5.3 ■ Differential Patterns of Independence from Party Affiliation by Age Group

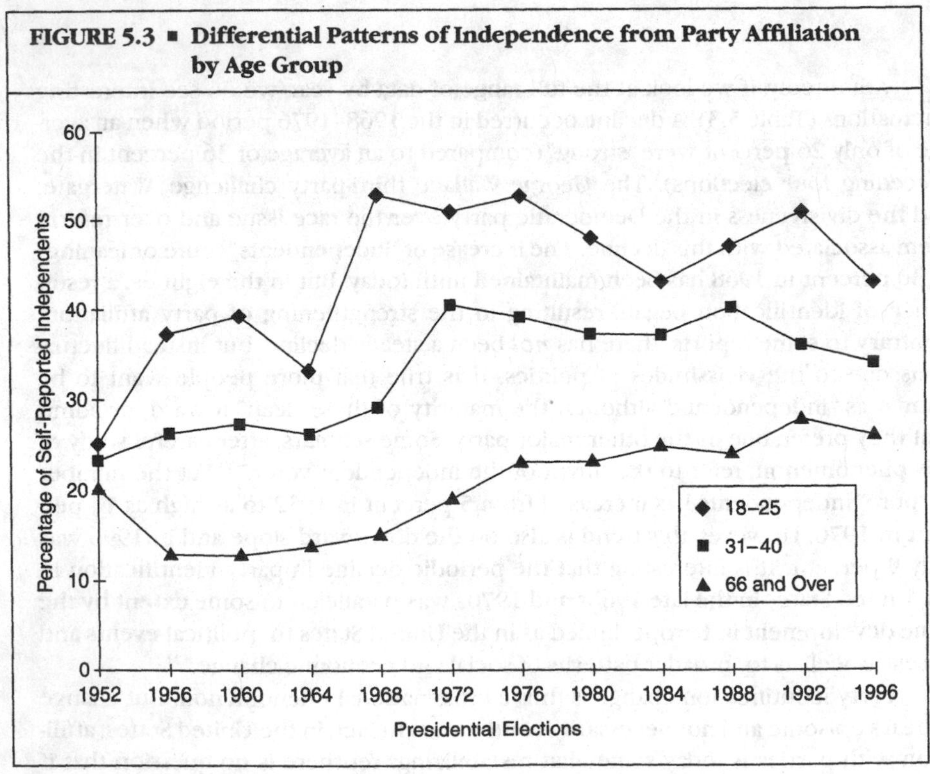

TABLE 5.4 ▪ Extent of Independence from Partisanship by Age Groups, 1952–1996 (in percent)

Age Group	*1952*	*1956*	*1960*	*1964*	*1968*	*1972*	*1976*	*1980*	*1984*	*1988*	*1992*	*1996*	*Change 1952-96*
18-25*	25	37	39	33	53	51	53	48	43	47	51	43	18
26-30	32	31	26	29	41	50	42	41	40	42	46	44	12
31-40	23	26	27	26	29	40	39	37	37	40	36	34	11
41-50	26	21	25	24	31	30	37	31	31	36	42	34	8
51-65	19	24	21	18	24	26	32	29	29	29	35	31	12
66 and Over	20	13	13	14	15	19	23	23	25	24	28	26	6

* For 1952-1972, age group is 21-25.

SOURCE: Eldersveld, *Political Parties in American Society*, 1982; National Election Studies.

drawal of commitment to the major parties (except for 1964) to the present day, leveling off at the 50 percent level from 1968 on (Table 5.4). In the meantime, the oldest age cohort (age 66+) reported a very low level of independence for some time (actually decreasing 1956-1968 and only in 1992 increasing to 28 percent).

Thus, if one distinguishes party identification by social groups in the electorate, it is clear that it is dangerous to generalize across the public. Party identification decline and party independence increase were not phenomena that were experienced by all social groups, and they did not occur to the same extent nor at the same time.

THE SPECIAL CASE OF AFRICAN AMERICAN PARTY LOYALTIES

One of the most radical cases in the transformation of party support is that of African Americans between Eisenhower's reelection in 1956 and the Kennedy-Johnson era of 1960 to 1968. This was the period of the election of many new Democratic liberals to Congress in 1958, the movement within the Democratic party to take a strong stand on civil rights (culminating in the battle at the Philadelphia national Democratic convention for a strong civil rights plank in the platform and the walkout of certain southern Democrats from the convention). It was the time of the push during Kennedy's and Johnson's presidencies for the strong Civil Rights Act of 1964 (followed by the Voting Rights Act of 1965). It was the period of the Goldwater candidacy in 1964, which repudiated the need for such civil rights legislation and aligned the Republicans with the conservative South. Throughout these ten years the parties changed sides on civil rights, the Democrats becoming the new champions of civil rights. As a result, a major "transformation of American politics" took place. What happened to the African American party orientations during this period is summarized in Tables 5.5 and 5.6.

TABLE 5.5 ▪ African American Patterns of Party Identification and Independence (in percent)

Party Identification	*1952*	*1960*	*1968*	*1972*	*1980*	*1992*	*1996*	*Change 1952-96*
Strong Identifiers								
(Democratic and Republican)								
African American	35	31	57	40	49	44	45	+10
Total electorate	35	35	34	25	27	29	31	-4
Independents								
(3 categories)								
African American	18	27	11	23	19	28	30	+12
Total electorate	22	23	30	35	34	38	33	+11
Democratic Identifiers								
(strong to weak)								
African Americans	52	43	85	67	72	64	66	+14
Total electorate	47	46	45	40	41	35	39	-6
Republican Identifiers								
(strong to weak)	13	16	2	8	5	5	3	-10
African American								
Total electorate	27	27	24	23	23	25	27	-0

SOURCE: National Election Studies.

TABLE 5.6 ▪ Critical Period of Transformation of African American Party Loyalties, 1956–1968 (in percent)

	1956	*1958*	*1960*	*1962*	*1964*	*1966*	*1968*
A. *Democratic Identifiers*							
(strong or weak)	50	51	43	58	72	59	85
Republican Identifiers							
(strong or weak)	19	18	16	13	7	9	2
B. Turnout:							
Democrats	42		65		70		71
Republican	36		61		73*		50*
Independents	60*		63		56*		50*
C. Presidential Vote							
Democratic identifiers voting Democratic	35		61		70		68
Republican identifiers voting Republican	33*		35		0		25*

* Few cases for analysis.

SOURCE: National Election Studies.

In the past forty years African Americans have reversed the tide of decline in party identification by increasing their commitment to the Democrats (a 14 percent gain since 1952). But what is particularly important to notice is the sharp increase which occurred 1960-1968 in Democratic identification (42 percent) and the simultaneous drop in Republican identification (from 16 percent to 2 percent). During "the critical period" of transformation, 1958-1968, not only did party identification (Democratic and total strong identification) increase, but so did voting turnout (Table 5.6). Yet the 66 percent Democratic identification in 1996 is a considerable decline from 85 percent in 1968. African Americans today also want to be considered "independents" more than was the case in 1980 (from 19 percent to 30 percent). Here is a special case, then, of both irregularity in patterns and sustained increase in party loyalty. This development again requires the generalization about persistent party decline to be severely questioned and qualified.

THE LINK OF PARTY IDENTIFICATION TO THE PUBLIC'S ISSUE POSITIONS

People are much more preoccupied with issues today. When they are asked to discuss what they like or dislike about parties, they place more emphasis on issues than on anything else. In the past, particularly in the 1950s and early 1960s, they talked more about "the candidate." But issues now rank highest in saliency — an increase from 49 percent in 1952 to 78 percent by 1980, and even more so in recent elections.[23] The question is, What does party identification have to do with *the way* the public thinks about the issues of the day? The answer is that party identification seems to have a considerable linkage, today as in recent years. First, if a person identifies with a party, he or she is more likely to see differences between the parties on issues than if the person calls himself/herself an "independent" (Table 5.7). Most independents think the parties are the same on issues, and that is part of the reason they want to insist that they are attached to no party. But the majority of citizens are partisans who consistently say that the parties are distinctive on the issues.

TABLE 5.7 ▪ Perceived Party Differences by Party Identification

"Are there important differences between Republicans and Democrats?" (percent who say "yes")

Party Identification	*1972*	*1988*	*1996*
Democrats	46	62	65
Pure Independents	30	28	47
Republicans	48	66	73

SOURCE: National Election Studies. See also Bruce Keith et al., *The Myth of the Independent Voter* (Berkeley: University of California Press, 1992), 148.

Second, we find that Democratic identifiers differ from Republican identifiers in ideology and in their positions on specific issues. When asked to classify themselves as "liberals," "moderates," or "conservatives" in ideology, the self-perceptions are quite a contrast, particularly in recent years. Here is a brief summary:

Ideological Self-Classification by Identifiers (in percent)

	Democrats			Republicans		
Ideology	*1980*	*1992*	*1996*	*1980*	*1992*	*1996*
Liberals	38	44	47	10	10	4
Moderates	34	33	32	23	27	21
Conservatives	28	24	21	67	64	75

Notice the ideology polarization by parties! On specific issues the differences also are fairly consistent. Table 5.8 provides some examples. These are not "all-or-none" differences. There are obviously considerable differences of opinion *within* parties. But on balance, the liberal-conservative difference about the role of the government in the society emerges.

There seems to be no doubt, therefore, that having a party identification means for many taking on opinions about public policy that are partisan. Many people do seem to get their cues from parties as to what their position, as a partisan Democrat or as a partisan Republican, should be. Party identification seems, then, to be functionally linked to issue positions and ideology.

THE LINK OF PARTY IDENTIFICATION TO VOTING

It has long been accepted that a high proportion of American voters are partisans with habitual voting preferences. Before the modern period of the Michigan national surveys, scholars like V. O. Key were speculating that "consistency of voter preferences" meant, by his estimate, that 75 to 85 percent of the voters voted for the same party in consecutive presidential elections.[24] The early study, *The American Voter,* confirmed this estimate and detailed at great length the determinative influence of party identification on the vote, as well as the many other consequences of partisanship.[25] In fact, "party identification" was considered a primary

TABLE 5.8 ▪ Differences in Issue Positions for Democratic and Republican Identifiers, 1992 (percent for)

Issue	*Democrats*	*Republicans*	*Difference*
1. Increase government services?	48	23	25
2. Increase federal spending on social security?	57	36	21
3. Need more progress to assist blacks?	35	14	21
4. Support the government health plan	64	34	30
5. Abortion available by law?	53	39	14

SOURCE: National Election Studies.

factor in explaining the vote in the 1950s. Since that time there has been more skepticism by scholars about the importance of party identification's relevance for the vote. An example is Fiorina's observation that "identification with the parties has declined in importance as a variable that influences — indeed, that at one time *structured* — the voting decision."[26]

Many other writers have followed suit or preceded this observation with their own negations. Yet when we review the evidence, the link between "party identification" and the vote appears strong. As Warren Miller says, "the big question is the stability (and the meaningfulness) of one's self-identification as a Democrat, Republican, or something else." The correlations of party identification to the vote reveal a "remarkable demonstration of stability." And the linkage has actually continued through the 1980s, although declining somewhat in the 1990s (Table 5.9).

With Perot on the ballot in the 1992 and 1996 elections, one might well expect a decline in the size of the correlation. And this was the finding for 1996 — a partial correlation of .58. However, considering that we have controlled for the effects of six other key factors, these are still robust correlations. Miller concluded in 1991: "there has been no across-time decrease in the extent to which the national presidential vote is a party vote.[27] There was a decline after 1991, it is true (a partial correlation of .58), but the linkage of party identification to the vote is still impressive.

Other studies have confirmed the same finding. The "party line defectors" (those who declare an identification for a party and then vote for the other party's candidate) have declined from 25 percent in 1972 and 24 percent in 1980 to 13 percent in 1984 and 12 percent in 1988.[28] Recent studies of the 1996 elections discover that 84 percent of the self-declared Democrats voted for Clinton and 80 percent of the self-declared Republicans voted for Dole.[29]

Now we all know that the American electorate is not tightly bound for all time to party affiliations. It is estimated that 22 percent of Americans do change parties during their lifetime, and about one-fifth "have left the political faith of their fathers."[30] The electorate is not in a straitjacket. Not only do 30 percent to 35 percent wish to be considered "independents," but even strong party identifiers will stray from the party fold. It is then perhaps all the more remarkable that Miller's correlations (see Table 5.9) stand up today, and that in an election like that in 1996, 80 percent of the self-declared partisans vote as predicted. Particular elections, such as those in 1964 and 1972, may cause a higher rate of defections, and it is true

TABLE 5.9 ▪ Correlations of Party Identification and the Presidential Vote

Correlation	*1952*	*1964*	*1972*	*1976*	*1980*	*1984*	*1988*	*1996*
Bivariate	.69	.57	.52	.69	.73	.74	.75	.54
Partials	.64	.48	.43	.64	.63	.65	.68	.58

Note: The partial correlations control for race, education, gender, religion, income, and union membership.

SOURCE: Warren Miller, "Party Identification, Realignment, and Party Voting: Back to the Basics," *American Political Review* 85 (June 1991): 565. National Election Studies, University of Michigan.

that Perot's candidacy in 1992 saw a decline in party voting for both Democrats and Republicans (77 percent of Democrats were loyal, 73 percent of the Republicans). But in 1996 both sets of partisans had returned to the high level of commitment to their own presidential candidates. Here are the 1996 results on loyalty and switching in the vote by partisans:[31]

Congressional Vote	Presidential Vote (in percent)		
in 1996	*Democrat*	*Republican*	*Perot*
Democratic	84	8	7
Republican	15	76	8

One concern scholars have is the increasing propensity for voters to split their ballots (supporting a presidential candidate of one party and a congressional candidate of the other party). In the past, the classic partisan allegedly voted a "straight ticket" from president down to county sheriff and was proud of this loyalty. Although certainly there is still much of this, many more citizens rebel at such automatic voting loyalty. Research has revealed a rise in the proportion of congressional districts that were split — won by a presidential candidate of one party and a congressional candidate of another party. The proportion fluctuates by years, but increased from 3.2 percent in 1920 to 44.4 percent in 1972, dropping to 28.5 percent in 1976.[32] In particular elections, we are aware of considerable defections in the presidential vote. A majority (58 percent) of the Democrats in 1952 split their ballots for Eisenhower. The same is true for 40 percent of the Republicans who defected to vote for Johnson in 1964. And 60 percent of the Democrats deserted McGovern in 1972. Kennedy won in 1960 by getting 30 percent of his votes from Republicans.[33] On the other hand, identifiers switching their vote to congressional candidates of the other party is even more common.

The important question for us here is, To what extent is this pattern of defection in voting linked to strength of party identification? We summarize these data over time in Table 5.10. Three observations stand out in these data: (1) vote defection by identifiers has been happening for a long time, back to 1952 and certainly before; it is not just a recent phenomenon; (2) such crossover voting for the other party is particularly associated with the type of candidate put up by the parties; (3) "strong" identifiers are much less likely to defect than weak or independent identifiers; only an average of 6 percent of strong partisans deserted their own presidential candidates in the last three elections.

Thus party identification still means something, still has a "braking effect" on inclinations to defect, even among those who call themselves "independents" but admit to being closer to one of the parties. Split-ticket voting and defection behavior are further manifestations of the non-rigidity of party affiliations in the United States. In many senses there are many "floating voters" out there, and as a result, our politics has the potential for dynamic change. The parties have to be aware of that and constantly adapt to the "independence" of the American public — a healthy state of affairs, in actuality.

What is the record of voting participation for party identifiers? Actually the record is relatively good for those who are *strong* identifiers. One must remember

TABLE 5.10 ■ Strength of Party Identification and Vote Defections, Voters Only (percentage of identifiers not voting for their own party)

Party Identification	*1952*	*1960*	*1964*	*1972*	*1980*	*1988*	*1992*	*1996*
Strong Democrats								
Presidential vote defectors	16	10	5	27	14	7	7	4
Congressional vote defectors	11	7	6	9	15	12	14	13
Weak Democrats								
Presidential vote defectors	38	28	18	52	40	30	31	18
Congressional vote defectors	23	14	16	20	31	20	18	30
Independent Democrats								
Presidential vote defectors	40	12	10	40	55	12	29	24
Congressional vote defectors	36	16	21	20	32	14	26	31
Independent Republicans								
Presidential vote defectors	7	13	25	14	24	16	38	32
Congressional vote defectors	19	26	28	27	32	36	35	22
Weak Republicans								
Presidential vote defectors	6	13	44	10	14	17	40	30
Congressional vote defectors	10	16	36	25	26	30	37	22
Strong Republicans								
Presidential vote defectors	2	2	10	3	8	2	13	6
Congressional vote defectors	5	10	8	15	23	23	18	5

SOURCE: National Election Studies, University of Michigan. Percentages exclude nonvoters; also, see Beck, *Party Politics in America* (New York: Longman, 1997), 157-8.

that from 1952 to the present, the percent of eligible voters who cast ballots in presidential elections fluctuated greatly: 53.5 percent in 1952, 64 percent in 1960, then a steady decline with slight variations, to 48.5 percent in 1996 (the lowest since the 1920s). When we ask a sample of adults whether or not they voted, there is always "overreporting." Allowing for that, however, we find that strong identifiers have, by their own report, the best record of turnout, and independents the worst. On the average, weak identifiers have a 12 percentage point lower turnout than strong identifiers in both parties, since 1952. But note: there is really no major, continuous trend of decline in turnout by strong partisans. The decline is really more observable for "pure independents" (see Table 5.11).

TABLE 5.11 ■ Turnout Rates of Identifiers and Independents in Presidential Elections (in percent)

Party Identification	*1952*	*1960*	*1980*	*1992*	*1996*
Strong Partisans					
Democrats	75	85	84	85	77
Republicans	93	89	88	88	89
Weak Partisans					
Democrats	69	79	65	73	62
Republicans	77	88	77	77	70
Independents					
Democrats	74	72	69	73	58
Republicans	78	87	75	72	66
"Pure" Independents	74	77	54	60	44

SOURCE: National Election Studies. For the earlier years, see Herbert Asher, *Presidential Elections and American Politics* (Homewood, IL: Dorsey, 1976), 80.

PUBLIC SUPPORT FOR REALIGNMENT IN THE SOUTH

Up to this point, we have analyzed and explored public support for political parties at the individual and collective levels. There is also a regional basis for public support. The South — defined here as the eleven states of the old Confederacy — has not only been a distinct region in the nation since its inception due to its culture of slavery and segregation and ideology of white supremacy. This region more than any other has exhibited a uniqueness in its party behavior throughout its history. Hence any discussion of the public support of parties would not be complete without an assessment of public support in this region of the country.

V. O. Key, Jr., in his article, "The Future of the Democratic Party" in 1952, made a major forecast about public support in this region when he raised and attempted to answer the question: "What about the South? How should it be weighted in a set of calculations on the long-run balance of power between the parties?" His answer is indeed interesting. He asserted: "The Democratic party enjoys the great advantage of the ingrained loyalty of most Southern voters . . . (yet) . . . conditions are conceivable under which a part of the South might be detached from the Democratic party at least temporarily."[34] At this point, Key asked the question "why"? Then he offered an answer that appraised both parties. He wrote: "In truth, the imagination demonstrated by the national Democratic leadership in coping with the South and the race question scarcely impresses." And as for the Republicans, he averred: "Republican improvisation in the near future of a policy appeal powerful enough to knock Democrats from control for more than one or two terms at a time seems unlikely, if one assigns considerable strength to the tempo and direction of Republican policy of recent years."[35] Thus in 1952, when the Republican party, with General Eisenhower as its standard bearer, won the presidency, the preeminent southern scholar, Key, predicted that the victorious party could only win in this region at best one or two elections, given the ingrained nature of Democratic partisanship there.

Philip Converse attempted his own forecast, nearly a decade later. He observed that "the convergence process will lead to a more vital Republican Party in larger urban areas, and a reshaping of the southern political map quite generally with a new kind of political differentiation setting city apart from exclusive suburb and both apart from rural counties fifty miles distant, in the manner which has become familiar in the north."[36] For Converse, a political realignment in the region was coming with public support that was changing its moorings.

Another scholar of southern politics, Samuel DuBois Cook, made a unique forecast, in the same time frame. "Events of the fifties threatened traditional southern values and hence mothered several movements aimed at the preservation of the traditional framework. The forces of conservatism and reaction took a variety of forms: massive resistance . . . (and) . . . overnight conversion to Republicanism as a method of protest."[37] According to Cook's forecast, the South would drop its Democratic partisanship and realign with the Republican party if that party would embrace race and use it as a wedge issue in party competition. Here, we have three different and unique forecasts about the potentials for changing public support of parties in the region.

But it was not just a simple matter of scholarly forecasting about altering the bases of public support for parties in the South. Party strategists got into the game. After the 1968 election, Nixon aide Kevin Phillips wrote *The Emerging Republican Majority,* in which he predicted a Republican electoral realignment based on the election outcomes of 1968. Here is how Phillips worked out his strategy for the Republican capture of public support in the Democratic South:

> Summing the Nixon and Wallace totals in 1968, Phillips predicted that the region would become predominantly Republican as the Wallace constituency of traditional southern whites eventually aligned themselves with the GOP as the only viable alternative to a Democratic party increasingly dominated by "new politics liberalism" on economic, racial, cultural and foreign policy issues. . . . Events in the 1972 presidential election worked out almost exactly as Phillips had predicted.[38]

In fact, numerous party scholars, pundits and political commentators have noted that the geopolitical strategy of Phillips "which argued for the eventual merging of the Nixon and Wallace constituencies after the 1968 election, served as the electoral handbook for the Republican Right during the 1970's."[39] However, this "southern strategy" of the Republican party was developed in 1964 by their candidate, Barry Goldwater, and after that election, it was retooled, refined, and polished by Phillips.[40] And his advice led to the party's penetration of the region and new victories everywhere in the region. The party was reborn in the South via new sources of public support.

By 1988, after seven major Republican presidential victories (Eisenhower, Nixon, Reagan, and Bush), another southerner, Charles Bullock, revisited the southern electorate and declared that there was a "creeping realignment" taking place in the region.[41] Slowly and gradually, Republican partisanship was displacing the traditional Democratic partisanship. Historian Nicol Rae commented on this fate that had befallen the Democratic party just before the 1992 elections: "With the rise of ideological politics, the Democratic party lost 'middle America'— the lower-middle and working class, white, northern ethnics and white southerners."[42] Having made this observation, he sought to explain why and the answer for him was simple: "The civil rights revolution isolated the traditional Democratic South . . . and assisted the creation of a viable southern Republican party in which the most conservative elements of the old Democratic South found a comfortable home."[43] Elements of the public in the region had now shifted their support.

However, in 1992 a southern native son at the head of the Democratic ticket won the White House, and this gave some of the analysts reason to pause. This native son was reelected in 1996 and his victory caused one observer to comment: "although the South has won back the national Democratic party, the Democrats are gradually becoming weaker in the South."[44] In the 1994 midterm elections, the Republican party captured the majority of the region's congressional seats, its gubernatorial positions, and state legislative posts. Table 5.12 permits one to compare across time the number of seats held by Republicans in southern state legislatures in 1939 to what it was after the 1994 realignment election. Republican ability to alter public support in their favor has been significant.

Beyond this changing of public support at the state level, there is the resurgence of the party at the congressional and gubernatorial levels. Table 5.13 tells us that in U.S. Senate contests the party was only winning 20 percent of the time in 1962, and by the 1990s that had risen to 59 percent of the time. In electoral contests for the House of Representatives, the party moved from capturing 23 percent

TABLE 5.12 ▪ Percentage of Republican Seats Held in the State Legislatures of the Eleven Southern States, 1939–1994

	1939	*1966–1971*	*1988–1989*	*1994*
State House	3	11	23	30
State Senate	2	8	20	31

SOURCE: Richard Scher, *Politics in the New South: Republicanism, Race and Leadership in the Twentieth Century*, 2nd ed. (New York: M. E. Sharpe, 1997), 142, Tables 5A and 5B.

TABLE 5.13 ▪ Percentage of Republican Victories in U.S. Senate, House, and State Gubernatorial Races in the South

Race	*1962–1978*	*1980–1988*	*1990–1994*
U.S. Senate	20	35	59
U.S. House of Representatives	23	34	40
Gubernatorial Race	19	30	47

SOURCE: Adapted from Richard Scher, *Politics in the New South: Republicanism, Race and Leadership in the Twentieth Century*, 2nd ed. (New York: M. E. Sharpe, 1997), 133–35.

of the seats to 40 percent. And in the gubernatorial contests, it jumped from 19 percent to 47 percent. Public support continued to shift toward the Republicans.

At the presidential level, changing public support for the party won the region for the Republican party in 1964, 1968, 1972, and 1980–1988. And with the 1994 election, a southern Republican became Speaker of the House and majority leader in the Senate. This major sea change in the American political process caused Bullock, who saw only a "creeping" realignment a decade earlier, to declare a sweeping, swift, and nearly complete realignment of the region just prior to the 1998 midterm elections.[45] This new forecast about the region may have been more than premature because in the 1998 elections, African American voters turn out in record numbers in Alabama, South Carolina, and Georgia and put the Democrats in those states back in power.

Thus, for those scholars who were forecasting and describing the changing nature of public support for parties in the South, it can now be said and shown that "while there is considerable variation across the South . . . in no state does the one-party system continue to exist. Republicans appear on the ballot, are competitive, and frequently win major elections.[46] And while this is true, whether a true two-party system has developed in the South is much more problematic."[47] Surely, as can readily be seen, the Republican National Committee, its activists, and racially conservative presidential candidates have altered the nature and scope of public support for the parties at a regional level. However, just how far a race-based realignment of public support can go remains to be seen.

CONCLUSIONS

Returning to the national picture of support for the parties, certain conclusions seem clear. The gloomy and alarmist critique of the state of the public's commitment to party in the United States by some writers seems misconceived and very exaggerated. The charges that parties are less relevant for citizens and no longer are functional and important institutions for the public do *not* seem to reflect a careful examination of the empirical evidence. True, in the United States as in other western democracies, citizens are not as closely bonded to one party as in the past. And the public's concern about their society and economy are manifest also in their concerns about politics. Strong partisanship has declined somewhat, but still today as previously, 30 percent to 35 percent of the electorate are "hard core" partisans, and another 30 percent are weak partisans, most of whom stay with their parties on election day (less than 25 percent normally defect). That is a large body of loyal followers! More people want to be known as "independents" today — 33 percent in 1996 compared to 22 percent 40 years ago. But many of these people are really fairly loyal partisans ("closet partisans," some would say). One real worry is that young people ages 18 to 25 are most likely to want to be "independents," and they do tend to support third parties, such as Perot's in 1992. But by 1996 they were back in the party fold — only 10 percent supported Perot. Yet their "floating voter" status has implications for the support structures of parties and the way parties have to develop strategies to win the young vote.

The major empirically based findings which stand out are these. First, voters are no less positive about parties today than previously, although they are more neutral and less adversarial. Second, groups vary greatly in their party identification patterns; there is not one pattern of decline for all groups. Third, there are periodic fluctuations in the strength of party identification — low in the 1968-1976 period, but an increase in the 1980-1996 period. This strongly suggests "period effects" and rebuts the claim of continuous, universal, and monotonic decline. Fourth, parties do provide cues on issues for party identifiers, who are much more likely to *see* differences on issues between the parties and more likely to *take positions* which are distinctive by partisan category. Fifth, party identification *is* linked to the vote, as before. Sixth, so far as the phenomenon of voter defection is concerned, it has been going on for a long time, varies by elections, and is greatly reduced by strength of party identification. It does not signal the disappearance of parties or their reduction to irrelevance. It does signal the emergence of a more politically sophisticated public that is more closely evaluative of the parties than ever before, but still possessed of strong partisan predispositions and partisan loyalties.

NOTES

1. Roper Center for Public Opinion Research, *America at the Polls 1996* (Storrs, CT: Roper Center, 1997), 5. Gallup survey, May, 1996; National Election Studies, University of Michigan, earlier.
2. National Opinion Research Center's General Social Surveys, reported in Roper Center, *The Public Perspective* 8 (February/March 1997): 3-5.
3. Ibid., 35.
4. Ibid.
5. Roper Center, *Polls 1996,* 12.
6. Richard Hofstadter, *The Idea of a Party System* (Berkeley: University of California Press, 1969), 2.
7. Austin Ranney, *Curing the Mischiefs of Faction* (Berkeley: University of California Press, 1976).
8. Quoted by columnist Donald M. Rothberg, Associated Press, *Ann Arbor News,* (May 12, 1997), A5.
9. Herbert McClosky and John Zaller, *The American Ethos: Public Attitudes toward Capitalism and Democracy* (Cambridge, MA: Twentieth Century Fund, Inc., 1994), 293.
10. Roper Center, "Americans Rate their Society and Chart Its Values," *Public Perspective* 8 (February/March 1997): 35.
11. Samuel J. Eldersveld, *Political Affiliation in Metropolitan Detroit* (Ann Arbor: University of Michigan, Bureau of Government, 1957), 127.
12. National Election Studies, 1980.
13. Steven J. Rosenstone et al., *Third Parties in America* (Princeton: Princeton University Press, 1984), 248.
14. National Election Studies, 1980.
15. Roper Center, *Polls 1996,* 12.
16. In 1952, 29 percent made this statement. Band on CPS/NES data.
17. Roper Center, *Polls 1996,* 90.
18. Rosenstone et al., *Third Parties,* 248.
19. Roper Center, *Polls 1996,* 52-53.
20. Bruce E. Keith et al., *The Myth of the Independent Voter* (Berkeley: University of California Press, 1992).
21. Russell J. Dalton, *Citizen Politics; Public Opinion and Political Parties in Advanced Western Democracies* (Chatham, NJ: Chatham House, 1996), 208-210.
22. See the early analysis of group variations in Norman H. Nie, Sidney Verba, and John Petrocik, *The Changing American Voter* (Cambridge: Harvard University Press, 1976).

23. See Nie et al., ibid., 142. NES data from 1952 to the present.
24. V. O. Key, Jr., *Politics, Parties and Pressure Groups* (New York: Crowell, 1952), 586.
25. Angus Campbell, et al., *The American Voter* (New York: Wiley, 1960), chap. 6.
26. Morris P. Fiorina, *Retrospective Voting in American National Elections* (New Haven: Yale University Press, 1981), 203.
27. Miller, "Party Identification, Realignment, and Party Voting: Back to the Basics," *American Political Science Review* 85 (June 1991): 565.
28. Keith et al. *Myth*, 202.
29. Roper Center, *Polls 1996*, 85–90.
30. Gerald M. Pomper, *Elections in America* (New York: Dodd, Mead, 1968), 85.
31. Roper Center, *Polls 1996*, 90.
32. Walter de Vries and V. Lance Tarrance, *The Ticket Splitters: A New Force in American Politics* (Grand Rapids, MI: Eerdmans, 1972), 30–33. Also, see William J. Keefe, *Congress and the American People* (Englewood Cliffs, NJ: Prentice Hall, 1980), 50.
33. Cited in Pomper, *Elections*, 84, attributed to V. O. Key, *The Responsible Electorate* (Cambridge: Harvard University Press, 1966), 20, 27.
34. V. O. Key, Jr., "The Future of the Democratic Party," *Virginia Quarterly Review* 28 (Spring 1952): 172–173.
35. Ibid., 173, 175.
36. Philip Converse, "On the Possibility of Major Political Realignment in the South," in Angus Campbell et al., eds., *Elections and the Political Order* (New York: Wiley, 1966), 242. See also Hanes Walton, Jr., *The Native-Son Presidential Candidate: The Carter Vote in Georgia* (New York: Praeger, 1992), 123–125.
37. Samuel DuBois Cook, "Political Movements and Organizations," *Journal of Politics* 26 (February 1964): 130.
38. Nicol Rae, *Southern Democrats* (New York: Oxford University Press, 1994), 47–48.
39. Ibid., 163.
40. Dan T. Carter, *From George Wallace to Newt Gingrich: Race in the Conservative Counterrevolution 1963–1994* (Baton Rouge: Louisiana State University Press, 1995).
41. Charles Bullock III, "Creeping Realignment in the South," in Robert Swanbrough and David Brodsky, eds., *The South's New Politics: Realignment and Dealignment* (Columbia: University of South Carolina Press, 1988), 220–37.
42. Rae, *Southern Democrats*, 126.
43. Ibid., 154.
44. Ibid., 152.
45. Charles S. Bullock III and Mark J. Rozell, eds., *The New Politics of the Old South: An Introduction to Southern Politics* (Lanham, MD: Rowman and Littlefield, 1998). In fact, to show the supposedly extensive shift in public support for the Republican party in the region, the book adds Oklahoma, which was not a part of the old South and its confederacy.
46. Richard Scher, *Politics in the New South* (New York: M. E. Sharpe, 1997), 157.
47. Ibid.

CHAPTER 6

The National Organization of Our Parties: Revived and Relevant

American major parties, as serious contenders for power, had to develop their organizations in due time if they were to endure. This is true for every party system. Parties rarely can survive for long without some type of organization. In democracies party organizations come into existence to fulfill two basic needs. Party leaders need organizations to communicate with, and mobilize support from, the public. The public, on the other hand, needs party organizations as channels of protest, guides for voting, and ultimately means to control elites.

The organization attests to the institutional reality and durability of the party. Like other political groups, it is the collectivity of individuals working together to perform the critical tasks necessary to achieve common goals in the political arena. But the organization of a party is a special system of interpersonal relationships, roles, and activities. It is usually collectively committed to competing for public, government, offices, and thus with such political power to obtain the opportunity and right to participate in the formulation of public policy. The party organization, then, is *theoretically* central to the public life and to political decision making of a nation. It can *actually* be central or peripheral, depending on the effectiveness of its performance.

American major party organizations have fluctuated greatly in this century, in centrality, relevance, and effectiveness. We are currently in a period, beginning in the late seventies, which is considered by many scholars to be a time of party organizational resurgence and revival. Thirty years ago (even in the early seventies) our party organizations were considered weak, in a state of decline, even decomposition. One scholar argued that "parties are more important as labels than as organizations."[1] They presumably had lost their monopoly of running campaigns, their nomination function, their policy roles, and the loyal support of their followers. The critiques were many, and severe, although not all criticisms were factually accurate.

By the late 1970s, party organizations in Congress and the national conventions were reformed, by the Democrats particularly. By the 1980s we begin to see a revival of the party organizations at all levels. The Republicans took the lead in nonlegislative national reform of organizations when Reagan was president, and the Democrats soon followed. The national parties rebuilt their financial resources, increased their national staffs, coordinated and financed many more candidates for Congress, established new linkages with state and local organizations, and played a major role again in campaigns in the '80s and '90s. This was so impressive that scholars commented that the "nationalization of parties is rapidly developing,"[2] that "both parties are moving toward the model of highly structured programmatic parties,"[3] and that "political parties are alive and well and the parties' national organizations are more powerful than ever before."[4] Others advocated a more cautious interpretation, suggesting that we should not exaggerate revival and await further developments.[5] Yet there was a striking turnaround in the evaluation by experts after 1980.

A HISTORICAL PERSPECTIVE ON PARTY ORGANIZATIONAL DEVELOPMENT

The early history of our national party organizations is somewhat uncertain and obscure. In the early days of party formation, 1790–1800, the two emerging parties, the Federalists (under Hamilton) and the Anti-Federalists (under Jefferson) soon developed intense rivalry over a variety of issues (see Chapter 3) and met together in the first party "caucuses" to discuss policy: the Federalist caucus in 1790, Jefferson's party in 1795. These legislative party caucuses were used by the leaders for a variety of purposes, including the nomination of candidates for the presidency and vice presidency. Through informal contacts, correspondence, and personal encouragement these national leaders contributed to the formation of political societies, committees of correspondence, and the associational forms which represented the embryonic development of parties at the state and local level.

Aside from these early caucuses and informal associational efforts, national organization was slow to develop. In 1831 the National Republicans, who opposed President Jackson, decided to call a national nominating convention to select Henry Clay as their nominee. The Democrats followed in 1832 to renominate Jackson. This was the first organizational step to break the Congressional leaders' hold over presidential nominations. Much later both parties created national committees to run the party between conventions — the Democrats in 1848 and the Republicans in 1854.[6]

In Congress, very little party organization appeared in the nineteenth century, aside from these infrequent caucuses. The Democrats did establish a congressional campaign committee in 1842 and the Republicans in 1862, but we know little about what these committees did. Nor do we see any evidence that the parties had other congressional organs. Woodrow Wilson was concerned in 1885, when his *Congressional Government* was published, that little responsible organization existed in Congress. He wrote then:

> There are in Congress no authoritative leaders who are recognized spokesmen of their parties. . . . there is within Congress no visible, and therefore no controllable party organization. . . . the legislation of a session is simply an aggregate of the bills recommended by Committees composed of members of both sides of the House. . . .[7]

The amount of national party organization in the latter half the nineteenth century, in Congress or outside of Congress, was apparently minimal. Nor is there evidence that the national organs that existed did much — the major exception to this is of course the quadrennial national conventions for nominating the candidates for President and Vice President, for which the national committees had to plan and coordinate. At the national level, then, it was a "shadow" organization, not assertive or dominant. It was at the local level that party organization began to develop real strength, with each state and city and county having its own Republican or Democratic party (or both parties) organs. This was what V. O. Key called a "confederative" system, by which he meant there was very little national organization independent of the states.[8] There was no emerging tradition of national party control or nationally oriented party discipline, as had developed in England in the nineteenth century.

Local party organizations and political machines, particularly in the big cities, dominated our party politics in the latter half of the nineteenth century and into the twentieth century. They played central roles in the selection of candidates for office at all levels, and in the mobilization of votes in all elections. The new immigrants from Europe and elsewhere and the urban poor generally were welcomed, converted, educated, and absorbed into politics by these big city machines. Politics was very partisan, but also very local. The national party organizations did cooperate occasionally with, and assist, local party officials, but they were more dependent *on* them than supervisors *of* them.[9] But in the early part of the twentieth century a variety of factors and forces began to emerge which resulted in the decline of these party machines, partially the result of the progressive movement, partially due to reforms initiated by the parties themselves, partially due to social trends. The direct primary took away the easy control of nominations by the bosses; civil service reform took away their control over political jobs; and with social and demographic change the hold of the machines on their publics diminished. What remained was the local party organizations without boss control, without control over nominations, without control over patronage, with diminished resources, relying heavily on volunteer efforts in campaigns. It is argued that these developments forced candidates to be more self-reliant and campaigns to become *candidate centered.*

The twin developments of egalitarianism and populism led to the eventual decentralization of party organizations — that is, the transfer of power from the legislative wing of the party to the local "delegate convention." By the late 1830s the parties began to take on this organizational form. It emphasized the election of delegates from local party units to a representative assembly called the county or district convention. At the national level the first such conventions were held in 1831–1832 by the Anti-Masonic party and then by the Democrats and Whigs. By 1844 they were well established.

Thus, between 1832 and 1860, there occurred a basic change in two senses in party organization: (1) the power to direct and control the organization was taken from the leaders at the national level and given more and more to leaders with local status; (2) the power of the party organization was taken from the party's legislative leaders and given to those leaders who were largely individuals with no status in either the Congress or the state legislatures. This populist or decentralizing orientation continued in the post-Civil War period and was pushed to its next logical step with the adoption of the direct primary starting in 1900. The primary, in a sense, takes the power of the party (particularly to nominate candidates) out of the hands of the delegate convention (which in the nineteenth century had been criticized as undemocratic and corrupt) and theoretically places that power in the hands of the loyal party supporters. Since 1900 we have moved steadily in the direction of expanding the opportunities for direct popular decision making in party affairs with less emphasis on making decisions through representative party institutions. Today the primary has been adopted in every state, the presidential primary in forty states. Figure 6.1 presents a diagram of these organizational changes in the party system.

From 1790 to the twentieth century, our party organizational system was basically transformed. The national organization lost control to lower levels of the party and to the party voters. And legislative leaders lost control to a variety of other actors in the system. The direct primary in a sense took over a major function of the party, the nominating function. Our system was decentralized and made more populist. This was quite contrary to developments in other systems. In Britain the national party units retained control throughout this century of change, and the parliamentary leaders of British parties still have a dominant role. Each party system in each country develops its own party organizational style.

FIGURE 6.1 ■ Transformation of Party Organizational Control Patterns since 1790

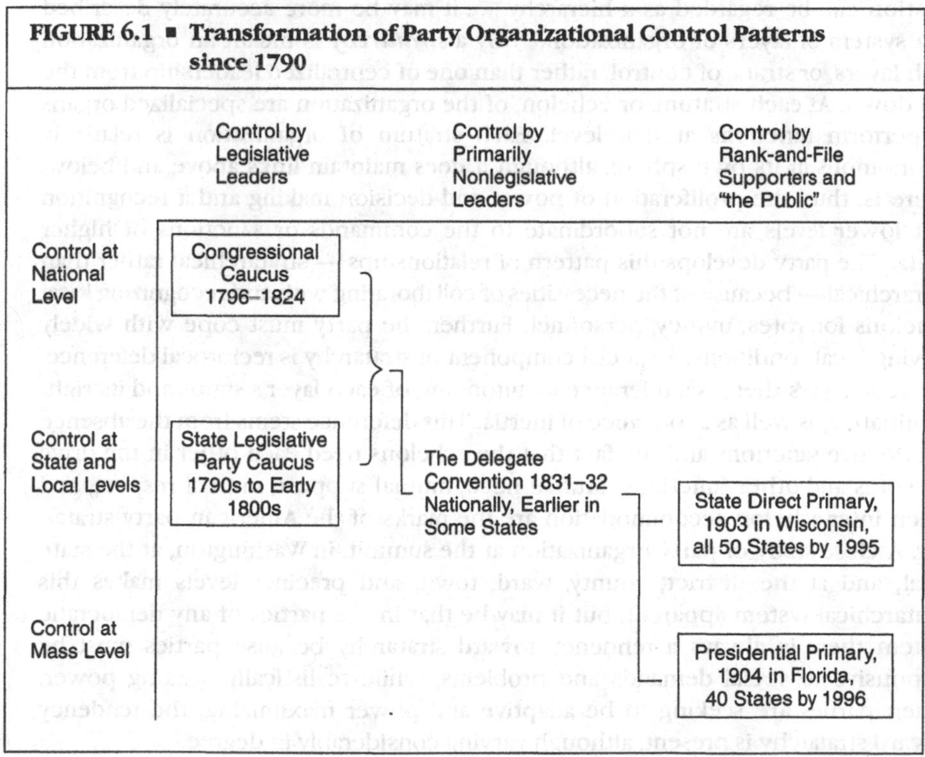

There are several major characteristics of American party organization today to keep in mind. There is a great variety of units in that organization from the precincts and wards and clubs, at the base, to the national committees at the top. Much of this organization is regulated by state law, particularly as to how state and county units are to operate — how they hold their meetings, how candidates for office are nominated, and how they collect and spend their funds. While there is much regulation of parties, the courts have also, particularly recently, permitted the party organization to act on certain matters in its own best interest. Another characteristic of our party organization is that it is loosely structured, and control is not integrated from the top to the bottom. There is usually no central leadership organ which commands the entire party organization. This is traceable to the historical development of our parties, the decentralization of power, the insistence by local units on autonomy, and hence the dispersion of authority at all levels of the system. We do not have a party hierarchy or strong discipline within our parties. It is difficult, then, for the rank-and-file party workers to hold higher-level leaders accountable through organizational mechanisms of control. Top organizational leaders are never subjected to a rank-and-file check by loyal supporters within the organization.

American party structures are in many respects essentially *stratarchies.*[10] This is apparent from a look at how the parties are organized from the national to the local level. V. O. Key long ago recognized this, saying, "Although the party organization can be regarded as a hierarchy . . . it may be more accurately described as a system of layers of organization."[11] By a *stratarchy* is meant an organization with layers, or strata, of control, rather than one of centralized leadership from the top down. At each stratum, or echelon, of the organization are specialized organs to perform functions at that level. Each stratum of organization is relatively autonomous in its own sphere, although it does maintain links above and below. There is, thus, the proliferation of power and decision making and a recognition that lower levels are not subordinate to the commands or sanctions of higher strata. The party develops this pattern of relationships — stratarchical rather than hierarchical — because of the necessities of collaborating with and recognizing local echelons for votes, money, personnel. Further, the party must cope with widely varying local conditions. A special component of stratarchy is reciprocal deference. Between layers there is a tolerance of autonomy, of each layer's status and its right to initiative, as well as a tolerance of inertia. This deference stems from the absence of effective sanctions and the fact that the echelons need each other in the drive for votes and other functions. Mutual need, mutual support, mutual respect, and much inter-echelon accommodation are the marks of the American party stratarchy. A close study of party organization at the summit, in Washington, at the state level, and at the district, county, ward, town, and precinct levels makes this stratarchical system apparent, but it may be that in the parties of any democratic system there is always a tendency toward stratarchy because parties must be responsive to social demands and problems, while realistically seeking power. When parties are seeking to be adaptive and power maximizing, the tendency toward stratarchy is present, although varying considerably in degree.

PARTIES AS UMBRELLA STRUCTURES

A major question which emerges from this description of the American party organization in overall terms is: What holds such a disparate, diffuse, decentralized party organization together? Why don't the Democratic and Republican organizations throughout the country disintegrate, since there is no central leadership organ and no hierarchical pattern of relationships? This will be discussed in greater detail later. But here it should be recognized that perhaps *the* Republican and *the* Democratic nationwide organizations really consist of a set of party organs at all levels loosely held together by a common party name, common symbols of organization, a common history, a sense of identity or loyalty, and, to a certain extent, similar beliefs and philosophies. The Republicans in Racine, Wisconsin, and the Republicans in Los Angeles County, California, in Springfield, Illinois, and in Portland, Maine, are not subordinate in any real organizational sense to a Republican party hierarchy. But they recognize most of the same leaders as Republicans, and they think of themselves as Republicans in terms of basic policy orientations, historical traditions, and sense of loyalty. And once in four years they or their representatives get together in a national convention with other Republicans from throughout the country to launch their campaigns for the presidency. The Republican organization, therefore (and the same can be said for the Democrats), is a loose-knit set of semiautonomous units at all levels, which may be very efficient or very lax, very liberal or very conservative, very competitive or very subdued — but they are part of an umbrella organization called the Republican party.

There is no power elite directing the Republican and Democratic parties. Nor is there an ideological straitjacket into which the party units, national and state and local, are forced. There is great diversity in the units in the Republican and Democratic party organization in terms of their goals, interests, incentives, techniques, ideologies, and efficiency, but there is a cement holding them together. They *are* viable structures, however diverse, diffuse, decentralized, and centrifugal they may be. Perhaps this is the reason for their survival. For almost 140 years the Republican and Democratic organizations have survived. Their longevity is truly remarkable and can only be understood through an analysis of the structure of these diffuse "umbrella organizations" and how they have adapted to the changing American social and political environment.

We will now look in detail at the party organizations at all three levels — national, state, and local. With the important facts in hand, we need to evaluate the state of our party organizations today. There are at least five major questions to ask:

1 *How well structured* internally are our parties? Do they have adequate vertical depth from top to bottom as well as geographical breadth? Is there a logical organizational framework for allocation of tasks, good lines of accountability, good communication basis?
2. *How well led* are our party organizations? Do they have able personnel at the apex and lower echelons of the organization? Do they work well together, at the same level and between levels? Are national leaders in touch with their party's organizational base?

3. *How well linked* are they to the governmental leaders of the party in Congress, state legislatures, and county commissions? Do they consult together on strategy, policy, etc., or are they isolated from each other?
4. *How well integrated* are they with civil society, interest groups, the mass media, and the public? Is their relationship with any of these imbalanced, antagonistic, nonexistent? Conversely, have they developed strategies to incorporate the interests of these external forces in party planning and activities?
5. *How well are they performing* their key functions of mobilizing resources (financial, personnel, etc.), recruiting and training activists for mobilizing support, developing critical strategies for campaigns and implementing these strategies, articulating and communicating the ideology, goals, and issue positions of the party, adapting to the changing environment (technological, social, economic, and political) with innovative organizational changes, in order to compete effectively with the opposition?

These are critical requisites which party organizations have to meet in any system if the party's goals are to be achieved and, indeed, if the party is to survive. The above basic questions suggest at minimum what those requisites are.

A DESCRIPTION OF OUR NATIONAL PARTY ORGANS: AUTONOMOUS BUT INCREASINGLY INFLUENTIAL

Normally citizens do not hear or read much about activities of the Republican and Democratic national organizations and their leaders. Unless there is a dramatic fight for the chairmanship of the national committee, or a conflict among the party leaders in the House or the Senate, or the somewhat staged spectacle of the national convention once in four years. Then we become aware that such party organs exist nationally, but we still have only a vague idea what they do. A basic description of each is thus in order.

The *national chair* has become a significant, even powerful, party leader in recent years. Ron Brown, who was elected Democratic chair in 1989, after much debate in the Democratic National Committee (DNC), and who was a dynamic figure in Clinton's victory in 1992, is an example of this. He had been active in Ted Kennedy's campaign earlier and campaign chair for Jesse Jackson when he ran for the presidency in 1988. Talented, experienced, and truly committed to winning, he decided when elected as DNC chair to transform the DNC into "a tough, aggressive, professional campaign organization." His actions demonstrated that he worked hard to do exactly that, for example, raising in 1992 $20 million in nonfederal funds (compared to $12.8 million by the Republicans) and developing more advanced and more systematic techniques in polling, research, constituency targeting, and relationships to state and local organization than ever before.[12] Lee Atwater was a similar type of campaign chair (1988) and national chair for President Bush, 1985–1991. And, of course, Haley Barbour, after Bush's defeat, has been a strong national chair 1993–1996. These recent instances of strong national chairs were

Party Chairs

For both parties the selection of the national chair has always been important. This is particularly true today as the national organization takes a greater role in party affairs. For a party holding the presidency the chair is usually approved or dictated by the president, and ratified by the national committee. For the party opposition the chair is selected by the national committee. At its national meeting on January 17, 1997, the RNC, unhappy about its failure to capture the presidency and strengthen its majority margin in the House of Representatives, replaced Haley Barbour from Mississippi with Jim Nicholson from Colorado.

Nicholson, who had been Colorado's national committeeman for the past ten years and a twenty-year party activist and organizational man, is known throughout the party as a grassroots specialist and fund-raiser. He has also served as chair of the RNC's Rules Committee, a member of the Budget Committee and vice chair of the Western States. In 1996, Nicholson chaired the RNC's Task Force on Presidential Primaries and Caucuses and developed for the party's 2000 effort a less hectic and compressed schedule of primaries. Moreover, the 1996 Convention included the Task Force's recommendations in the RNC's rules. Although he won in a multi-ballot election, his election was in part driven by the party's concern to capture and keep the western states, as was Barbour's election to hold on to the solid Republican South.

Over at the DNC, President Clinton in January 1999 proposed to restructure the national organization so that Vice President Gore will have his own people in place before the 2000 campaign gets under way. Former Colorado Governor Roy Romer will continue to serve as general chairman and the party's leading spokesman, but the co-chairman's position now held by Steve Grossman will be abolished and replaced by a triumvirate of three individuals. Of the triumvirate one will be the African American mayor of Detroit, Dennis Archer; another will be a Hispanic woman, State Representative Loretta Sanches, (D. Calif.); and the third one will be the successful state party chair of Indiana, Joe Andrew. If this proposal is finalized, this triumvirate structure will not only put Gore's people into place, but it will address the race issue. This is a new strategy the Democrats are adopting, and it will be interesting to see how it works out. Notable about these chairs is that both are from the same state, which is also a new development!

SOURCE: Ceci Connolly, "Archer is tapped to help lead Democratic National Committee," *Detroit Free Press* (January 11, 1999).

preceded by the very effective reform-minded Senator Brock of Tennessee in 1977, who really developed the foundations for the modern, revitalized, Republican party. And Charles Manatt was the Democratic chair who performed the same type of party-rebuilding role for the Democrats after 1980.

The national chair has thus become a full-time, well-paid, prestigious position, working behind the scenes of national party politics. Selected by the national committee, with the presidential candidate often having considerable influence in this decision, he or she (there have been two women national chairs in the past

25 years) has a demanding set of tasks. Among these is presiding over a national committee which meets up to four times a year, a large body representing state, local, and auxiliary party officials. As important is the directive responsibility for a fairly large staff, which has grown over the years to 300 or more persons, with very specialized campaign, organizational, and research responsibilities. In addition, as the 1996 party finance investigations revealed so clearly, a major function of the chair is taking a leading role in fundraising, particularly in identifying possible (hopefully legal) sources for large contributions to pay the bills at the national headquarters as well as to assist congressional and Senatorial candidates, or to allocate "soft money" for generic party-centered campaign advertising. As part of this operation, the national chair's staff works with state and local party leaders. Planning for the national convention, and keeping within budget for the convention costs, is a major job. In 1996, each party was allocated $30 million from public funds for this purpose. Above all, the national chair, is in a real sense, the ceremonial organizational head of the party, proof that a national party exists.

We used to scoff at the position of national chair, claiming it was an ineffective office. One chair, Kenneth Curtis, when asked to step down by President Carter, called it a "lousy job."[13] In a sense it does appear powerless, inevitably frustrating, and very contingent on many others for success. The chair is dependent for operational leeway on the president, or the national committee, or the willingness of other major actors, within and outside Congress, as well as state and local leaders. While seemingly a perilous job, it can also be a creative and effectual position. There is great potential for leadership. And recently, with the types of national chairs we have had, that potential has been converted into striking organizational performances at the national level.

The national committees of the two parties are large — 165 for the Republicans and over 400 for the Democrats. The Republicans allow three representatives from each state (usually top state party leaders), plus a few additional choices; the Democrats allocate representation on the basis of the party vote and population size of the state, and then add state chairs, congressional party leaders, a few governors, officers of the national committee, some mayors and state officers, and additional members to represent the Young Democrats and other auxiliary groups. They are selected in different ways by the state parties (at state conventions, or state central committees, or direct primaries). Thus one must remember that these people come out of the state and local organizations, usually after much experience in many campaigns, and theoretically at least beholden to their state constituencies. One must also remember that election to the national committee is very prestigious, the job is really sought after, and very serious battles occur in state committees over these appointments!

These committees have in the past been derided as powerless, unwieldy, and fragmented due to the widely varying types of people who emerge through this decentralized representative process. Superficially much of this criticism may still seem true. But one must ask what is their role, and how well is this achieved? Originally when set up in 1844 by the Democrats, it was supposed to be the campaign committee for the presidential candidate. That is not what it does, at least not explicitly, although many of its activities today are linked to that objective. Both Clinton and Dole had their own special set of strategies for their campaigns, and yet consulted continuously with the national committee staffs and leaders on the implementation of strategy.

Occasionally, there are discussions of actual policy, foreign and domestic. But this does not seem to be their primary interest. The national committee's major role is organizational policy making in the party. Many decisions have to be made in planning the national conventions once in four years: Where and when it will be held, how the delegates will be apportioned to the states, what rules the state parties must follow in selection of delegates to the national convention, determination of the convention agenda and the selection of key speakers — all of these can engage party leaders in considerable controversy. But the national committee has the authority to set the rules under which the conventions will operate. As discussed earlier, the recent Supreme Court decisions in cases in which state parties (or state governments) oppose national committee decisions have regularly upheld the power of the national committee. This applies particularly to the authority of the DNC which has in the past 25 years, since 1972, sought to impose on state parties precise and rigid rules for selecting national convention delegates.

While these procedural reforms initiated by the national committees are important, it is the expansion of the functional role of these committees in financing and helping manage campaigns that particularly must be kept in mind. In 1996 the DNC raised $99 million and distributed 56 percent of it to state and local party candidates. In 1996 the RNC raised $110 million and also distributed a large proportion of it (43 percent) to local and state parties and candidates. This provides the clue to the real story about the significance of the national committees today. Their large staffs and very successful fundraising efforts, combined with the expansion of their organizational and management functions, including newly developed technologies for campaigns, polling, and working systematically with political action committees, candidates, their own media resources, and state operatives, make these national committees critical actors in national party politics. As decision-making bodies the national committees are cumbersome and often incapable of more than broad and general decisions, indeed on occasion internally dissensual. But as broadly representative forums of leaders from all parts of the country, working with, as well as devolving authority on, their national committee chairs, officers, executive committees, and staffs, they have become much more influential and relevant in the presidential and congressional campaign operations than ever before.

We used to say that the quadrennial *national party convention* was "the basic element of national party apparatus."[14] No longer. It is the only national party structure that has declined in function and probably in importance in the past thirty years. If we know American party history, we remember the dramas of many earlier national conventions as places where the critical decision on the party's candidate for the presidency was made. There were many historic conventions. In 1932 FDR battled Al Smith and others in Chicago, winning on the fourth ballot after all-night voting and negotiations, when Speaker Garner of Texas released his delegates to Roosevelt. Eisenhower battled Taft and the conservative Republicans in 1952. Kennedy confronted Lyndon Johnson in Los Angeles in 1960, winning with barely 50.1 percent of the delegate votes. Goldwater was nominated by the Republicans in San Francisco in 1964, when the conservatives took over the party convention.

These great moments in our party history have faded somewhat, even though today we still have the national convention every four years. But now the convention has lost its power or function to nominate the presidential and vice presidential candidates. The pre-convention state caucuses and primaries have preempted this function. The convention merely ratifies the results of the primaries. In 1996 Dole clinched the Republican nomination with 51 percent of the delegates on March 26, after winning the primaries of California, Nevada, and Washington (twelve states still had to hold their primaries). In 1992, Clinton clinched the nomination at the completion of the primaries on June 1 when he had won 59 percent of delegates.

We will discuss the nomination process in detail later. Here it is important to see the convention as a party structure which still has an important place in the party system. The conventions are large bodies (in 1996, 4,291 Democratic delegates and 1,990 Republican delegate votes) which bring representatives elected from state and local party units from all over the country. It thus has a representative function, some would say a unifying and integrating function. Delegates see themselves as part of a large political party community.

The convention today also has other roles and uses. Technically the convention still retains its status as the plenary body of the party organization power; it can pass rules and adopt resolutions on how the party should be organized. It has done this in the past, for example, in specifying that state delegations shall reasonably reflect minority groups and that one-half of the delegates selected shall be women. Normally, however, the convention does not engage much in such restructuring of the party.

The convention is very important as a body which adopts a party platform stating the party's policy goals and direction. In the process of doing that the convention's platform-writing committee holds extensive hearings before the convention meets at which interest groups, or individuals in the party, can present their proposals for planks in that platform. For scholars of parties, these party platforms are considered significant indicators of party ideology or party positions. Fights do occur at conventions on party plank content, a recent example being the Republican continued controversy over the phrasing of its abortion plank.

The convention must be seen today as a gathering of the party faithful from all over the nation — leaders, activists, followers — selected by party voters in local primaries or caucuses or conventions, and representing a great heterogeneity of interests and beliefs. It has a learning function; party activists in California and Utah mix with those from Illinois, Louisiana, New Jersey, and Maine. These interactions are important, as is the excitement for the individual delegate as a participant in this periodic spectacle. It *is* a spectacle, which today is perhaps more important for launching the presidential campaign, giving the rank-and-file of the party an opportunity to see and hear its old and budding leaders, and reaching out via television with all the party's communication resources to American citizens, hoping thereby to mobilize new supporters and activate previous supporters. Recent conventions, according to some research, have not maintained the high pitch of public enthusiasm as in an earlier period.[15] Nevertheless, the convention, despite its loss of the presidential nominating function and despite its sometimes overdone and superficial celebratory character, remains a key part of the national operation. In America's diverse and centrifugal politics, it is important once in four years to try to bring the disparate elements of the party together from across the nation and try to convey the appearance at least that there is a "national presidential party."

THE CURRENT STATE OF NATIONAL PARTY ORGANIZATION

We have come a long way in the past twenty years in the building of more effective national party organs. The research of many students of our parties demonstrates that this is so. One of the most careful scholars, Paul Herrnson, concludes that the once weak national party organizations "are now financially secure, and highly influential in election campaigns and in their relations with state and local committees."[16] There are two basic ways to look at, and evaluate, the national party entities, in terms of (1) their functions, activities, roles and influence, and (2) their structural character, and the organizational relationships between the diverse party organs, at the Washington level.

We have a variety of national party units for each party, but they are not part of a neat pyramid of power, with the national chair at the top and the other committees and units filling subordinate positions in a party hierarchy. Rather, it is a segmented and pluralized assortment of party units with no clear and precise linkages to each other. We try to suggest this pattern in Figure 6.2.

FIGURE 6.2 ▪ Components of a National Party

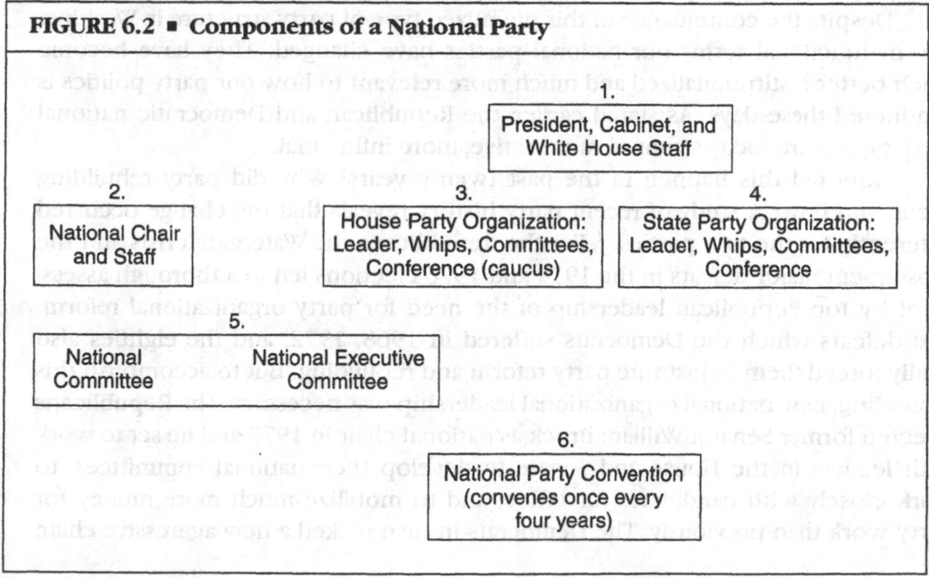

Note that in Figure 6.2 there are no lines suggesting the relationships of these units to each other. They are fairly autonomous centers of party activity. While the president may have a close relationship with, and some control over, the national chair of his party, this has not always been true. And the only generalization one could make about the influence of the national committee over the House and Senate committees, or vice versa, is that there is no such influence relationship! Thus we have a "co-archy:" a "structure of equal power relationships."[17]

The conception of our national party organs as separate units which jealously guard their autonomy was the way we looked at the national party years ago. And there was much evidence to support this image. For example, the rules of the Democratic party specified that the congressional campaign committee "has no organic connection with either the Democratic National Convention or the Democratic National Committee."[18] This does not mean that these committees did not work together, consult together, or informally collaborate from time to time. But they maintained their autonomy. Recent research suggests this is still true. As Herrnson reports, "each Hill committee is autonomous and self-governing . . . committee staffs scoff at the idea (they) take orders from party leaders in the White House or at the national committees."[19] He adds that the members of these committees do interact, have common interests, and cooperate in a variety of ways in campaigns — sharing voter lists, research, get-out-the-vote drives, et cetera. But they particularly protect their separate status in financial matters. As one member put it, "relations are usually friendly, cooperative, but fiercely competitive."[20]

What this means is that there is no central managerial control of the national party, no singular body of party authority and responsibility. Rather it is a structure of several centers of power and decision making. This U.S. model is quite different from political parties in Europe, where there is usually a national executive committee that is usually closely linked to the parliamentary party, which is the focus of national party leadership.

Despite the continuance of this pluralized type of party structure in Washington, in functional terms our national parties have changed. They have become much better institutionalized and much more relevant to how our party politics is conducted these days. As stated earlier, the Republican and Democratic national party organs are today stronger, more active, more influential.

How did this happen in the past twenty years? Why did party rebuilding occur, and how? A study of recent party history reveals that the change occurred differently for the two parties.[21] For the Republicans the Watergate crisis and the subsequent major defeats in the 1974 and 1976 elections led to a thorough assessment by top Republican leadership of the need for party organizational reform. The defeats which the Democrats suffered in 1968, 1972, and the eighties also finally forced them to institute party reform and rebuilding. But to accomplish this rebuilding, new national organizational leadership was necessary. The Republicans selected former Senator William Brock as national chair in 1977 and he set to work with leaders in the House and Senate to develop their national committees, to work closely with candidates for office, and to mobilize much more money for party work than previously. The Democrats in turn picked a new aggressive chair,

Charles Manatt, former finance chair of the Democratic National Committee, who also worked with the House and Senate leaders to revitalize the Democratic organization after 1980. The new leadership of the parties realized that they had to make the national party organs more dominant and useful in campaigns, as well as in between elections. Above all, they realized that they had to adapt to the new developments in American society and the political system. They had to face the reality of technological changes in campaigning, the increased role of political action committees, the high costs of media advertising, the need to coordinate state and local campaigns with national campaigns, and the trend toward a very candidate-centered politics. The Republicans took the lead in the late seventies in rebuilding their national party, and the Democrats after 1980 realized they were falling far behind and began to rebuild as their rivals had already done. Thus election crises, new leadership, with realistic insights about the "new American politics," led to creative and adaptive organizational rebuilding — these were the conditions responsible for the "reemergent national party organizations."[22]

NATIONAL PARTY ACTIVITIES IN RECENT YEARS

Much detailed research has been done on what the Republican and Democratic parties have done to make their national organizations more effective. We can only summarize briefly here the main features of these efforts. First, both parties have greatly increased their national committee staffs. In 1972 the Democratic National Committee had a staff of 30, by 1992 it was 270; for the Republicans it was also 30 in 1972, reached 600 by 1984, 425 by 1988, and was at 300 in 1992.[23] Both parties now own impressive headquarters in Washington, DC.

Second, both parties today have the ability to mobilize large amounts of money and to use this money to finance their own requirements in Washington, as well as to contribute to Congressional and Senatorial candidates and also to state organizations. Later in this book, we will describe party finance in detail. But as an indicator of change, the following data may illustrate what has happened. The amount of "soft money" that the national committees have mobilized (by "soft money" we mean funds given to the national party that are exempt from federal limitations on spending) was as follows in recent elections:[24]

Soft Money Collected by National Parties (in millions of dollars)

Party	*1992*	*1994*	*1996*	*Compared to 1980*
Democrats	36.3	49.1	122.3	.3
Republicans	49.8	52.5	141.2	2.4

These were funds national parties could distribute as they wished. In 1996, the Democratic National Committee raised $99.4 million and gave 56 percent ($53.9 million) to state and local committees; the Republican National Committee raised $110.3 million and gave 43 percent ($47.8 million) to state and local committees.[25]

This is a basically new development in our politics and attests to the new financial capacity and influence of the national organizations, including the national committees which in the past have been considered relatively impotent financially (as the 1980 data reveal was true.)

Third, as indicated above, the national committees are now deeply involved in working closely with state and local organizations. Transfer of funds is one way. But other assistance is also provided, such as assistance in organizational management, voter registration drives, candidate training, polling, and data processing. The Republicans appear to do more of this active assistance than the Democrats are able to, although recent reports suggest the Democrats are catching up.[26] While in the past state and local parties were considered completely on their own in campaigns, in our very decentralized system today there seems to be much closer interaction between the national and state levels and more efforts at integrating the two levels of the system.

A fourth type of national involvement is with the recruitment of candidates for Congress. The national committees (DNC and RNC) as well as the congressional and senatorial committees are today focusing considerably on this function. As one official with the Republican Senatorial Campaign Committee put it: "Candidate recruitment is the biggest thing we do."[27] The national parties let potential candidates know the election opportunities and financial assistance available, and they conduct "training colleges" and to some extent develop programs focused on the recruitment of certain types of persons, such as women candidates. The parties also go beyond this general encouragement to "active recruitment" efforts, particularly to find candidates for competitive districts. These efforts can (and have) caused conflicts with local organizations that may have their own preferred candidates. But the pressure to find able candidates is obviously very real, and the extensive work to identify such candidates, meet with them, even bring them to Washington for discussions and help them plan their campaign strategies, is clearly what is going on today. And both parties are engaging in this, the Democrats perhaps less extensively than the Republicans. As Herrnson describes it, the new approach can be called the "national party as intermediary" model.[28]

The national committees have developed cooperative relationships with the political action committees (PACs) as a fifth way of playing a significant role. In the past it has been argued that PACs have weakened parties by taking over the funding of candidates. Today the relationship is much different. The national party committees work with PACs; mobilize PAC money, identifying competitive candidates who need funds and who have a chance to win; prepare detailed reports for PACs on the latest developments regarding particular campaigns in states and districts; and try to deter PACs from supporting the candidate of the opposing party. On the other hand, the national committees' representatives work with congressional candidates to advise them on PAC funds available and on strategies for PAC fundraising. Thus the arms-length and hostile relationship between the party organizations and PACs of the past has been transformed into a set of mutually beneficial help relationships which are sometimes informal, and often even formal, PAC party linkages.[29]

Sixth, the national parties are deeply involved in the preparation of media materials and ads, for dissemination nationally or targeted for particular regions and states. This again runs counter to the image of the role of the media in our parties, an image that emphasized that the media had displaced the parties as the major force in campaign communication. What has happened is that the national parties play a major role in developing issues, in planning media strategy, in preparing their own media ads, and in working directly with candidates in the preparation of "media packages" tailored to a particular campaign. The national party congressional committees have their own television and radio production and broadcasting facilities. These are extensive facilities. After 1994 the Republican National Committee was given $2.5 million from Amway Corporation to set up its own broadcasting center. The Democratic congressional campaign committee has its Harriman Communications Center. Thus both parties are able to produce media ads in-house, or by agreement with outside specialists. Much of these services to candidates is selectively done, favoring the most competitive candidates of the party. As Herrnson reports, this means that there has been both a nationalization and professionalization of political communication, with the national parties again playing a greatly expanded role.[30]

Finally, the national party organizations have for some time now provided a variety of "public opinion" and voter mobilization assistance and advice to local party organizations. In a presidential election there are three types of such local organizations: the presidential activists, the workers committed to a local congressional candidate (as well as other candidates), and the regular local party organization. The national parties, of course, have their own polling organizations and thus are provided general information about the state of public opinion and polling findings sometimes targeted to particular subgroups in the electorate or to particular states and constituencies. That is, specific information on voters can be provided to targeted local campaigns. And candidates can be sold survey results or can arrange to receive them as "in-kind contributions." In addition, the Hill committees advise on local campaign strategy based on survey results (national, state, or even district surveys). Also voter registration drives can be (and have been) mounted successfully, working with local activists. And there is some attention given even to training local party leaders in how to recruit and train party workers.[31] The relatively recent argument, therefore, that the local party activist is irrelevant and has virtually vanished is not supported by the evidence of how much importance the national organizations place on the local party work and how they seek to improve it. A study by Paul Beck and his colleagues of the impact of local activists working in the 1992 presidential campaigns reveals empirically again (as many scholars have done in the past) that such local activity can be very significant. As Beck and his co-authors conclude, "our study has demonstrated that there is a vibrant party campaign at the grass roots."[32] National committees are today aiding and abetting these local activists.

We have, then, clear evidence from recent research on the national party organizations that they are today major actors in our party system. Their roles and relationships seem strikingly different from two or three decades ago. They work on all major facets of the campaign, from fundraising to media ads to close cooperation with PACs to voter mobilization. One can say that these national organizations have developed closer *functional* integration with, and to some extent control over, the communication and the media, recruitment, campaign funding, interest groups (PACs), and voter mobilization aspects of politics than ever before. Further, they have better *territorial* integration, with state and local party organizations. These national organs are not neatly centralized or combined into a single structure, but they do cooperate better than before. This is not to say that our decentralized system with more autonomy for local organizations has now been reversed. We still have a rather "federal" party system. Rather, a new balance has been developed between national, state, and local party organs, with the national party structures playing a more important, and intermediary role, working effectively, reciprocally, with state and local party structures and candidates. As a result of all of this development in recent years, party organization has become more important than it has been for some time, at all levels of the system.

REFORM OF PARTY ORGANIZATION IN CONGRESS

The lack of leadership in Congress, which Woodrow Wilson deplored in his 1885 book, was reversed rather abruptly toward the end of the nineteenth century by Speakers Reed and Cannon who imposed strong, almost dictatorial, control in the House from 1889 to 1910. But the revolt of 1910 returned the House to its previous state of "committee government," with committee chairs selected under the seniority system dominating. This remained the situation until long after World War II. Only in the late sixties did Congress begin reform efforts, first leading to the Legislative Reorganization Act of 1970, and then the major efforts by the Democrats to decentralize their party organization, followed by more limited efforts at reform by the Republicans. The basic reason for such change, it is argued, was electoral. The 1950s and 1960s brought a new group of liberal Democrats to Washington who were upset by the control which conservative leaders, Republicans and southern Democrats, had in Congress. They set to work to make institutional changes. The Democratic Study Group after 1968 discussed how to better work as a party to adopt liberal legislation and to make the people who had positions of power responsible to rank-and-file Democrats.[33] This group then activated the Democratic Caucus in the House and pushed for a whole series of reforms.

When we think of the party organization in the Congress up to November 1998, we may remember certain top legislative leaders, such as the Republican floor leaders Trent Lott in the Senate and Dick Armey in the House, or Dick Gephardt, the Democratic House floor leader, and Tom Daschle, the Senate Democratic leader, as well as Speaker Newt Gingrich. These leaders are important because they hold key party positions in Congress. What these positions are and the units which make up the party organization in the House are presented in Figure 6.3.

The relationships between all of these units is not always clear, in the diagram or in actuality. It is no hierarchy or pyramid of authority from the Speaker or

FIGURE 6.3 ▪ Party Organization in Congress

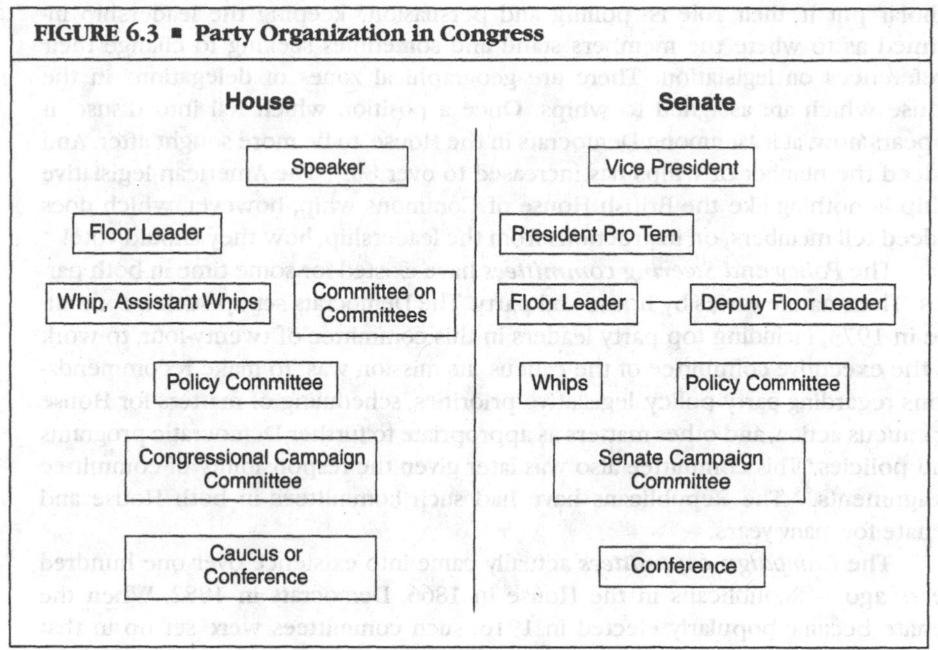

majority/minority leader downward. A brief explanation of these units may be helpful.

The *party caucus or conference* is theoretically the plenary body of ultimate power, since it includes for each party all of the elected members. At the beginning of each Congress it selects its own party leaders, subject to self-imposed constraints. It can, and often does, discuss policy and take policy positions, but usually does not bind its members to vote a certain way on a roll call, as is done in the British parliamentary system. The caucus can be the instigator of reform, adopting new procedures, setting up new units, changing leadership — which is what the Democrats did in the reforms of the seventies. The chair of the caucus, if assertive, can play a major role in making that body influential.

The *floor leader* of each party is selected by the party caucus and plays a major role in legislative planning. For the party in control of the House or Senate, the floor leader works closely with the Speaker and other leaders to manage the legislative schedule. He or she is a person of considerable congressional experience and commands respect. He sits on important committees, such as the Committee on Committees, which assigns members to the standing committees of the House, and the Steering or Policy Committee, which is the executive body for the caucus or conference.

The *whips* work with the floor leaders (under an assistant floor leader) to maintain contact with members, informing them of the legislative schedule as well as, on certain bills, the position of the party leadership on these bills. As one scholar put it, their role is "polling and persuasion," keeping the leadership informed as to where the members stand and sometimes seeking to change their preferences on legislation. There are geographical zones of delegations in the House which are assigned to whips. Once a position which fell into disuse, it appears now, at least among Democrats in the House, to be more sought after. And indeed the number of whips has increased to over 80.[34] The American legislative whip is nothing like the British House of Commons whip, however, which does indeed tell members, on instructions from the leadership, how they should vote!

The *Policy and Steering committees* have existed for some time in both parties. Their nature varies by house and party. The Democrats set up such a committee in 1973, including top party leaders in this committee of twenty-four, to work as the executive committee of the caucus. Its mission was "to make recommendations regarding party policy, legislative priorities, scheduling of matters for House or caucus action, and other matters as appropriate to further Democratic programs and policies." This committee also was later given the responsibility of committee assignments.[35] The Republicans have had such committees in both House and Senate for many years.

The *Campaign committees* actually came into existence over one hundred years ago — Republicans in the House in 1866, Democrats in 1882. When the Senate became popularly elected in 1916, such committees were set up in that body also. Their goal today is essentially the same as earlier. But now they have become the focal points for activities such as fund-raising in campaigns, working closely with candidates at the state and district levels, providing strategic device, assisting with polling and research, and media advertising. As indicated in our previous descriptions of their work, with the funds at their disposal congressional campaign committees can have great influence. Here is an indication of their disbursements in 1996:[36]

Soft Money Expenditures, 1996 (in millions of dollars)

Expenditure	*Democrats*	*Republicans*
Senatorial		
Campaign Committee	13.9	25.6
House		
Campaign Committee	10.3	27.4

These committees clearly are very important as part of the party organization in Congress.

WHAT REFORMS SINCE 1970?

The objectives of the reforms, as indicated earlier, were to reduce the power of committee chairs, and in the process to increase the relevance of the party caucus (or conference), as well as to enhance the role and participation of the individual member in the legislative process. One consequence, it is argued, is that party cohesion, as revealed in roll call votes, was increased; indeed, that "the party" became more important in Congress.

A great many specific reforms emerged in the seventies. Of primary importance was the loss of automatic tenure for standing committee chairs or, to put it differently, the radical modification of the seniority system. The Democrats adopted the rule that at the beginning of each Congress the caucus would vote on the chairs for all committees (by secret ballot if 20 percent of the members requested). As a result several chairs have been challenged and deposed. Committee chairs suffered power reduction in other ways; for example, a majority of the committee could move for floor consideration of a bill even if the chair was opposed. In addition, a "subcommittee bill of rights" was adopted, permitting a member to chair only one subcommittee, taking away the power of chairs of large committees (such as Appropriations) to appoint subcommittee chairs unilaterally, and guaranteeing subcommittee chairs their own budgets and staff (which they could appoint).

In reforming the party organization itself, the caucus was revived as a major forum for discussion of policy as well as party strategy. Not only did it acquire the power to seat or unseat committee chairs, but it was given the power to elect the whip and the chair of the campaign committee as well as the selection of its own Steering and Policy Committee. Above all, the caucus was convened more frequently (and could be called on petition of fifty members). Its policy role was enhanced greatly, particularly after Gillis Long, of Louisiana, became Democratic caucus chair in 1980, resulting in much caucus debate over issues in the Reagan period. As Sinclair puts it, "the mid- and late-1980s witnessed the emergence of strong policy-oriented leadership in the House of Representatives."[37]

The Republicans also instituted a series of reforms. They gave their conference the right to vote on the ranking members of committees, particularly the ranking Republican on the Budget Committee. They revived their Policy Committee. They set up a new, more representative Committee on Committees and attempted to instill more discipline by the conference's adoption in 1988 of rules permitting the party leader to identify certain "leadership issues" on which it was expected that all Republican leaders would support conference positions.[38] In 1994, when the Republicans took over control of both houses of Congress, they instituted the same basic reforms providing for party leadership and control as the Democrats (discussed in detail in our later chapter on the party in Congress).

There has been much critical and supportive discussion of these reform actions by the parties, within Congress when they were adopted, and among academicians in evaluating the meaning and impact. Not all scholars feel the reforms have been effective; some feel the reforms are contradictory, increasing democratic control but also in some respects increasing the power of the top party leaders. Nevertheless, on balance the view is positive — that a more coherent concept of "party" has emerged in Congress, that members can participate more meaningfully than before, and that leaders are more accountable. As Rohde concludes, "While in an earlier era it may have been possible for scholars accurately to assert that political parties were of little theoretical importance in explaining political behavior and legislative results in the House, that is certainly not true now."[39]

EVALUATION OF THE NATIONAL ORGANIZATIONS TODAY

Our description and assessment of the national party organizations has stressed certain key points. First, there has been great change in the functioning of these national party units, the product of reforms which began really in the early seventies. Second, obviously today these national organizations have become very important, thus confounding those who predicted that party organization would increasingly deteriorate and prove hopelessly decentralized and dysfunctional. From a structural standpoint the national party organs or units are almost as disjoined and disconnected as ever. It is truly a co-archal structural pattern. There is no central leadership entity, no focal center of party power. It is a collection of diverse bodies. But individually and collectively these party organs have reformed themselves, usually under able, aggressive, perceptive leadership, to improve their *party roles,* to assert themselves as major actors in the party system. It is true that the function and nature of the national convention has certainly been modified and has probably declined. But for the other components in the national party arena one sees only an increase in the level of critical performance. Thus, the national party has enhanced its prestige, its powers, and its dominant role in the American party system. In the past it was perhaps mainly a mere symbol of national party legitimacy; today it is much more than a symbol. It is taking the lead in transforming the parties into more effective competitors in elections and for governance than has been true for a long time. The long-predicted deinstitutionalization of party structures has been reversed; instead today we see a revitalization of American party structures.

NOTES

1. James Q. Wilson, *Political Organizations* (New York: Basic Books, 1973), 95.
2. Gerald Pomper, quoted in James W. Ceaser, "Political Parties — Declining, Stabilizing, or Resurging?" in Anthony King, ed. *The New American Political System* (Washington, DC: AEI Press, 1997), 129.
3. James Reichley, quoted in Ceaser, "Political Parties."
4. Paul S. Herrnson, *Party Campaigning in the 1980s* (Cambridge: Harvard University Press, 1988), 130.
5. Ceaser, "Political Parties," 89.
6. Austin Ranney, *Curing the Mischiefs of Faction* (Berkeley: University of California Press, 1976), 16–17.
7. W. Wilson, *Congressional Government* (Boston: Houghton Mifflin, 1875) 76, 80. Quoted in David W. Rohde, *Parties and Leaders in the Post Reform House* (Chicago: University of Chicago Press, 1991), 3.
8. V. O. Key, Jr., *Political Parties and Pressure Groups* (New York: Crowell, 1964), 334. See also David E. Price, *Bringing Back the Parties* (Washington, DC: Congressional Quarterly Press, 1984), 36–46.
9. Herrnson, "Reemergent National Party Organizations," in L. S. Maisel, *The Parties Respond* (Boulder: Westview Press), 43–45.
10. Samuel J. Eldersveld, *Political Parties: A Behavioral Analysis* (Chicago: Rand McNally, 1964), 9–10, 98–117.
11. V. O. Key, Jr., *Politics, Parties and Pressure Groups,* 4th ed. (New York: Crowell, 1958), 347.
12. Anthony Corrado, in Daniel M. Shea and John C. Green, eds., *The State of the Parties* (London: Rowman and Littlefield, 1994), 65.
13. Referred to by William Crotty, ed., in *The Party Symbol* (San Francisco: W. H. Freeman, 1980), 40.

14. Key, *Political Parties,* 396.
15. Sabato, 1997, 118.
16. Herrnson, "Reemergent National Party Organizations," 41.
17. Harold Lasswell and Abraham Kaplan, *Power and Society* (New Haven: Yale University Press, 1950), 204.
18. Key, *Politics,* 356. (Based on Democratic National Convention *Proceedings,* 1948, 548–49.)
19. Herrnson, *Party Campaigning,* 40
20. Ibid., 42.
21. Price, *Parties,* 34–86; Herrnson, 46–66.
22. Herrnson, "Reemergent Parties," 41–48.
23. Herrnson, *Party Campaigning,* 1994.
24. Herbert E. Alexander, "Financing the 1996 Election," in Roper Center for Public Opinion Research, *America at the Polls 1996* (Storrs, CT: Roper Center, 1997), 154.
25. Anthony Corrado, "Financing the 1996 Election," in Gerald Pomper, ed., *The Election of 1996* (Chatham, NJ: Chatham House, 1997), 152–53.
26. Herrnson, *Party Campaigning,* 42–44.
27. Ibid., 44–46, 74–77; Frank J. Sorauf, *Party Politics in America* (Boston: Little, Brown, 1980), 445-64.
28. Herrnson, 51.
29. Ibid., 56.
30. Ibid., 60-65.
31. Ibid., 77-81.
32. Paul A. Beck, Audrey Haynes, Russel J. Dalton, and Robert Huckfeldt, "Party Effort at the Grass Roots" (unpublished manuscript, 1994).
33. Rohde, 1991, 18.
34. Barbara Sinclair, "The Congressional Party," in Maisel, *The Parties Respond,* 223.
35. Rohde, *Parties and Leaders,* 24.
36. Corrado, "Financing the 1996 Election," 152.
37. Sinclair, "The Congressional Party," 227. See also Rohde, *Parties and Leaders,* 66-69.
38. Rohde, *Parties and Leaders,* 136–37.
39. Ibid., 192.

CHAPTER 7

State and Local Party Structures: Strengthened and Still Relevant

Implicit in our discussion of the national party organizations was the significance of the state and local levels of the party. Since the activity and the powers of a party are only partially exercised at the national level, with heavy emphasis on local representation and support, it is necessary to study party organization below the national level to discover what an American party really is. Recent reforms, as we have shown, have made the national organizations much more effective. But it is clear they have to work with the state and local organizations and indeed are dependent on them. If one asks who has the power and who has the responsibility, in vote mobilization terms, the answer for the American system is that the state and local units of the parties share both power and responsibility with the national.

There has been considerable controversy recently over the question of the actual effectiveness of state and local structures. There are those who argued twenty-five years ago that a disintegration of lower-level party organization had taken place. David S. Broder of the *Washington Post* wrote:

> The condition of the political parties at the state and local level is so varied as to defy safe generalization. For the most part, however, they are plagued by inadequate finances and the lack of a trained, stable cadre of personnel to man the headquarters and provide essential services for the party's office holders and candidates. . . . Someone has said that the political parties may well be the worst-managed large enterprises in America.[1]

Broder goes on to argue that some states from time to time have strong organizations and these are tied to strong executive leadership in the mayors' and governors' offices. However, the bulk of the "task [of reform] is still ahead of us," he says.

Along with this theory of deinstitutionalization was the view of displacement of the local organizations by other institutions, forces, and groups. Many scholars maintained that the new techniques of mass persuasion in campaigns had replaced the local party organizational efforts at communication. The voting public presumably secured its campaign information and stimuli from the specialized media campaigns. Others argued that state and local organizations were being displaced by other forms of group action. Single-issue groups, often representing new interests and causes, had appeared. It was argued that, on one hand, people participated much more in politics through such groups and, on the other hand, the public listened to the appeals of these groups more than to the parties.

The argument that these developments led to the demise of state and local party organization, if once true, can no longer be accepted. There is considerable evidence from recent research on state party organizations that they have become larger, more professionally structured, and better financed. State and local parties in many ways are important infrastructures undergirding the electoral process. We will present evidence of this a bit later.

BASIC FORM OF THE STATE AND LOCAL ORGANIZATION: A STRATARCHY

To understand parties within the states, it is necessary to comprehend the basic structural form and the essential nature of organizational relationships. It is like a series of layers of organization, or strata, superimposed figuratively one on the other but with no indication that it is a neat pyramid of authority. Figure 7.1 attempts to represent the strata in this nonhierarchical sense. There are no dependency connections between these layers. It is not usually a hierarchy of command from the state chair to the district chair to the county chair and on down to the precinct chair. That does not mean that there are no formal connections whatsoever, because surely there are. Lower-level conventions select delegates to upper-level conventions, for example, and some city chairs may appoint ward chairs. But the key elements in a stratarchy in the states are:

1. No command from the state level down to the base of the organization
2. Limited control (if any) by intermediate echelons of authority over another
3. Autonomy in decision making at each stratum and, thus, dispersion of decision-making centers
4. Reciprocal deference, or the tolerance by each level of authority of other levels and a recognition of the independent status of each level in the system.

This means, in addition, that (1) there is limited interference in the operations of each level from above or below and thus freedom to develop indigenous approaches to party work; (2) the clear lines of communication found in a hierarchy are not found here, and leaders rely on informal associations and persuasion, rather than directives; and (3) holding leaders and committees accountable for the functioning of the total structure is virtually impossible, although individual chairs and committee members are, indeed, periodically elected. There is no centralized leadership locus for the entire state party organization in such a stratarchy.

Recent research on the strengthening of state and local party roles in campaign operation, especially in cooperation with the national committees, suggests more integration in national-state-local relationships. While this is so, it does not mean the disappearance of the type of stratarchy discussed here. Now, to say the system tends to stratarchy is to suggest a model with certain characteristics. Not all of these elements of the model may be equally true in each of the fifty states, however. Two alternative models emphasize other characteristics of the state and local

FIGURE 7.1 ▪ Party Stratarchy

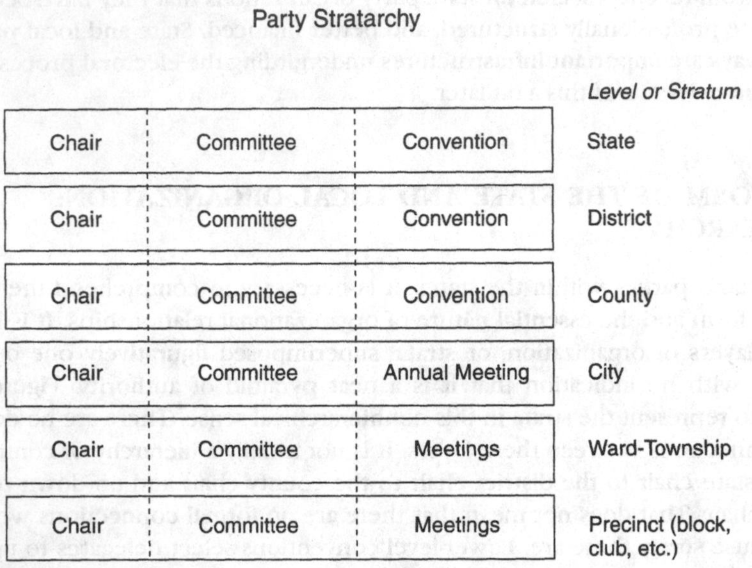

Party Stratarchy

			Level or Stratum
Chair	Committee	Convention	State
Chair	Committee	Convention	District
Chair	Committee	Convention	County
Chair	Committee	Annual Meeting	City
Chair	Committee	Meetings	Ward-Township
Chair	Committee	Meetings	Precinct (block, club, etc.)

parties. The two alternative models highlight the striking contrasts which theoretically could exist in state structures. The command model emphasizes a chain of command and a flow of directions from the state chairperson on down through the five levels. Committees are subordinate to the chairs, who if controllable at all are controlled through election and reelection by the state convention. The echelons are certainly not autonomous under this model, and the deference is strictly upward!

In the accountability model, there are various possibilities. But its essence is that those at the base of the organization have opportunities to elect, control, and hold accountable their own local chairpersons and committees as well as, through indirect or direct electoral processes, district, county, and state chairpersons and committees. There may still be considerable autonomy in each stratum, but the deference in the organization is certainly downward.

While the state organization is usually more of a stratarchy than a conformity to either of the other two models, from time to time in certain state and local party structures there is a tendency away from stratarchy to either of these other two models. Certainly the city boss and his machine, exemplified by former Mayor Richard J. Daley and his party in Chicago, came close to the command model, although studies of his organization revealed much evidence of reciprocal deference.[2] On the other hand, in some states attempts have been made to at least implement the accountability model. Thus in about one-fourth of the states there is direct election of state level committees by party supporters at the base of the system.[3] Caution, therefore, should be taken before jumping to the conclusion that all state and local party structures are complete stratarchies. Obviously this is not so. However, if parties are to respond to changing social problems, adapt to the needs of citizens, and maximize their power in elections by mobilizing votes effectively, there must be considerable tolerance by upper-echelon party leaders of the lower-level activists, diffusion of control, and much autonomy in decisions. There must be respect for local initiatives, local strategic improvisation, and recognition of the need for adaptation at the base of the organization. In organization relationships, parties must accommodate to actors at all levels of the organization. There also must be deference by the local leaders to those at the top and respect for their leadership prerogatives, but certainly there must be respect by the leaders at the top for the local organization activists, for the latter theoretically have the power to mobilize the vote. As Elmer E. Schattschneider said long ago, noting the decentralization of power in American parties, "The strongest and most stable organization within a city may be a ward organization rather than a city-wide machine."[4]

OTHER MAJOR FEATURES OF STATE ORGANIZATIONS

There are other important characteristics of American state party organizations. First, the indirect method of election and representation from one level to the next is found in many states. Precincts, wards, and townships (or clubs, assembly district, etc.) elect delegates once in two years to county or district conventions, which in turn elect delegates to the state convention. Therefore, if an interested and attentive voter wishes to influence the party organization, he must be certain to vote in that one election opportunity, the party primary election when local delegates are selected. In terms of party control, that is the crucial election.

Second, in most state party structures there is separatism or dualism, which is the tendency for the legislative party organization to exist pretty much by itself. Some states do give seats on the state central committee to legislative party leaders, or legislators may seek control over the state organization, but often they are separate party leadership systems. This is true also at the county level, where often the party representatives on the county board or commission keep organizationally separate although they may maintain informal relationships with the county party chairman and the county committee. This separation tendency parallels identically the national party level, with Senate and House party organization generally keeping aloof from involvement with the national committees.

Third, in most state organizations there is a preoccupation with organizational specialization. Thus, many states set up a professional financial organization specializing in fund raising. Similarly, there are other specialized organizational subunits and divisions dealing with such matters as patronage, race and ethnic relations, women's affairs, and coordination of publicity.

STATE-LEVEL PARTY UNITS

The same types of party organs are at the state level as at the national: state chairperson, state central committee, state convention, and party organizations in the state House and Senate. The state chairperson is usually selected by the state convention to a full-time position and is often well paid. The tenure is, however, rather brief, about two years. The job of the state chairperson is variable from state to state, but generally it is a relatively important position in the state organization. The tasks and prerogatives include being a member of the national committee, presiding officer of the state central committee, manager of the state party headquarters, patronage clearer if the party is in power, spokesperson for the party in the state, and above all, if the chairperson wishes to assume the roles, factional broker and organizational liaison with the critical groups in the party. Of all the tasks, research reveals that state chairpersons consider the following as very important: building of the party organization, fund raising, campaign activity, and candidate recruitment. In addition they see themselves as liaisons with lower party leaders. State chairpersons often maintain that they participate in primaries and are actively in contact with county party organizations. What this means, however, in terms of types of activities and the extent to which county organizations can be influenced is another matter.

The role of the state chairperson is, again, often highly contingent on the extent of support from the governor. One study identifies a particular type of chairperson as the "political agent" of the governor, contrasted to the one who is independent of, and possibly in conflict with, the governor as well as the chairperson of the party out of power, who has to act more on his or her own authority. The state chairperson thus has to act within the context of powerful other people — governor, House and Senate leaders, big city leaders, interest group leaders, and entrenched county bosses. The chair's role is to take care of the organization, encourage party workers, and resolve conflicts. The chair is better financed today and has a larger permanent staff than before. He or she works through a state central committee varying in size from fifty to five hundred and is usually accountable to a state convention that reviews the chair's mandate at least every two years.

The other state party organs vary in form in each state, but there are similarities. The state central committee seems to be a buffer group. It is strictly representative of the district and county parties, often on a proportional basis of representation, and is a prestigious position for an upwardly mobile party activist. It can be important for making certain organizational policy decisions, such as planning for state conventions. It is not a key policy leadership body, although it may debate policy and thus provide a certain catharsis for those party activists who are ideologues. In most states it usually makes no critical policy decisions for the party.

The state party convention is the analogue of the national convention, including the reality that its nominating function is usually also atrophied. The state convention used to be of great importance in selecting the party's candidates for governor, lieutenant governor, and a host of other state officer. Today it is still used in some states for that purpose. In Michigan, for example, it selects the candidates for lieutenant governor, attorney general, secretary of state, state treasurer, superintendent of public instruction, and members of the boards of regents of the three main universities, as well as state supreme court justices. In several states the convention is a pre-primary endorsement forum. In Colorado any candidate who gets 20 percent of the convention vote automatically qualifies for the primary. In New York a candidate needs 25 percent of the convention vote to enter the primary. In Utah a convention endorsement is necessary to get into the primary. Thus, the convention is still relevant to the nominating process in several states, but it does not have the final authority. The emphasis in the state convention, as in the national, is on the representation of lower units of the party; the opportunity to discuss mutual experiences and problems; involvement in discussions (usually off the convention floor) of party policy; and, periodically, participation in the selection of (1) the state chairperson, (2) the state committee, (3) the delegates to the national convention, and (4) the members of the national committee. As an important intermediate personnel selection body, as a buffer group, and as a symbolically important representative assembly, therefore, the state convention has a significant role. Theoretically it is the plenary body in the state party organization; actually, it does not often assert its authority. It must be seen as a key integrative party institution that brings together people of heterogeneous backgrounds and interests to share a feeling of like-mindedness while performing organizational tasks.

Although state party units are not powerful in a hierarchical control sense, they are important nevertheless. They are indispensable for the maintenance of a viable Republican and Democratic party. They operate collaboratively with district and local organizations, and above all, they work closely these days with the representatives of the national committees in planning and financing campaigns. They are infrastructures of coordination, accommodation, and liaison, working on state problems with state leaders to resolve conflicts and promote party harmony.

VARIATIONS IN STATE PARTIES AND THEIR POLITICAL ENVIRONMENTS

It has been said that the United States has one hundred different state party systems, and a study of state organizations indicates this characterization may be true. State organizations live in different social and political environments and thus seem to develop differently. To begin with, state parties reveal great differences in the extent of organizational unity. The analysis of contests within the parties for state office reveals this. In some states party factions fight regularly for leadership positions; in others there are rarely such primary contests.[5] At national conventions some state party delegations may be very unified in their voting while other delegations seem very divided. Further, state parties differ in their liberalness or conservativeness. Thus, two pictures emerge — one of considerable factionalism and disunity, the other of variations within the parties, in terms of ideological differences. The effect of these conditions on the way politics is conducted in a state may be great.

Further, the political competition and turnout environment in each state within which the parties have to compete varies greatly. The proportion of voters going to the polls can range from 70 percent to 30 percent. The extent of party competition in a state can also vary greatly. The two (competition and turnout) may be related, but not always. Parties will function quite differently in a situation of voter apathy than of greater voter involvement. In one-party areas the task of the opposition party to persuade candidates to run is much more difficult.

No doubt other environmental conditions under which state and local parties must function — such as the type of direct primaries states have, the nature of the ballot, and similar legal regulatory conditions — limit the party organization's effectiveness or at least require extra efforts by the party to mobilize support.

State party organizations have changed over the past two or three decades, in some respects similarly to the national organizations. We have evidence from a variety of special studies that document this.[6] Surveys of state party chairs from 1960 to 1992, as well as data provided by the Federal Election Commission, reveal what these changes are. First, the state parties have built up their own staffs, headquarters, and funds for their own operations. Almost all now have their own buildings, full-time chairs or executive directors, and some full-time salaried staff. The average budget increased from under $200,000 annually in the sixties to double that by the early eighties, and much over that today. A careful analysis of the financial situation of state party committees in the 1992 election based on FEC reports reveals that these committees are in excellent financial health. In that year, 80 percent of the Democratic state parties raised over $500,000 each, as was true of 72 percent of the Republican state committees. The total amounts raised in 1992 that state parties reported to the FEC were, for the Democrats, $89 million and for the Republicans $86 million. The high was California for the Democrats ($12.6 million) and Texas for the Republicans ($7.0 million). The Democratic and Republican national committees contributed in nonfederal "soft money" to state and local parties considerable funds: $35 million to the Democratic state and local committees, $29 million to the Republicans.[7] In addition, local organizations raised funds which, as reported to the FEC, amounted to $4.5 million (Democrats) and $7.1 million (Republicans). As a result of these resources, state (and also local) parties can assist candidates for offices at all levels with a variety of services such as polling, media advertising, recruiting and training workers, campaign managing, and conducting seminars on campaign planning and strategy. Previously state organizations have not had such extensive resources, although there is evidence that in 1978 the RNC began to assist state parties and the Democrats followed in the mid-eighties. By 1988 each party was committing at least $20 million to state and local activities. In addition, LCCs (state legislative campaign committees) appeared after 1970 and are found now in almost all states, raising considerable funds on their own, apart from the state party committees.[8] Thus, state and local party organizational finance has become much more complex, affluent, and influential than before.

V. O. Key said that the strength of the state party organizations and their competitiveness determined the type of policies the citizens in a state would see adopted and hence determined their social and economic welfare. The organization interacts with its political environment, reflected in the level of voter interest and the nature of its opposition. Strong organizations in a healthy political environment should mean more effective governmental action. In the last analysis this is the critical linkage.

DISTRICT AND COUNTY LEVELS OF ORGANIZATION

District and county units can be, depending on the locality, key cogs in the state party structure. The obvious reason for this is that they are the geographical electoral areas from which congressional representatives, state legislative personnel, and county governmental personnel are selected. Theoretically, therefore, the party organization at these levels could have an important role to play in electing these individuals. This may be the case, depending on the type of campaign organization the party candidates for Congress, state senate, county sheriff, et cetera, wish to put together. Often they will indeed depend, at least in part, on the district and county organizations; often they prefer to set up their own ad hoc organizations.

The county, or district, chairperson was traditionally considered to have a powerful position. In earlier days, before the primary, the recruitment and nomination of candidates was an important task for the local organization, but this has declined, although county and district chairs may in certain areas still spend a great deal of time persuading party regulars or outsiders to run for office — or not to run. Although this function may be of less importance today, the county or district chair can still be influential. If the party is in power in the state or the county or both, there may be considerable patronage to dispense, and this is ordinarily funneled through the county or district chair's office. As the coordinator and planner of party activities in the area (not only the planning of the county convention), the chairperson has a major opportunity to play an important role. He or she can build the organization from the precinct and township up, particularly if there is no strong city organization in the area. Building a strong party machine in the area is a goal which may be either a satisfaction in itself or a stepping-stone to a higher career. The county and district chairpersons are often kingpins in the state organization, playing major roles at the state convention and being consulted by the state leaders before any critical decisions are made. While the potential for a significant role is there, the county and district chairpersons often operate in dangerous conditions. They have to maintain harmony among the many warring factions in the organization, some of whose leaders may aspire to the very position the leader is seeking to preserve. An early study of Detroit congressional district party leaders found that dealing with factions was the overwhelming priority. Here are quotations from two district chairmen:

> The district chairman has to be a Sherlock Holmes in the Republican party. We have, as you know, many splits in our party, and there is a constant battle of wits as to who is for whom.

This from a Democratic district chair:

> Let's take this district here. There are two major racial groups, the Negroes and the Jews. But it is not only nationality groups, but also political groups — for example, we have the Young Democrats and the CIO. We do not have a right wing faction, but we have small businessmen's interests and labor interests. Our trade unions are split into two separate units, even though they are externally united. The AFL is conservative, the CIO is the liberal group. . . .[9]

Clearly these leaders are aware of the linkage of the organization to socioeconomic interest groups. They also are realists and know how fragile their hold on power is! These observations still hold true today even though the types of factional groups in the parties may be somewhat different.

County and district committees can be important adjuncts to the chairperson's activities as well as important representatives of all the diverse activists in the area. They also can be merely social get-togethers, relatively quiet, and perform only intermittently. Yet to get on a county or district committee may be an important goal for the party worker. The county or district convention, which used to hold the all-important nominating function (now taken over by the direct primary), is still usually the basic representative link of the party with the local activists. It is the first stepping-stone upward in a party career, and from this convention the aspiring careerist can go to the state convention as a delegate or assume other responsibilities in the party. But the heavy emphasis in the county and district arenas is *organizational* influence.

Today we have much better knowledge of the level and type of activities of county organizations than in the past. Excellent systematic surveys of county party leaders in 1964, 1979–80, and 1992 give us much of the information we need to assess county party organizations.[10] The county leaders reported in 1980 that in their opinion the county organizations had become stronger (since the sixties): 54 percent stronger, 22 percent little change, 23 percent weaker. Actually the 1964 National Election Study (based on a national probability sample of counties) revealed a higher level of county party activity than most scholars anticipated — for example, 60 percent conducted voter registration drives and engaged in fundraising and publicity for local candidates while 80 percent distributed campaign literature. The later studies collected much more data and permit a comparison of activity by the parties. Although the level of activity in 1992 was on average (for eleven types of activities) only slightly higher than in 1980, there were certain types of work where both parties seemed to be working harder. See Table 7.1 for a few examples.

It is clear from these data that, for the past thirty years at least, a majority of counties have parties which do work in campaigns, and in some of these activities they are more active in the nineties than previously. The real point is that there is strong evidence here of substantial county party activity. So far as the function of recruitment of candidates is concerned, a majority of leaders in these studies say they are "very" or "somewhat" involved in finding, persuading, and supporting possible candidates for a variety of offices — 90 percent say they did this in 1992 for

TABLE 7.1 ▪ Level of County Party Activity (percent of county chairs responding affirmatively)

	Republicans		Democrats	
Activity	*1980*	*1992*	*1980*	*1992*
Distributing literature	79.0	88.0	79.0	90.0
Fundraising	68.0	74.0	71.0	76.0
Organizing campaign events	65.0	77.0	68.0	61.0
Contributing money to candidates	70.0	75.0	62.0	67.0
Coordinating county level campaigns	56.0	57.0	57.0	59.0
Organizing door-to-door canvassing for votes	48.0	52.0	49.0	55.0
Average for all eleven activities	54.6	56.3	54.1	58.2

SOURCE: James L. Gibson, et al., *American Journal of Political Science,* 29 (1985): 139–60; John Frendries, et al., in Daniel Shea and John Green, eds. *The State of the Parties* (Lanham, MD: Rowman and Littlefield, 1994), 138–41.

county offices, over 80 percent for state legislative offices, and 66 percent for congressional offices. All of these are increases over the reports of leaders in the 1980 study. Such data constitute a strong rebuttal to those who are skeptical of local party activity.

THE CITY, WARD, AND PRECINCT ORGANIZATIONAL LEVELS

This, in many respects, is the most fascinating level of the party. However, to generalize about party organization at the local level (below the county) is difficult. Local party organization varies by region, by state, and by communities within states. In some states there may be only town committees or clubs and no wards or precincts. The form of organization and the functions performed by these lower units can vary greatly. Such differences are linked to political traditions that influence party practices.

The style of "grassroots" local party organization is actually bewildering in its variety. Although the day of the city boss and his party machine is allegedly behind us, the signs of machine politics of the old style can still be found in certain areas. On the other hand, a majority, indeed over 70 percent, of cities with a population of 25,000 and over has *nonpartisan* elections, which doesn't mean the party organization is completely quiescent but does contribute to a different type of party environment. Partisan politics differs at the local level and leads to great diversity in party organizational form. At the risk of oversimplification, all that can be done here is to summarize from the available literature some of the major characteristics of American local party organization.

First, local party organization is very voluntaristic. The organization is usually wide open to anyone who is interested in participating. Positions are usually not rewarded by pay or by patronage or other material rewards, although this was true in the days of the big-city machines. Individuals are not trained but often learn on the job and eventually do achieve some real competence for party work. Self-starters and those who are not recruited by the party leadership make up a large proportion of the workers and those who hold precinct and ward positions. Turnover is considerable each year. Studies reveal that even for a majority party up to a third of the lower-level positions (such as precinct positions) in an area may be unfilled, and many more are filled by untrained personnel. Further, many lower-level leaders are brought into the organization by friends, family members, and working associates, or join by themselves. Thus, the need for workers is continuous, recruitment is haphazard, with no careful screening of available personnel, and people are appointed often by default, because there is no one else for the position.

In certain big-city machines, however, even very recently, there was much more control over the selection of party personnel than the above description would indicate. Patronage was (and, to some extent in some areas, still is) important as the incentive and reward for local party personnel. Mayor Daley is reputed to have had thirty thousand to thirty-five thousand jobs to give out through the Cook County Democratic organization.[11] A 1970 study claimed that three-fourths of Pittsburgh's Democratic committeemen held public jobs. During John Lindsay's administration in New York, there were thirteen thousand personnel on "temporary" appointments and another seventy-five thousand noncompetitive jobs not under civil service.[12] There are no systematic studies to indicate how much patronage is still available as payment to party personnel, but the above illustrations are the exceptions rather than the rule. In many large cities, no more than 10 percent of precinct leaders are on the public payroll. For the large majority of local party personnel, therefore, it is basically a voluntary and avocational role.

A second characteristic is the diversity in the backgrounds, interests, expertise, and incentives of people who work at the local level. It is obvious that with such little control over entry in party activism and into precinct and ward positions, it is inevitable that the workers will not be a homogeneous, like-minded group. Again, local organizations will vary a great deal in the class, racial, ethnic, and other demographic characteristics of their activists. In some areas there is a tendency toward more middle-class backgrounds; in others, more working-class. Some wards may be almost entirely lower-class blacks, others may be middle-class Polish Catholics, others upper-class professional Protestant whites. Although there will be clusters of socially homogeneous neighborhoods with their socially homogeneous party activists, the local party organization for the entire community will be a great mixture. And the potential for differences in perspective, for controversy over ideology and policy, and hence for disruptive factionalism is considerable. Above all, such heterogeneity makes it difficult to achieve party harmony.

A third characteristic of the local party organization is its improvising mode of operation: there is a looseness, flexibility, and informality in how local party cadres operate. They are not controlled much and resort to their own ingenuity. Since there is much autonomy at the local level, the local ward and precinct leaders and their coterie of supporters and workers (often family and friends) can decide by themselves how much and what to do. Communication is often poor, either from the city level downward or laterally from ward to ward and precinct to precinct. Thus, leaders often define their own tasks and roles. This may mean that nothing is done or the precinct work is done poorly, but it may also mean that it is done well and done in a way adapted to the conditions of the people living in a particular neighborhood. In a sense, the local level of party operations is a small army or company of party activists who have somehow wandered into the party operations and then very often are left to themselves to improvise as best they can in order to maximize the party's goals, however they conceive those goals. This, again, is to a great extent the derivative of a centrifugal and stratarchical system.

A fourth characteristic of the local structures is that they often thrive on and reflect apathy about politics. Few citizens take an active and continuous interest in local elections. The percentage of eligible voters turning out for city elections is usually 25 percent or less. Fewer of them will take the time to vote for precinct delegates to county conventions or participate in the election of precinct, ward, and city leaders at annual meetings. As a consequence, local party elections can often be controlled by a small number of votes, precinct leadership and delegate positions are won by default, and reform of the local party organization usually does not occur. When attempted, however, it can be easier than originally predicted. The local party establishment survives often because of inert opposition. The organization is in the hands of activists whose interest is fleeting or careerists who are not innovators in policy, indeed who may not be very well informed on policy matters. The pressure for policy leadership so badly needed at the city, county, state, and national levels of the party organizations is minimal and infrequent. Status quo politics flourishes under these conditions, for state and national leaders hesitate to risk policy innovation in the absence of pressure from below.

A fifth and final basic characteristic of local parties is organizational slack, that is, a tendency for the local activists to perform at a minimal level of efficiency, without too much system, in a rather hit-and-miss mode of operation. In an efficiently functioning local organization, able personnel staff the key posts in the wards and precincts with sufficient workers who have been adequately trained and know their jobs. In a well-run local party organization there is a plan of operation, a campaign strategy, an effective communication system.

The "grassroots" local party organization ideally should be involved in voter registration drives, fundraising, public rallies (and smaller social events such as coffee hours, barbecues, etc.). In mobilizing the vote the grassroots activists should be engaged in telephone canvassing, house-to-house canvassing, literature distribution, and election day get-out-the-vote activities (last minute contacting of voters and, if necessary, transporting them to the polls).

Recent studies again suggest to what extent these types of activities "on the ground" are actually being performed. The study by Beck, Dalton, Haynes, and Huckfeldt in 1992, which interviewed party leaders and campaign coordinators in forty representative counties and focused on the activities of the local party organizations on behalf of the presidential campaigns, is very indicative of grassroots party efforts.[13] The presidential local parties differed considerably in the extent of their involvement in the presidential campaign, as the data in Table 7.2 reveal.

TABLE 7.2 ▪ Types and Levels of Local Presidential Party Activities in 1992 (percentages reported by campaign coordinators)

Activity	*Republicans*	*Democrats*
Register voters	45	80
Transport voters to polls	21	72
Distribute literature	50	73
Telephone voters	58	87
Distribute news releases	32	54
Plan media advertising	20	24
Organize campaign events	60	73
Average for eleven activities	40	62.2

SOURCE: Paul A. Beck et al., "Party Effort at the Grass Roots: Local Presidential Campaigning in 1992" (paper presented at Midwest Political Science Association, Chicago, April 1994).

They concluded from the study that "the local parties were active players in the 1992 presidential campaign. Despite the fixation of reporters and analysts on the national campaign, the tentacles of the presidential parties clearly reached down to the grass roots, where they were locked in 'heavy combat.'"

Many studies of precinct and ward leaders and activists have corroborated these findings, studies done in particular cities across the nation.[14] One such study was done in Detroit in 1956 and repeated frequently until 1992. The level of activity varied from election to election and by party (the Democrats in recent years were more active than the Republicans). What is impressive, however, is the continuance of the local party activity throughout these ten presidential election campaigns. Illustrative findings are presented in Table 7.3.

TABLE 7.3 ▪ Activity Level of Local Party Activists in Detroit, 1956–1992 (in percent)

	Democrats			Republicans		
Basic Activity	*1956*	*1980*	*1992*	*1956*	*1980*	*1992*
1. Registered voters	93	42	71	80	19	16
2. Did house-to-house, or telephone, canvassing	46	60	63	32	61	46
3. Got out the vote on election day	68	69	88	80	62	49
4. Performed all three of these critical tasks	17	30	56	25	16	11
Performed two of these tasks	38	35	22	22	34	19
Total for Number 4	55	65	78	47	50	30

SOURCE: Samuel J. Eldersveld, *Political Parties: A Behavioral Analysis* (Chicago: Rand McNally, 1964), 74–75; more recent data from special studies by Eldersveld.

Studies reveal also that precinct leaders are relatively active in party work between elections — holding rallies and social events, giving advice to and aiding constituents. Further, over half usually are joiners and involved in community life by belonging to social groups. These contacts may be of considerable value in vote mobilization work.

The local party has tremendous potential organizational power. This is obvious from the evidence of this great organizational slack. If the local party is activated and mobilized properly, it can be extremely effective. Why isn't the potential mobilizational power of the ward and precinct realized more frequently? A combination of factors is responsible: activists are often ignorant of what to do and are poorly trained, advised, and led. The motivation to put a great deal of time into the precinct work doesn't exist often, either because the candidates are not attractive or the issues not clear-cut. Conflicts among factions destroy the sense of organizational unity and purpose and also result in the loss of a tremendous amount of time spent in promoting harmony. Local leaders may be engaged in a holding operation, feeling no need to exert themselves in order to hold to their position and its prerogatives. Because the turnover rate is high, the local effort is too sporadic and not a continuous operation; hence, start-up costs for each campaign are considerable. The system of rewards for local activists may be inadequate to keep them performing at a high level of efficiency, offering little patronage, limited choices for upward mobility, and inadequate recognition. All these are important reasons, but there is a more basic condition. What will activate a local political party is an organization which is exciting to be a part of, which is working effectively for community projects, which involves people with the party in a social, as well as a political, sense. If this element of excitement and involvement in worthwhile projects is present and the local party has able, dedicated leadership, the local organization can be dynamically involved, produce the vote, demonstrate its organizational power, and in the process have a great effect on the community.

BOSSES AND POLITICAL MACHINES: ARE THEY WITHERING AWAY?

In March 1979, an event occurred which some considered impossible: Jane Byrne defeated Mayor Michael Bilandic and the Chicago Democratic machine in the primary, thus being virtually assured of election as the next Chicago mayor. *Time* magazine quoted Alderman Vito Marzullo as saying before the election, "The Machine may not be well oiled but it will never break down. Mayor Bilandic is going to swamp them."[15] But the Daley-Bilandic machine did break down, to the surprise of many. The question raised by many now is, Have we seen the last of the political boss and his machine?

American party politics is often considered by foreigners to be unique because of the phenomenon of big city bosses with their machines. James Bryce in *The American Commonwealth* described bosses and rings at the turn of the century and while cautioning his European readers not to take too dim a view of "the

The Philosophy of a Political Boss

Beginning at the end of the nineteenth and lasting into the twentieth century, many cities had political machines and bosses. New York City had its Tammany Hall, and George Washington Plunkitt was the boss.

Plunkitt imparted his political wisdom and lore to a newspaperman who wrote these insights up for posterity in the book *Plunkitt of Tammany Hall.* He was asked: Why did bosses emerge and how did they build political machines? Here was part of his answer:

> These days . . . nobody thinks of drawin' the distinction between honest graft and dishonest graft. There's all the difference in the world between the two.
>
> There's an honest graft, and I'm an example of how it works. I might sum up the whole thing by sayin': I seen my opportunities and I took 'em.
>
> First let me explain by example. My party's in power in the city, and it's goin' to undertake a lot of public improvements. Well, I'm tipped off, say, that they're going to lay out a new park at a certain place.
>
> I see my opportunity and I take it. I go to that place and buy up all the land I can in the neighborhood. Then the board of this or that makes its plan public, and there is a rush to get my land, which nobody cared particular for before.
>
> Ain't it perfectly honest to charge a good price and make a profit on my investment and foresight? Of course, it is. Well, that's honest graft.

To many bosses in other cities, these sage insights by Plunkitt made a lot of sense.

SOURCE: William L. Riordon, *Plunkitt of Tammany Hall.* Edited with an Introduction by Terrence J. McDonald (New York: Bedford Books of St. Martin's Press, 1994).

Boss," said, "He is a leader to whom certain peculiar social and political conditions have given a character dissimilar from the party leaders whom Europe knows."[16]

It may be argued that this was not so in Europe in Bryce's time nor today, but the machine is still considered one of our unique contributions to government. To think of machine politics is to think of organizations led by bosses such as Tom Platt and William Tweed of New York, Frank Hague of Jersey City, Edward Crump of Memphis, Tennessee, Huey Long of Louisiana, J. Henry Roraback of Connecticut, Anton Cermak of Chicago, Tom Pendergast of Kansas City, Missouri, Jim Curley of Boston. What was comprehended in the conception of these earlier, classic machines? Essentially they were considered to be authoritarian structures dominated by a boss, carefully organized as command structures with authority exercised from the top down and responsibility from the bottom up, and a cohesive group of individuals that systematically and efficiently (like a machine) performed the tasks of the party (in particular electing leaders, mobilizing votes, and distributing patronage) while exploiting the political system for private purposes. In the classic type the emphasis was on:

1. Hierarchical control of the organization and usually also centralization of authority over the entire urban political community
2. Vote mobilization linked to the provision of a great range of services to the public
3. A material reward system with patronage and financial success coming to those who performed efficiently in the hierarchy
4. Iron discipline and use of a variety of sanctions in maintaining the highest level of efficiency and in replacing those who were not continuously productive
5. Very little emphasis on ideology, although the machine often did promote particular programs of action
6. Manipulation, through the use of the public power of the machine, of other groups and sectors of the community, including the media, labor, business, et cetera
7. Utilization of threats, intimidation, surveillance, violence, and harassment to make organizing an opposition difficult
8. Tolerance of illicit and corrupt actions and relationships as the rewards of power
9. Fraud on election day in vote-getting activities and election administration
10. A very limited ethic of day-to-day responsibility of the machine to the public, except in the special sense of providing services to maximize votes

These were the attributes of the old-style machines, most of which in this extreme form had disappeared by the 1960s. A conception of political organization which was considered peculiarly American, it came into existence in the latter part of the nineteenth century and thrived in the early twentieth century as the response to the needs of the immigrants who were swarming into our cities, the needs of the poor which were not ministered to properly by government, the dissatisfaction of citizens with the poor government they were getting, and the emergence of a type of personal leader who saw the possibility of accumulating personal power by building a special type of political organization combining vote-getting with social welfare. Because it performed critical political functions and fulfilled many human needs and, as most studies of this early period show, because voters were satisfied with the kind of government they got, these machines survived. The machine was a cohesive structure which, whatever else its shortcomings might have been, provided leadership.

There are still many machines with somewhat different characteristics in the United States (and in Europe) today. But this old-style machine has been disappearing for some time. Perhaps the only one to survive into the 1970s in this classic form is the Chicago machine, created by Mayor Anton Cermak in 1931 and continued by mayor Edward Kelly and Patrick Nash from 1933 to 1947, by Mayor Martin Kennelly (despite his attempts at reforms) from 1947 to 1955, and then flourished in classic form under Mayor Richard Daley until he died in late 1976. Mayor Bilandic carried on after him until his defeat by Jane Byrne in March 1979. Harold

Washington became mayor in 1985, winning against the old machine, and was reelected in 1987. His death led to the election of the son of "Boss" Daley, Richard M. Daley, Jr. But while the new Daley has his own effective organization, the old Daley machine was not resurrected.

That the Daley machine was a classic cannot be denied. The journalistic accounts and analyses of writers like Mike Royko of the *Chicago Sun-Times* and Milton L. Rakove have described its recent nature in detail. The 3,500 precincts in Chicago in the fifty wards each had a captain and most had an assistant captain, making upwards of $15,000 a year on patronage jobs.[17] Below them were those who had lower jobs in government (swinging mops at the county hospital, digging ditches, etc.) and who helped out at election time, contributing part of their salaries (estimated at 2 percent) and engaging in other party activities such as attending political fund-raising dinners. Above these personnel were the ward committeemen (of which there were fifty): "the clout," "the chinaman," "the guy," and "our beloved leader." It was a well-articulated structure with power delegated and responsibility specified. Power was delegated from the Cook County central committee of the party to the fifty ward committeemen and from them to the precinct captains. "The entire system operates on the principle of autonomy of authority at each level in the political pyramid." Daley was at the top of this pyramid. Vote production was a responsibility and failure meant discipline, loss of a job. Daley personally took an interest in the disciplinary action. "The leverage a political boss can exert on his lieutenants is directly in ratio to the patronage and other benefits that he can grant or withhold from them. Daley absolutely controlled all such matters in Chicago."

The description of the actual functioning of the Daley machine as Rakove describes it is fascinating reading. The patronage clearance system was precise. For example, a young Chicago attorney in the Democratic party (a precinct captain of Daley's machine) who was being considered in Washington for an appointment to a federal regulatory commission had to get Daley's approval, but this was not possible without the ward committeeman's approval, who in turn checked with his other precinct captains before endorsing him and informing Daley of his approval. An estimated thirty thousand jobs locally were distributed through this system. Precinct captains usually held down these patronage jobs as long as they produced. Before elections they had to estimate what their precinct's vote would be. If they could not do this fairly accurately (estimating too low was as bad as estimating too high), their reputations suffered and it was assumed they were not in touch with the public. A captain who had lost his precinct and abjectly reported to his superior was asked by his ward committeeman bluntly, "What kind of job are you going to look for now?" The ward committeeman has considerable power in such a system. The famous leader of the Twenty-fifth Ward, Vito Marzullo, at one election said, "The mayor don't run the Twenty-Fifth Ward. Neither does the news media or the do-gooders. Me, Vito Marzullo. That's who runs the Twenty-Fifth Ward, and on election day everybody does what Vito Marzullo tells them."

Marzullo was an Italian immigrant, seventy-seven years old, who came to Chicago at the age of twelve and spent over fifty-five years in politics there after only a fourth-grade education.

The services performed by precinct captains were extremely diverse: legal service for the poor, repair of broken street lights, more police squad patrols in a neighborhood, special antirodent cleanups, new garbage cans for tenants provided by landlords, and talks with probation officers for youngsters in trouble. In middle-class neighborhoods, precinct captains could help get tax bills appealed, curbs and gutters repaired, scholarships for students to the University of Illinois, summer jobs for students, and help with the bureaucracy. Marzullo said he had

> the most cosmopolitan ward in Chicago — 30 percent blacks, 20 percent Polish, 12 to 15 percent Mexican, 5 percent Italian, Slovenians, Lithuanians, Bohemians . . . but I take care of all my people. . . . My home is open 24 hours a day. . . . I'll go to a wake, a wedding, whatever. I never ask anything in return. On election day, I tell my people, "Let your conscience be your guide."[18]

Marzullo told his captains, "Mingle with the people. Learn their way of life. Work and give service to your people." According to Rakove, on election day every captain received between $50 and $200 for expenses in his precinct. What the precinct leader did was adapted to the individual needs of people. A top Democratic leader in Cook County maintained that there had been a considerable change in the types of services provided: "The political organization today [in the 1970s] is a service organization, an ombudsman, and an inquiry department. I consider my ward has fifty-seven community organizations doing public service [by his fifty-seven precinct captains, his 'community representatives']."

Another leader agreed, saying, "To a great extent the service we offer now is referral." Daley's language, in March 1975, to describe his precinct captain's services and work is interesting:

> He never gets in the newspapers unless he is criticized. He's as honest as the rest of us and he's a better neighbor than most of us. . . . He has solicitude for the welfare of the family in his block. . . . He gets your broken-down uncle into the county hospital, if he lives in the slums. He's always available when you're in trouble. . . . The people who make up the Democratic party of the city of Chicago are always and will always be a reflection of the people in their respective neighborhoods. Until someone devises a better method of Democratic political organization that has a broad base of precinct captains who punch doorbells and recite the virtues of their party and candidates, we ought to be more careful in our moral judgment about a political party. . . . as long as we have the kind of organization we have in the city of Chicago, we will time and time again be victorious.[19]

Rakove's overall image of the classic Chicago machine is worth keeping in mind:

> As an organism, the Chicago machine is a hydra-headed monster. It encompasses elements of every major political, economic, racial, ethnic, religious, governmental, and paramilitary power group within the city. It recognizes the reality of all forms of power . . . and it understands the need to subordinate all. . . . forms of power to the political. . . . The machine believes with Machiavelli that men in politics are greedy, emotional, and passionate and are not governed by reason, morality, or concern for their fellowman. It believes that men can be coopted, bought, persuaded or frightened

into subservience to or cooperation with the machine. Every man has his price, according to the machine.[20]

This, then, is the image of the classic machine. Its decline has long been celebrated by many scholars. The old-style machine has been disappearing because, it is argued, the adoption of the merit system has led to the decline of patronage. Further, it is claimed that the absence for some time of the conditions that produced this type of organization has led to its demise: the need to take care of immigrants, ignorant of living conditions in the big city, has disappeared, as has the need for social welfare services, since presumably these are now taken care of through governmental and other private agencies. It is also argued that the public is not willing to tolerate the machine today because of its manipulative and illicit activities and techniques.

Has the machine really disappeared from American politics? Those who take this position are thinking only of the old-style boss and his very specialized authoritarian structure. Other types of party machines are conceivable. And there certainly is still a need for political organizations in American cities (and in cities everywhere) to provide certain services and fill certain human needs. Raymond Wolfinger has raised the question in his article entitled, "Why Political Machines Have Not Withered Away." He concludes both that the need for machines continues to exist and that although these needs may be left unfulfilled in certain areas, certain types of machines still do exist, but that the types of constituent services they perform now are different. Direct material assistance to the poor is still necessary, but new types of welfare assistance are required (such as helping people get into public housing units), while new types of help (for small businessmen, for example) and for those dealing with the bureaucracy may be necessary. There are basic social conditions in most American cities of medium and large size which require and will reward (with votes) a political organization with some of the characteristics of a machine.[21]

The basic point is that political machines can be of many different types. The Daley machine with its emphasis on discipline, payoffs, patronage, and loyalty to a leader is not the only type. Other types of machines can be suggested by utilizing two of the key dimensions of the classical machine and conceiving of organizations with opposite characteristics (see Figure 7.2).

The Daley machine was an authoritarian, materialist-oriented structure. But theoretically there can be authoritarian, ideological structures (as the fascist type of organization), decentralized materialist structures, and nonmaterialist, decentralized structures. In fact, some of the local party organizations today tend to be either decentralized and materialist, or decentralized and nonmaterialist, or hybrid structures with a mixture of these characteristics. Yet, they may still be machines in the sense of being relatively efficient organizations for the performance of particular types of political functions. The workings of the party in American municipalities show that many of the conditions of the classic political machine no longer exist — the intimidation, the blatant manipulation, the fraudulent control of votes, the use of the payroll to reward all party personnel, the unquestioning loyalty to a

FIGURE 7.2 ▪ A Two-Dimensional Classification of Party Machine Types

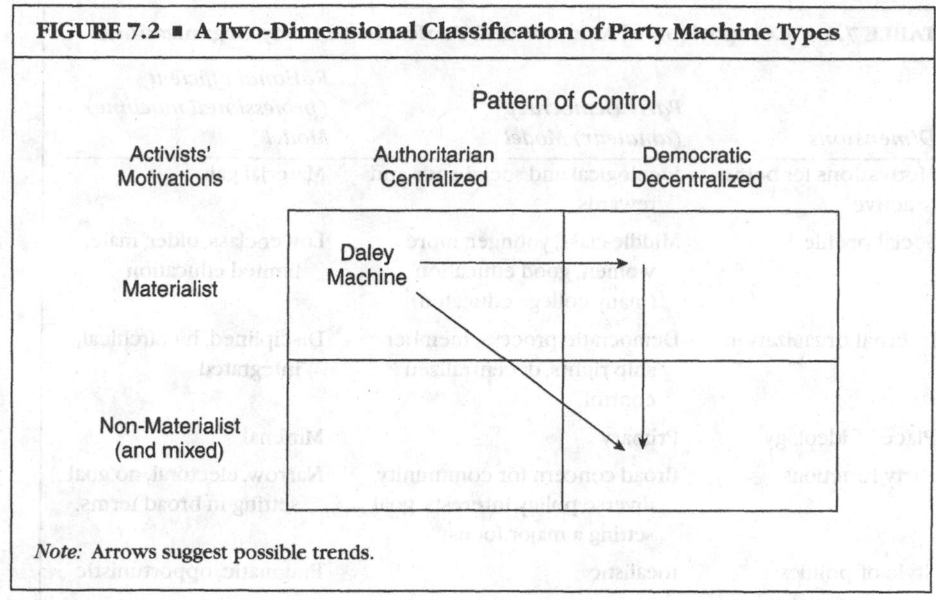

Note: Arrows suggest possible trends.

leader. But machines — in the sense of a rather cohesive group of committed partisans, both inside and outside the formal organization of the party, doggedly and tirelessly performing the tasks of the party (screening candidates, raising money, recruiting workers, registering voters, getting out the vote) and being rewarded on other than purely materialistic grounds — are often a reality. It is easy for people to shout at Chicago on election night as they did when Jane Byrne defeated Mayor Bilandic, "There is no more machine. There will never be a machine again," but surely this is a misconception. Where there is local politics there will be machines; otherwise the work of politics at the local level would never be completed.

PARTY ORGANIZATION AS A HYBRID TYPE

An attempt can be made to contrast the oligarchical or professional machine model and the amateur model on certain key dimensions. William Wright has suggested that there are two extreme or "polar types" of party organizations — the "party democracy" type, which is close to the amateur organization, and the "rational efficient" type, which is close to the boss and machine model (see Table 7.4).[22]

It is important to remember that the "pure" party organization, one which fulfills perfectly all the particular characteristics of a type, is rarely found. Thus these dimensions are continua, with party organizations ranging from one extreme to another and most likely tending not to be extreme in all characteristics. Any party organization consists of a heterogeneous group of activists who will have a variety of motivations, social backgrounds, and orientations to politics. Rarely is an organization made up of activists, precinct leaders, and ward leaders who are completely

TABLE 7.4 ▪ Comparison of Amateur and Professional Party Organizations

Dimensions	*Party Democracy (amateur) Model*	*Rational Efficient (professional machine) Model*
Motivations for being active	Ideological and social ends and rewards	Material gain
Social profile	Middle-class, younger, more women, good education (many college educated)	Lower-class, older, male, limited education
Internal organization	Democratic process, membership rights, decentralized control	Disciplined, hierarchical, integrated
Place of ideology	Primary	Minimal
Party functions	Broad concern for community, diverse policy interests, goal setting a major focus	Narrow, electoral, no goal setting in broad terms.
Style of politics	Idealistic	Pragmatic, opportunistic

ideological about politics and about their motivations for party work, as the amateur model suggests. On the other hand, rarely is the party organization populated exclusively by those seeking to plunder the public treasury for private gain. The world of reality differs considerably from the world suggested by these models of party organization. It is not a question of whether the local organization matches either this or that model. The world of actual party politics is too complex for such simplistic formulations. A close look at the world of the activists (presented in the next chapter) is necessary to understand party organizational life at the grass roots.

EVALUATION OF THE LOCAL ORGANIZATIONS

What basic images should there be of the parties as local structures? First, they are the structures which maintain the closest personal relationships to the public and, therefore, can be repositories of important powers of the party: power to select candidates, to produce the vote, to raise money, to recruit personnel, to determine issue positions (at least at the local level), to reward the activists, to seek to control elected candidates, to develop party membership, even to discipline their members. They can be influential and powerful if they are well organized and wish to assert their powers. Here is where much of the party's power can lie. The concerting together or coordinating a large number of effective local precinct and ward operations can mean a successful county, district, state, and national party.

The problem with most local party structures is, again, the looseness and diffuseness of control (aside from the few boss-run machines today). At the local party level are the same maladies found at state and national levels: fragmentation in authority, no centralized leadership, no clear lines of responsibility. The local organizations are not undemocratic oligarchies, but they also are not strongly democratic structures in which members have rights and duties and participate effectively and regularly in making party decisions. A key problem is that there is no membership concept, as most European parties have, the idea of the card-carrying, dues-paying member who has voting and other participatory rights and roles in the organization. The closed-direct-primary states attempted to move in that direction but very ineffectually. The idea of a mass-membership-based party is something rarely found in the United States at the local level.

Another problem with most local party structures is that they are preoccupied with organizational tasks more than with deliberation on policy and utilization of the organization to achieve policy goals. Local party politics is often just not very issue-oriented, partly because in over 70 percent of the cities the local elections are nonpartisan but also because there is no local tradition to see the party as basically a policy instrument. The idea is to elect candidates to office and then give them free rein unless their behavior is so outrageous as to require opposition in the succeeding party primary. Strengthening the policy deliberative role of the local party would make involvement with the local party infinitely more attractive. When the local party has been activated, it is usually due to policy controversies.

All local organizations are not of one type, however. They differ in several basic dimensions:

1. They can tend to be hierarchies, or they can be stratarchy systems with much autonomy and democratic control.
2. They can be wide-open systems which anyone can join, with great diversity of personnel, or they can be relatively closed membership systems.
3. They can be voluntaristic structures with a haphazard incentive system, or structures which have developed a reward system based on recognition of work done, patronage, or other material rewards.
4. They can place considerable emphasis on the role of ideology in recruiting members, selecting candidates, and running campaigns, or they can ignore the role of ideology.
5. They can be highly efficient structures with well-trained cadres performing the critical tasks of maintaining contact with the public and getting out the vote, or they can be quiescent and utterly inefficient and status quo structures.

Local parties in a democratic society such as ours, if they are to maximize their power and role, are inclined to be nonhierarchical, open, voluntaristic, nonideological, and inefficient, given the conditions of American society and political culture. There are, however, community differences and the possibility, indeed the capability, of local parties to move in other directions.

Despite the problems of American parties at the local level, local parties do exist, are more active in some localities than before, and do perform important functions. They are the entry point for the member of the public who wishes to do more than just vote and wants to act through the instrument of party. They are usually his or her initial contact with the party system. Local parties are often structures through which other social-interest groups work or with which they work in order to influence local governmental decision making. They are important links in the new national committee-state committee-local committee networks of fundraising and candidate support which have developed and which were discussed earlier. They can represent the public's concerns (or that segment of the public which they see as their constituent group) upward when their delegates attend county, state, and national conventions or sit on county, state, or national party committees. In all this the local party organization can play a vital integrative role in the political system. Local parties are much more than merely symbolic. Even in their relatively ineffectual organizational condition, they can perform vital functions as critical links between the public and the government. When activated, they can make the difference between stagnant consensus and policy innovation.

American local parties operate differently from the way that local European parties usually do, except in Great Britain. On the Continent rarely do parties engage in any house-to-house canvassing, because it is considered contrary to their political traditions and is believed to be resented by the public and an invasion of the privacy of the home.[23] Telephone canvassing is equally unheard of in Europe. Of course, many European parties are organized as mass-membership parties, may have professional agents directing party work (as in England), and may have different types of units at the base of the organization than in the United States. Without the payment of membership dues, these parties would not be able to function. Parties in Europe are organized differently, and their roles and techniques in getting out the vote are quite different from those of American parties. There is much more reliance on the large rally, the sound truck, the distribution of literature, and candidate speeches and appearances. The local organization mobilizes votes differently. It usually has a reliable clientele who vote regularly. In the United States the task of the local organization to get out the vote is more demanding.

Finally, in evaluating local party structures it must be remembered that the local party is a social group as much as it is a political group. Fulfilling social as well as political functions for those whom it attracts and to whom it appeals, the party penetrates deeply into the social structure. To many, politics at the local level is social fun and provides opportunities for social interaction as well as chances for social mobility. The party is thus a critical social institution for many Americans.

NOTES

1. David S. Broder, 1978, "The Case for Responsible Party Government," in Jeff Fishel, ed., *Parties and Elections in an Anti-Party Age* (Bloomington: Indiana University Press), 26.
2. Edward C. Banfield, *Political Influence* (New York: Free Press, 1965).
3. Malcolm E. Jewell and David M. Olson, *American State Parties and Elections* (Homewood, IL: Dorsey, 1978), 56.
4. Elmer E. Schattschneider, *Party Government* (New York: Rinehart, 1942), 147.
5. Jewell and Olson, *American Parties,* 134-36.
6. James L. Gibson, Cornelius Cotter, John Bibby, and Robert Huckshorn, "Assessing Party Organizational Strength," *American Journal of Political Science* 27 (1983): 193-222; Malcolm E. Jewell

and David Olson, *American State Political Parties and Elections* (Homewood, IL: Dorsey Press, 1982, 1988); John F. Bibby, "Party Organization at the State Level," in S. L. Maisel, ed., *The Parties Respond* (Boulder: Westview Press, 1990), 21-40; Robert Biersach, "Hard Facts and Soft Money: State Party Finance in the 1992 Federal Elections," in Daniel M. Shea and John C. Green, eds., *The State of the Parties* (Lanham, MD: Rowman and Littlefield, 1994), 107-32; Shea, ibid., 219-36.

7. Biersach, "Hard Facts and Soft Money," 117-19, 111-12.

8. Bibby, "Party Organization," 31-33; Shea, "State Legislative Campaign Committees," in Shea and Green.

9. Samuel J. Eldersveld, *Political Parties: A Behavioral Analysis* (Chicago: Rand McNally, 1964), 74-75.

10. A 1964 study by the Center for Political Studies was done in 208 counties. See Paul Beck, "Environment and Party," *American Political Science Review* 68 (1974): 1229-44; the 1979-80 study was done by a group of scholars who mailed surveys to over 7,300 county-level party leaders with a 52.7 percent response rate. See James L. Gibson et al., "Assessing Party Organizational Strength," 139-60; two 1992 studies were done, one of 659 county leaders in eight states (John Frendreis, Daniel M. Shea, and John C. Green, 1994, 133-45) and the other by Paul Beck, et al., of 40 county chairs "Party Effort at the Grass Roots" (unpublished manuscript, 1994).

11. Mike Royko, *Boss: Richard J. Daley of Chicago* (New York: New American Library, 1971).

12. M. Margaret Conway and Frank B. Feigert, "Motivation, Incentive Systems, and the Political Party Organization," *American Political Science Review* 62 (1968): 151.

13. Paul A. Beck et al., "Party Effort at the Grass Roots: Local Presidential Campaigning in 1992," paper delivered at the annual meeting of the Midwest Political Science Association, Chicago, April 14-16, 1994.

14. See, for example, William Crotty, ed., *Political Parties in Local Areas* (Knoxville: University of Tennessee Press, 1986).

15. *Time*, March 12, 1979, 22.

16. James Bryce, *The American Commonwealth* (New York: Macmillan, 1916), 118.

17. Royko, *Boss*, 60-61. Much of this summary is based on Royko's accounts and that of Milton Rakove, *Don't Make No Waves, Don't Back No Losers* (Bloomington: Indiana University Press, 1975), chap. 4.

18. Rakove, *Waves*, 119-20.

19. Ibid., 129-30.

20. Ibid., 3-4.

21. Raymond Wolfinger, "Why Political Machines Have Not Withered Away," *Journal of Politics* 34 (1972): 365-98. The dimensions and types of machines used here were stimulated by this article.

22. Adapted from William Wright, ed., *A Comparative Study of Party Organization* (Columbus, Ohio: Merrill, 1971), 53.

23. Leon D. Epstein, *Political Parties in Western Democracies* (New York: Praeger, 1967), 111-18.

CHAPTER 8

Party Activists: Working Elites with Organizational Links to the Masses

Parties in democratic societies need to mobilize their vote to survive, and thus they need to operate in close contact with the voters. As part of this function parties need to persuade the public to be involved in the electoral process and, hopefully, to understand and approve, at least in a general sense, the party's policies and leaders. Securing and maintaining mass support and participation is a continuous and complex set of tasks. There need to be actors in the party organization who assume these tasks and responsibilities. As Rosenstone and Hansen put it, "people participate in electoral politics because someone encourages or inspires them to take part."[1]

The actors who do this party work at the base of the party organization are the "party activists." They are those persons holding positions and/or doing the organizational work at the very bottom of the party structure (such as county, city, precinct, ward, and township leaders) and the campaign workers who work for party and/or candidates at the local level. With the technological advances in electoral politics, the activists may include campaign specialists in using mass media, in conducting polls and surveys or engaging in other types of party research, in fundraising, in conducting voter registration drives, and in canvassing for voters door-to-door or by telephone. The activists may work for the regular local party organization or in a particular geographic area, for particular candidates, from the president on down to candidates for local office. Party activists are frequently delegates to county, state, and national conventions. There is, thus, a great variety of party operatives at the local level fulfilling a great variety of tasks and functions. Sometimes they are called "grassroots workers," or "active party volunteers," or in other countries, "militants" (Europe), or "linkmen" (India), or other characterizations. But in essence they are the "party elites" at the base of the party system.

Each system develops its own strategy and concept for maintaining successful contacts with voters. In the past there were, and still are today, parties which are very centralized with, it is claimed, national control of vote mobilization efforts. But as Duverger pointed out over forty years ago, modern parties tend to be mass-based parties, even if in appearance controlled from above ("there are few purely cadre parties," i.e., parties without some mass base).[2] The parties need workers/members to meet the critical need for involving citizens in the electoral process.

In the United States from the early days there have been local party structures, although certainly not demonstrating the functional diversity of recent activists. As Chambers argued convincingly, the early national Federalists and Anti-Federalists (Republicans) extended their contacts to the countryside, encouraging the setting up of the "self-created societies" which Washington deplored. These leaders in states, counties, and towns began to express dissent against or support for Hamilton's and Washington's policies. Further, "national leaders and local leaders or cadres collaborated to develop increasingly standardized ways of performing 'functions'" such as "nominating candidates, electioneering and mobilizing voters, and shaping opinion in the states and localities."[3] Eventually these decentralized party activities led to the development of pressure for the "delegate convention," which turned the nominating function over to the delegates of the local party organizations, resulting in 1831-1832 in the first national delegate party conventions.

In the latter part of the nineteenth century and the early twentieth century, the emphasis in the study of local parties was on the big-city machines, as we discussed earlier. But with the decline of the machines after World War II, studies of parties seemed to have assumed that our major parties actually had very weak local organizations, if any at all. In the early editions of V. O. Key's basic book on parties (1948, 1952) one finds very little discussion of the local party organizations (aside from the big-city machines). It is almost as if local party organizations hardly existed, or if they existed were irrelevant. And this conception persisted for some time. Even in Sorauf's 1968 text we find this summary set of observations: "in truth, political party organization in many states and localities is weak, undermanned, even torpid. . . . there is often only sporadic activity by a handful of dispirited party regulars . . . virtual disorganization — the American parties remain largely skeletal . . . manned generally by small numbers of activists and involving the great masses of their supporters scarcely at all."[4] And for some students of American parties this interpretation is still accepted today.

THE REALITY OF LOCAL PARTY ACTIVISM IN THE POSTWAR PERIOD

It is important to look carefully at the empirical evidence on the state of local party effort and its relevance, particularly because of this early tendency to demean local party effort. Many observers have continued to say that local party activism is insignificant, if it exists at all. Scholars have claimed that "parties are more important as labels than as organization," "parties are empty vessels," "parties . . . just aren't important institutions for most Americans today . . . less relevant now than ever before," et cetera. The implication in much of this attack is evident: Perhaps parties and local party activism once were relevant, but certainly are not today. Is this actually the case? Some over-time evidence we have may correct these speculations.

One approach is through the national surveys conducted since 1952, which asked a representative sample of adults whether they participated in each of a list of possible activities during the election campaign. The image of American citizens as being much more active in the past disappears with a look at these over-time data (Table 8.1). In the fifties, as a matter of fact, the activity level was relatively

TABLE 8.1 ▪ Participation by American Adults in Presidential Election Campaigns (in percent)

Campaign Activity	*1952*	*1960*	*1972*	*1980*	*1988*	*1992*	*1996*
Contributed money to party or candidate	4.2	11.6	10.4	8.0	8.7	6.6	7.9
Tried to influence others in how to vote	27.5	33.5	31.6	36.1	28.9	33.9	28.9
Worked for the party or candidate	3.2	5.6	5.0	3.6	3.3	3.1	2.7
Active in 3 to 5 of the activities surveyed*	2.6	9.1	7.3	5.0	5.2	5.5	4.6

* These five activities include the three listed in the table plus "going to political meetings" and "wearing a campaign button, putting a campaign sticker on your car, or placing a sign in your window or in front of your house."

SOURCES: National Election Studies 1952–1996; see also Steven Rosenstone and John Hansen, *Mobilization, Participation, and Democracy in America* (New York: Macmillan, 1993), 56–63.

low; 1960 was a year in which involvement was high, a year when voting turnout was also high (64 percent). Actually, participation has fluctuated considerably due to the circumstances and stimuli of particular presidential contests. In the nineties the record is moderately good.

The actual reports of party activity at the local level given by county leaders provide another set of evidence on the continuity of fairly high level of local party performance (Table 8.2). Scholars have gone to great and painstaking trouble to collect the facts from county party chairs about local party activity, systematic research that began in 1964 and continued into the nineties. Several observations stand out from these data: (1) already in the sixties over 60 percent of the county leaders reported that their parties' activists were engaged in significant types of work in identifying voters through registration, funding campaigns, and working to "get out the vote"; (2) for most of these activities there has been an increase over the years, *not a decline;* (3) all of these studies emphasized the existence of a local "organization" that helped select candidates, worked with candidates, coordinated campaigns, and maintained itself over time as an effective activist group. All of the scholars involved in these studies claim that these local organizations exist, are active and effective, endure, and continue to have a role in their party's electoral success. The authors of the 1992 study concluded that "the local parties were active players in the 1992 presidential campaign."[5]

There have been many studies over the years of local activists and campaign workers in diverse urban environments, including Houston, Nashville, Chicago, Manhattan, Gary, and cities in North Carolina and Massachusetts.[6] In two large cities there have been studies of local party officials over a long enough period of time to provide us with considerable longitudinal perspective on the extent of

TABLE 8.2 ▪ Evidence of Thirty Years of County/Local Party Activity Based on Reports of National Samples of County Chairs (in percent)

Activity	*1964*	*1980*	*1984*	*1992*
Distributing campaign literature	78	93	90	79
Arranging campaign events/affairs	64	84	86	85
Conducting voter registration drives	64	76	78	92
Telephoning voters	N/A	68	77	92
Fund-raising	63	86 (78)	82	N/A
Number of cases	*110*	*121*	*199–321*	*40*

SOURCES: (1) The 1964 study was done by the Survey Research Center, University of Michigan, along with the National Election Study. Questionnaires were sent to the two major party chairs in 129 counties, with over a 90 percent response rate. See Paul Beck, *American Political Science Review* 68 (September 1974): 1229–44.
(2) The 1979/80 study was a survey of all county or county-equivalent party leaders, with a response rate of 52.7 percent. See James L. Gibson et al., *American Journal of Political Science* 29 (February 1985): 139–60. This is called the "Party Transformation Study" (PTS). The data here are for the common counties for 1964 and 1980.
(3) The 1984 study, called the "Party Dynamics Study," was a panel study of the leaders of a sample of the counties selected in the PTS in 1979/80. One thousand counties were selected; the response rate was 62 percent. See James L Gibson et al., *American Journal of Political Science* 33 (February 1989): 67–90.
(4) The 1992 study was a survey of 160 leaders in a probability sample of forty counties nationwide, including both regular county organization chairs and presidential campaign coordinators. Mail questionnaires and follow-up telephone interviews were conducted: the response rate was 99 percent! See Paul Beck et al., "Party Effort at the Grass Roots: Local Presidential Campaigning in 1992" (paper presented at the April 1994 American Political Science Association annual meeting in Chicago.)

local party activity, whether it has changed, and if so, in what ways. The two cities are Detroit and Los Angeles. We will be referring to these cities from time to time in this chapter to illustrate the changing character of local party activism. We can begin here with some basic data on party activity in these two cities to answer the question again, Is there evidence of decline in the activism of local party structures?

The Detroit study from 1956 to 1992 illustrates the variations in local activism. It is based on interviews with precinct leaders randomly selected from the universe of activists in the city of Detroit. Among the many questions asked were detailed inquiries about a variety of campaign activities. The Detroit study's findings have already been described in detail in Chapter 7. Here only a brief summary is presented in Table 8.3. A careful look at the findings from this research leads to certain observations for party activism in Detroit:

1. There has been an over-time average *increase* in Democratic activity, not a decline; there has been a *decline* in Republican activity.
2. Local party activism varies by election, declining for both parties in 1972, increasing in 1980, and decreasing for the Republicans in 1992.
3. Activity type varies; for example, voter registration efforts declined for both parties, while telephone canvassing increased for both parties.

TABLE 8.3 ▪ Extent of Activity by Detroit Precinct Leaders, 1956–1992 (in percent)

	Democrats				Republicans			
Activity	*1956*	*1972*	*1980*	*1992*	*1956*	*1972*	*1980*	*1992*
Voter registration drives	93	70	44	71	80	39	19	16
House-to-house canvassing	46	30	64	63	32	11	65	46
Fund-raising	—	—	41	62	—	—	29	29
Mean	*69*	*39*	*61*	*73*	*64*	*24*	*50*	*37*

Notes: The number of cases varied considerably for these studies: 281 (1956), 223 (1980), 141 (1988), 112 (1992), 95 (1984), 41 (1972).

SOURCE: All studies by S. J. Eldersveld. For publications presenting these data, see the author's *Political Parties: A Behavioral Analysis* (Chicago: Rand McNally, 1964); "The Party Activist in Detroit and Los Angeles: A Longitudinal View, 1956–1980," in William Crotty, ed., ***Political Parties in Local Areas*** (Knoxville: University of Tennessee Press, 1986), 89–120; the author's 1982 textbook, ***Political Parties in American Society*** (New York: Basic Books).

4. The Republican party activists were remarkably involved in a city which was increasingly more Democratic in presidential elections (62 percent in 1956, 88 percent in 1992).

There is no doubt that considerable "slack" in local party efforts exists and has existed for a long time. However, if one takes the three critical tasks in vote mobilization (registration, canvassing, and election-day get-out-the-vote efforts), one notes that the Democratic Detroit organization is doing slightly better (almost three-fourths are fulfilling these tasks) while the Republicans, though still involved, declined in such task performance in 1992.

In Los Angeles the study of the local party organization began also in 1956 and focused particularly on the work done by the county central committees as well as the political clubs which had been created.[7] Marvick's studies covered the period from 1956 to 1978.[8] In 1956 he found that already over half of these local party elites were active in campaigns; indeed, 70 percent were engaged in house-to-house canvassing. This level of activity declined only slightly over the years, with the Republicans doing at least as well as the Democrats. One key measure used by Marvick was the average hours per week worked during campaigns. He found that in 1968 this average was nineteen hours; a decade later the average was also nineteen! One final point: both the Detroit and Los Angeles studies reveal that these activists, even in the early years, were experienced in local party politics. In Detroit in 1956 we found that 40 percent of the precinct leaders had been active for ten years or more; in Los Angeles 64 percent said they had been active for more than twelve years.

SOCIAL BACKGROUNDS: HOW INCLUSIVE ARE LOCAL PARTIES?

Having established that local party structures and activists have had a significant presence in the postwar period and have demonstrated considerable vitality, we now discuss the characteristics of these activists: Who are they, what motivates them, what do they stand for, how well do they compete, and what impact do they have. Social backgrounds constitute a good entry point for that analysis. American parties are thought to be open groups in which anyone can get into party work and, indeed, move quite rapidly upward without impediment in the structure if he or she wants to. How true is this actually? How undiscriminatory are parties in encouraging people from diverse social backgrounds to participate? Parties might be open because of their wide appeals for support, or closed because of a desire for solidarity and efficiency.

The social-class status of activists, based on educational and occupational data, can give some idea of the openness of local organizations. In an early study of Chicago, Harold Gosnell found that the precinct captains included a large proportion from the lower class. In 1928, 58 percent of them had only a grade-school education and only 20 percent went beyond high school.[9] They were also primarily from lower-class occupational backgrounds. In 1981 the Chicago local party organizations were again studied, by Crotty, who found a tremendous change in social-class backgrounds, virtually the reverse of the prewar Chicago machine: 80 percent (Democrats) and 90 percent (Republicans) had more than a high school education, in fact 67 percent (Democrats) and 41 percent (Republicans) had college degrees or better.[10]

This tendency for party activists to have higher education today is apparent also in other studies, although it varies by level and by locale. The national sample of county leaders in 1980 found only 20 percent with a high-school education, and one-third had education beyond the bachelor's degree.[11] In Los Angeles in 1980, all the activists had some college and almost 50 percent had post-graduate training. Detroit precinct leaders in recent years are not quite that well educated — over 50 percent have completed college but, on the other hand, 15 percent (R) to 20 percent (D) have not gone beyond high school.

There is unquestionably a great "social-class bias" in party activity, even at this base level of the system. One sees it also in the types of occupations of activists, which are increasingly professional and business-managerial, as well as in their income levels. However, these local activist structures should not be considered extremely exclusive. There is evidence that local parties in many areas and for certain types of positions are still "opportunity structures," in socioeconomic status terms. In a city like Detroit, precinct leader positions are held by Democrats with blue-collar positions (21 percent in 1992), labor union members (52 percent), and those with limited education. (This is true also for Republicans but less frequently: 21 percent labor-union members, 19 percent blue-collar Blacks.) As noted earlier, the study of county party chairs reveals that leaders are not all well educated. And even delegates to the national conventions, studies reveal, include one-fifth or more with a low education.[12] Thus, while lower-class representatives in the local

Party Activists in the South: The Changing Political Context

Prior to the passage of the 1964 Civil Rights Act and the 1965 Voting Rights Act, the party activists in the South were white Southern Democrats. All that would start to change when the Goldwater-led Republican party in 1964 captured five states in the region. The Nixon victories in 1968 and 1972 made possible further inroads by Republicans into the region and the party activists in the region switched parties and became movers and shakers for the BOP. The evolving trend set into motion by Goldwater, enhanced by Nixon became a major flood tide with Reagan in 1980 and 1984 and Bush in 1988. The success of the Republican party in the region was fueled by the strategy of these leaders and the South's reaction to the civil rights revolution. Said efforts helped to change the nature and scope of party activists in the region. But there is more.

In 1984 and 1988 Reverend Jesse Jackson ran for the Democratic party's presidential nomination and on both occasions captured the majority of the southern states in the regional Super Tuesday presidential primaries. He did better than the native sons of the region. And in so doing, the white Democratic delegates from a variety of southern states were pledged to him and went to the National Democratic Convention to work in his behalf. For the first time in the history of the country and region, white party activists were supporting an African-American Democrat. Yet the Jackson phenomenon hastened the defection of southern party Democratic activists to the rising Republican party in the region.

Today, at national Democratic presidential conventions, African-American activists and delegates abound. One of the main catalysts for changing this region's party activism was the rise of the South Carolina Progressive Democratic Party in 1944, the Mississippi Freedom Democratic Party in 1964, and the National Democratic Party of Alabama in 1968. Each party caused the Democratic National Committee (DNC) to revise the rules and regulations governing delegate selections and require the next conventions to ensure fair representation in the delegate selection process. This brought newer and different activists into local, state, regional, and national organizations.

SOURCE: Hanes Walton, Jr., *Black Political Parties: An Historical and Political Analysis* (New York: Free Press, 1972).

parties are certainly outnumbered, the door is not shut and opportunities do exist. No doubt the parties should do more to attract those with lower-class credentials if the parties are to do a better job of reaching out to their potentially loyal supporters. After all, 60 percent of the American public does not have a college degree, and the median family income is much closer to $30,000 than $75,000.

The representation in the activist cadre of women, blacks, other minorities, and young people is perhaps even more critical than socioeconomic status. How well are these groups represented in local politics? In all studies women have been doing well, increasingly included in these local activist structures: over 20 percent among county chairs, over 30 percent among Los Angeles local leaders (1980), over 40 percent among Detroit precinct leaders (1980 to 1992), and among delegates to state and national conventions their proportion has increased to 50 percent (Democratic national convention, 1996), and 34 percent (Republicans 1996). Here and there one finds more exclusion of women; Chicago's parties in 1981 had only 6 percent women.

Blacks and other racial/ethnic minorities do much more poorly. But here again it depends on locality and party and level. Among county chairs less than 2 percent are "non white." In Los Angeles (1980) 10 percent of the Democratic cadres were black and only 2 percent of the Republicans (a decline from 7 percent earlier). At state and national conventions one can find a higher percentage of blacks — 20 percent at the 1996 Democratic national meeting but only 3 percent for Republicans.[13] In striking contrast, the steady increase in the assumption of precinct positions by blacks in Detroit resulted in their control of 83 percent of these positions in the Democratic organization by 1992 and 49 percent in the Republican organization (the population of Detroit was 76 percent black in 1990).

The key issue here is whether there is evidence of change in the inclusiveness/exclusiveness of local party activist structures. It is difficult to get good overtime data on the representation of these groups from the studies we have available. For national convention delegates we do have such data (Table 8.4).

The contrasts in Table 8.4 are striking! Women have achieved real parity or near parity to men in both parties. Blacks have gradually been more successful in the Democratic party, but the Republicans remain relatively exclusive political territory.

But in many large cities blacks do very well in achieving access to local party positions and indeed, in some cities, to dominance in the local parties. James Q. Wilson has written at length on this matter in an earlier book.[14] Detroit is an example of a city where this has occurred as the composition of the local activist cadres have been truly transformed (Table 8.5).

The Detroit data illustrate how two major parties adapted to populational, social, and economic change in their communities. One notices changes and continuities: a continuous surge in the mobilization of blacks, women, and the college educated, while maintaining the support of blue-collar workers, union members, and young voters. The parties do not respond to social environmental changes identically, but in their own special ways, mindful of their hard-core supporters who have to be "brought along" as the party engages in creative social renewal.

TABLE 8.4 ▪ Representation of Women and Blacks at National Conventions (in percent)

Social Category	*1968*	*1972*	*1976*	*1980*	*1984*	*1988*	*1992*	*1996*
Blacks: Democrats	5	12	11	15	18	23	16	20
Republicans	2	4	3	3	4	4	5	3
Women: Democrats	13	44	38	52	49	48	48	50
Republicans	16	39	34	31	44	33	43	34

TABLE 8.5 ▪ Social Transformation of Detroit Party Activist Cadres (in percent)

Social Group/ Category	*1956*	*1964*	*1972*	*1980*	*1984*	*1988*	*1992*	*Percent Change 1956-92*
Democrats								
Blacks	26	40	65	59	65	75	83	+59
Women	24	33	43	43	43	40	55	+31
College educated (plus graduate work)	20	28	35	51	52	65	58	+38
Blue-collar workers	56	37	39	29	18	19	21	-35
Professionals	16	29	26	34	27	46	45	+29
Labor union members	56	35	52	43	60	55	52	-4
Age below 40	28	40	39	33	18	29	13	-15
Republicans								
Blacks	20	36	22	33	45	43	49	+29
Women	11	30	39	38	44	48	45	+34
College educated (plus graduate work)	28	38	33	54	62	56	73	+45
Blue-collar workers	25	31	17	21	15	8	9	-16
Professionals	15	33	33	29	32	44	49	+34
Labor union members	22	15	44	18	21	18	21	-1
Age below 40	19	38	33	34	26	33	24	+5

SOURCE: Studies completed by Samuel J. Eldersveld.

In interpreting these Detroit data, keep in mind these basic changes in the census demographics for Detroit, as well as the electoral realities.

Census and Electoral Statistics for Detroit over Time

	1950	*1960*	*1970*	*1980*	*1990*
Blacks as a percentage of the population	16.2%	28.9%	43.7%	63.1%	76.0%
Size of population (in thousands)	1,850	1,670	1,511	1,203	1,028
Blue collar in the labor force		57.0%	58.8%	53.6%	
Unemployed		9.9%	7.2%	18.5%	
High School graduates		34.4%	57.2%	54.2%	

Voted Democratic for president	*1952*	*1956*	*1980*	*1984*	*1992*
	60.2%	61.7%	78.5%	81.0%	88.0%

One notices immediately that the parties were confronted with major sociological and economic changes in the city. The Democrats appear to have adapted very well if one looks at the changes in the characteristics of their activists and their progressively larger election majorities. A careful, detailed analysis, precinct-by-precinct, of how the Democrats responded by renewing and strengthening their precinct leadership over the years suggests that, indeed, the causal relationship *may be:* demographic and social changes ⟶ party organization social adaptation ⟶ electoral success.[15] But this is difficult to demonstrate statistically, although it seems logically and empirically convincing. The other major aspect of these Detroit data is the amazing, though more limited, transformation which the Republican activist organization went through in order to sustain itself in the face of this Democratic strength.

Women (and blacks to a certain extent), have achieved a greater recognition and role in local party activism, within the Democratic and Republican parties. In a third party such as Perot's Reform party in 1992 the activists were less inclusive — they were 96 percent white and 62 percent male.[16] However, this openness of the *local* parties stands in contrast to the opportunities for women and blacks (and lower-class persons generally) at the national level, in Congress. In 1997 there are only nine women and one black in the United States Senate. In the House 11.7 percent of the membership is female and 8.5 percent is black. (In 1989 only 26 percent of city council members were women, 10 percent African American).[17] Of course a college education is virtually a necessity for Congress (only about 5 percent or fewer have less than a college education). Our parties do not recruit those with lower social-class backgrounds as candidates for Congress to the extent that European parties do. In Britain, Germany, and other Western democracies, parties on the "left" particularly attract and select candidates for their parliaments who have lower social-class origins.[18] To move upward from the base of the U.S. party system to the top policy positions is much more difficult for women, blacks, and those with less education and lower social-class origins. We have been training these types in politics at the local level and relying on them to do the work of the party at the lower level, but we put a limit on how high these women, blacks, and lower-status activists can rise in the system.

PATHWAYS INTO PARTY WORK

How people become involved in party organization work is difficult to summarize. There are so many different stimuli and circumstances. Interviews with precinct activists reveal a multiplicity of conditions for involvement in party work.

"My father was a saloon keeper in Detroit and quite a politician. A lot of politics went on in the back room of the saloon. I heard a lot as a child. I was brought up in a political atmosphere and having had a taste of politics I wanted to know something about what went on in the inside. So I started to work for the party when I was first able to vote. I then ran for precinct delegate."

"When I was a kid I carried papers and worked for [James M.] Cox, the Democratic presidential candidate. Cox was a newspaperman and I always liked him. I also believed in the League of Nations. My family have always been Democratic and interested in politics."

"We were the only Republican family in our block and one day a ward chairman who was a friend of mine asked me whether I would like to be a precinct worker. He told me the job of the precinct worker was simple — to sell the candidate and get out the votes — not so simple! My parents were immigrants . . . and they came over to escape tyranny and build a new life. Therefore they were what you might call rugged individualists."

"I was active in the UAW as a shop steward. One night at a local educational meeting I heard Governor [Mennen] Williams speak. I don't remember what he said, but a state senator pointed out to me the position of the parties in Lansing. . . . I wanted to find someone else to run, but couldn't. The usual candidate was associated with the Communists so everybody was for me when I said I would run."[19]

Often the role of the family in discussing politics and in interesting family members in the possibility of party work is very important. Studies reveal that 30 percent to 40 percent of those who become active in politics did so through their families. There are additional conditions which, after this early introduction, lead to actual entrance into party activism:

1. The party leadership itself may play an important active role in recruiting (38 percent in Detroit, but as high as 54 percent in North Carolina and 69 percent in Massachusetts).
2. Other groups such as unions, business groups, or civic, racial, and ethnic groups may play a decisive role in recruitment (10 percent in Detroit, but usually less than 10 percent in other communities in which studies have been done).
3. Friends and relatives not in politics may also be the stimulus for getting people into the party apparatus (13 percent in Detroit, higher elsewhere).
4. Self-recruitment occurs for many of these activists, because of their interest in issues or causes and their independent decision to seek a career in politics (up to 32 percent in the Detroit, North Carolina, and Massachusetts studies).
5. Desire to work for a particular candidate (12 percent in Detroit — there is evidence that in other communities it is higher).
6. Accidental involvement — individuals who just happen into the party group even though they have never had much interest in party politics, a haphazard, unpremeditated social involvement.

Getting involved with the party thus is the result of quite different events in the life of a party activist and several different routes into party careers. There are three basic routes — the recruitment, self-starter, and accidental models (see Figure 8.1). The party organization by itself recruits only one-third or two-fifths of its activists; the remainder come in through other routes, with a fifth becoming accidentally involved. A minority gets involved because of an interest in a particu-

FIGURE 8.1 ■ Alternative Pathways to Party Activism

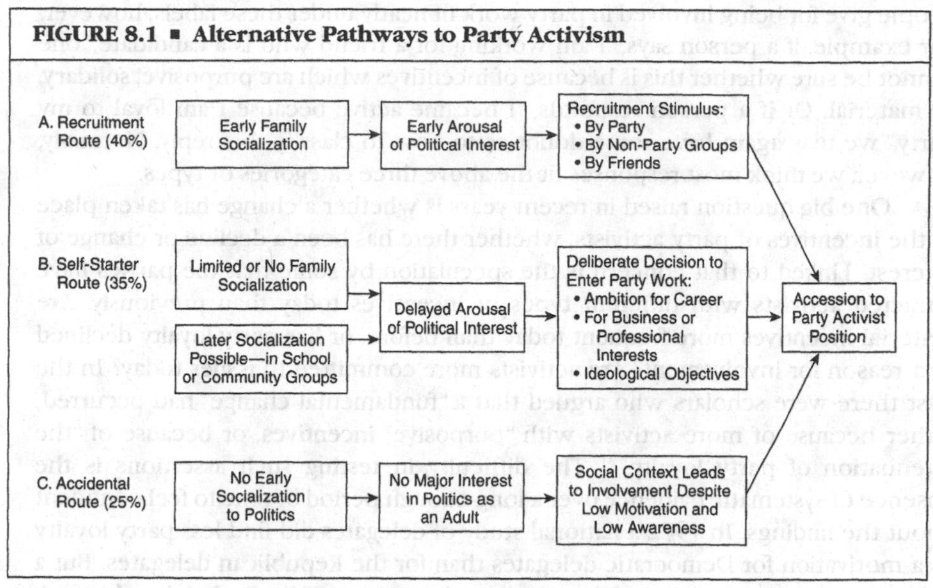

lar candidate. Persons arrive in the local party organization, therefore, after having traversed quite different paths, and the particular path they have traversed will probably have considerable influence on how they look at party work. Those from the labor unions will have a quite different perspective from those who became interested as a result of their work in an ethnic, racial, or business organization or who decided on their own volition to enter party work because of their beliefs and preferences concerning governmental policy.

MOTIVATIONS FOR PARTY ACTIVISM

People work for political parties for a variety of reasons. For the individual it satisfies certain needs, ranging from the enjoyment of group associations, to a desire for personal influence and perhaps a career in politics, to altruistic commitment to work through parties in order to contribute to the solution of social problems. Having a cadre of motivated activists is critical for a party if it is to develop and maintain the support of its followers. The party therefore has the task on the one hand of keeping its activists, with their very diverse motives, happy while utilizing their skills effectively to mobilize the vote. A party with a cadre of dissatisfied activists will have no collective drive potential and cannot survive for long.

Scholars have identified three major types of motivations, or incentives: (1) *material,* including the desire to develop a career in politics or to make business contacts; (2) *solidary,* which means an interest in social contacts and friendships and enjoyment of political group associations; and (3) *purposive,* or a commitment to work for causes, specific issues, or philosophical principles.[20] Not all reasons people give for being involved in party work fit neatly under these labels, however. For example, if a person says, "I am working for a friend who is a candidate," one cannot be sure whether this is because of incentives which are purposive, solidary, or material. Or if a person responds, "I became active because I am loyal to my party," we may again have some doubt as to how to classify this reply. Normally, however, we think most responses fit the above three categories or types.

One big question raised in recent years is whether a change has taken place in the incentives of party activists, whether there has been a decline or change of interest. Linked to that concern is the speculation by some that the parties have attracted activists with different types of incentives today than previously. Are material incentives more frequent today than before, or has party loyalty declined as a reason for involvement? Are activists more committed to issues today? In the past there were scholars who argued that a "fundamental change" had occurred, either because of more activists with "purposive" incentives, or because of "the attenuation of party loyalty."[21] The difficulty in testing such assertions is the absence of systematic evidence over a long-enough period of time to feel confident about the findings. In 1972 a national study of delegates did find less party loyalty as a motivation for Democratic delegates than for the Republican delegates. But a subsequent study of a panel of these same delegates in 1980 revealed that this finding was short-lived due to the fact that the 1972 Democratic newcomers were subsequently socialized, or they dropped out and were replaced by more loyal supporters, so that by 1980 the party difference was much smaller.[22]

To get a very long-term perspective on whether incentives have changed over the years we can present data from one study, in Detroit, collected from 1956 to 1992. And one other study can be referred to, in Los Angeles, which collected similar data from 1956 to 1980. The Detroit study data are presented in Tables 8.6 to 8.8.

What do these data reveal? It is clear that certain changes in incentives for party activism have been occurring, but one could not call these changes "revolutionary." First, there does seem to be some decline in party loyalty, that is, in its "importance" as a reason for *becoming* active. It is certainly not as strong a reason recently in Detroit as in 1956. (In Los Angeles the same is true: 69 percent cited loyalty as "very important" but by 1980 this declined to 50 percent). Second, the desire to work on issues and for "causes" has grown in importance in Detroit (in Los Angeles 70 percent of their activists have always reported that concern for issues was very important). Third, the relevance of social contacts as a particular reason for being active is less important today in Detroit (it was also on the decline in Los Angeles by 1980). But the "fun and excitement of party work" has become even more important, it seems, and a desire to work toward a career in politics is also a continuing ambition, even increasing somewhat. So "personalized" motivations are by no means on the decline. Above all, what appears to be true today as well as in the past is that the activists *as a group* have a diverse set of "drives" and "interests" that brings these activists into politics and which the parties need to appeal to if they are to secure their support as working cadres. The parties differ somewhat in motivations in Detroit (the Republicans more issue-driven than the Democrats recently; see Table 8.7), but the differences otherwise are not great.

TABLE 8.6 ▪ Reasons for Becoming Active in Parties, Detroit, 1956–1992

Percentage Saying the Reason Was "Very Important"	*1956*	*1972*	*1980*	*1992*	*Percent Change*
1. Influence on policies	58	54	57	68	+10
2. Party loyalty or attachment	53	44	33	44	–9
3. Social contacts or friendships	56	68	36	24	–32
4. Friendship for a candidate	23	10	24	35	+12
5. The fun and excitement of politics	34	49	38	47	+13
6. Personal career in politics	11	10	14	15	+4
Number of cases	*281*	*41*	*219*	*113*	

SOURCE: Samuel J. Eldersveld's surveys in Detroit.

TABLE 8.7 ▪ Do the Democratic and Republican Activists Differ in Motivations? (in percent)

"Best" or "Major" Reason for Becoming Active	*1956*	*1980*	*1984*	*1988*	*1992*	*Average Party Difference 1980–1992*
1. Party loyalty						
Democratic	11	4	7	5	2	
Republican	15	6	10	8	3	2.3
2. Influence on policies						
Democratic	19	40	27	27	25	
Republican	25	29	39	46	49	19.0
3. Social contacts						
Democratic	26	15	15	21	14	
Republican	15	25	22	20	12	5.0
4. Friendship for a candidate						
Democratic	12	5	2	6	8	
Republican	14	4	3	8	13	2.3

SOURCE: Samuel J. Eldersveld's surveys in Detroit.

What keeps these activists working, what satisfies them *now*, rather than what they thought were their incentives when they joined? We asked a key question to get at that: "If you had to drop out of politics tomorrow, what would you miss most?" Our findings are rather striking (Table 8.8). Many activists, when forced by such a question to decide on their *major* current satisfaction, report that the social contacts and friendships of party activism and the fun of the political game are most rewarding, or at least "what they would miss most." That is, they no longer see issue partisanship as most salient or primary, perhaps losing some of their philosophical fervor but admitting to the reality of social gratifications. In the jargon of motivational research, "purposive" incentives give way to

TABLE 8.8 ▪ Change in Motivational Orientations of Activists as They Work for the Party* (Detroit data, in percent)

Motivation	*1956*	*1980*	*1992*
Major motivation when they became active			
1. Influence on policies and/or a sense of community obligation	42	60	59
2. Social contacts and/or the fun of politics	21	15	13
Current major satisfaction			
1. Influence on policies/community obligation	13	28	43
2. Social contacts/the fun of politics	65	61	42

* The open-ended question used was: "If you had to drop out of political activity tomorrow, what things would you miss most from such work?"

SOURCE: Samuel J. Eldersveld's surveys in Detroit.

"solidary" incentives. But one must note the trend line in our data — the differential between these two incentives is much smaller in 1992 than in 1956 or 1980. Thus, in 1992, 59 percent were "purposive" when they began to work and

43 percent were "purposive" when interviewed, for a

16 percent differential.

In 1956 this was a 29 percent differential, in 1980 a 32 percent differential. Still, it appears there is a motivational "reorientation" which takes place, or a "resocialization" of the activist to more solidary incentives as a result of his or her exposure to the local party. This is a finding which has been confirmed in several other studies, American and foreign.[23]

If one wants to understand politicians, it is necessary to understand their motivations — what drives them, what keeps them involved? Obviously our data suggest that no one theory will suffice. Motivations are very diverse. But for local activists to be kept working, one must recognize that for many people the party is a social group providing social rewards and opportunities. Ideology and issues are also more important than ever today. It is this mix of high purpose and social enjoyment that demonstrates local party activism today.

THE IDEOLOGY AND ISSUE POSITIONS OF THE ACTIVISTS

As we noted in the analysis of activists' motivations, they are more interested than ever in party involvement because of their desire to work on issues. Thus, the interesting questions posed are: What are the views of the Republican and Democratic activists?; How do they differ?; How polarized are they?; and, how have they changed over time? Above all, since activists are in such close touch with the voters, especially with their loyal supporters as "opinion leaders" communicating their views on pending issues to these supporters, what is the ideological and issue content of their contact with the public?

The first analytical question is how distinctive ideologically are the two sets of local party activists? Are they "tweedledum and tweedledee," holding identical or very similar views, or are they distinctive? When we began a study of this matter back in 1956, the early studies showed that Republican and Democratic local leaders were strikingly different. One study at that time, of national convention delegates, revealed that on human welfare issues (aid to education, public housing, social security benefits, minimum wage, etc.), on average 64 percent of the Democratic activists took the liberal position while only 25 percent of the Republican leaders did.[24] Recently, a similar study in 1992 of the views of national convention delegates on the question of government action to provide more services for social welfare and health insurance found that 62 percent of the local Democratic leaders favored the liberal position (5 percent the conservative position) and 63 percent of the local Republican leaders favored the conservative position (fewer services and more use of private insurance companies) and only 5 percent the liberal position.[25] Thus today, as in the past, there is evidence that the local party activists really do differ strongly in their issue and ideological beliefs.

Of course, party activists' views do change, even radically, from time to time, and the level of disagreement varies by election campaign. One major example of this is in the views of the activists of the two parties on civil rights, particularly the rights of African Americans. A significant change took place between the late fifties and mid-sixties as a result of the passage of national civil rights legislation in 1964 and 1965, the Goldwater Republican candidacy for the presidency which questioned such legislation, and the decision of the Republican party to appeal for presidential election votes in the South and to oppose the extension of the civil rights legislation and its enforcement. As a result, the Republican party deserted the Blacks for the South and the Democratic party espoused the cause of the Blacks. And the attitudes of the party activists reflected this transformation in party ideology. A significant study documented this issue change.[26] It revealed a significant change from 1956 to 1972 in desegregation attitudes by Republicans, from over 50 percent pro-desegregation views to 65 percent anti-desegregation views, while Democratic activists changed from 46 percent to over 66 percent support for desegregation. The civil rights legislation and ensuing political campaigns had reversed and polarized the parties on this issue.[27] In our recent special study of Detroit activists we find a significant contrast in recent years between the parties:

Civil Rights Attitudes of Detroit Party Activists (percentage in favor of civil rights legislation)

Activists	*1980*	*1988*	*1992*
Democratic	62	78	68
Republican	27	34	30

For such to be the case in a large metropolitan area which is over 70 percent black is indeed significant.

These great differences between the activists of the two parties suggests an increasing polarization. Is this actually the case for most issues which the parties face today? Are the Republican and Democratic party activists farther apart on issues in the nineties? If we confine ourselves to these local leaders and look at their positions on social and economic issues over the years we can see what the trend has been. The 1956 study of party convention delegates found that on the average 64 percent of the Democrats took "liberal" positions on such issues (aid to education, social security, public housing, minimum wage) while only 25 percent of Republican delegates did — a considerable difference of thirty-nine percentage points.[28] In subsequent studies (1972–1984), also of convention delegates, scholars found increasing polarization between 1972 and 1980, and between 1980 and 1984. The average difference between the parties on such issues (busing, aid to women, abortion, etc.) revealed an increase of fifteen percentage points.[29] The actual percentage difference between the parties on social issues in 1984 was 45 percent (the Democrats 65 percent liberal and the Republicans 20 percent liberal). This appears to represent slightly more polarization than in 1956.

In recent years, studies show an even greater evidence of polarization. An analysis by Crotty for the years 1984–1992 provides the evidence of this polarization, based again on samples of local leaders (primarily national convention delegates and county chairs) (Table 8.9).[30]

These data strongly suggest a *net* increase in polarization between the parties. But if one inspects the data carefully, one notices that this increase in the ideological gap is not really because the Republicans were more conservative in 1992 than in 1984. On two of the issues the Republican activists actually became *more* liberal; but the increase in Democratic liberalism was greater — hence, the increased polarization.[31] National studies have shown that Democratic *identifiers* are more inclined to call themselves "liberals" (when asked the self-classification question on ideology) and Republicans are more likely today than previously to call themselves "conservatives" (in 1980, but slightly less so in 1984).

TABLE 8.9 ▪ Party Elite Differences on Key Social Issues in Presidential Election Years of 1984, 1988, and 1992 (percentage taking liberal position on the issue)

	1984		1988		1992	
Issue	*D*	*R*	*D*	*R*	*D*	*R*
1. Social welfare	71	34	76	31	76	21
2. National health insurance	72	19	74	23	81	19
3. Government aid to help blacks	60	31	63	36	75	35
Average difference between the parties	*39.7*		*41.0*		*52.3*	

Note: Based on surveys completed by John S. Jackson III, David Bositis, Denise Baer, in the American National Election Studies cumulative file 1952–1992. Elites interviewed were national convention delegates and county chairs, with a few state chairs and national committee members included in 1988 and 1984: *N*s were: 875 (1992), 1208 (1988), 1728 (1984).

Self-Declared Ideology over Time (in percent)

	Democratic Identifiers			Republican Identifiers		
See Themselves as:	*1972*	*1980*	*1992*	*1972*	*1980*	*1992*
Liberal	33	38	44	13	10	10
Moderate	41	34	33	37	23	27
Conservative	26	28	24	50	67	64

Source: National Election Studies.

Thus, in a sense the activists are moving to left and right along with a shift in the public's perceptions of their own ideological orientations. As a result, we have a more distinctive set of party activists and leaders today than before, although there have always been these striking contrasts in beliefs.

How representative are local party activists of the opinions of their followers? Is the ideological distance great or small, and has it been widening or narrowing? This question was posed already in the 1956 studies. The propositions suggested by the McCloskey findings were:

1. The followers of each party in the public (the party identifiers) are much closer to each other than are their local leaders; the "followers differ only moderately in their attitudes toward issues."
2. There is a closer congruence between Democratic leaders and their followers than between Republican leaders and their followers.
3. Republican leaders are much less liberal in their views on policy than their followers, while Democratic leaders are closer to their followers.
4. It follows, then, that "the views of the Republican rank and file are, on the whole, much closer to those of the Democratic leaders than to those of Republican leaders."[32]

This 1956 picture was, then, one of much *consensus* between followers of the two parties and much *conflict* between the leaders of the two parties. To what extent has that continued over the past forty years?

One of the most careful studies to explore these "elite-mass linkages" was the 1984 analysis by Miller. He found "a picture of national partisan politics that is sharply polarized, at both elite and mass levels, over questions of basic political philosophy."[33] The question we are interested in, however, is one of "representation" or "closeness in opinion" of elites and their mass loyal supporters. Adapting the findings on mean scores for policy preferences of party leaders and mass followers, and presenting them in summary form, we find this basic pictures in Miller's 1984 study:

Party	*Liberal*	*Difference*
Democratic leaders	73%	14%
Democratic followers	59%	
Republican leaders	34%	10%
Republican followers	44%	

Such findings suggest that the situation had changed considerably since 1956: (1) the followers were closer to their leaders than to each other; (2) there is fairly close intraparty congruence between leaders and followers for both parties; and (3) Republican followers seem to take their cues from their leaders (or, conceivably, vice versa), as do the Democratic followers. Further, this picture shows a fairly close representative relationship between leaders and followers for both parties. Yet Republican leaders continue to be more conservative than their followers, as in the past.

The final set of data brings us up to date to the eighties and nineties. The distance between activists and followers seems to have remained rather narrow for the Republicans on social issues — a difference below 10 percent. For the Democrats the same was true for 1988 (15 percent difference), but in 1992 the gap widened slightly (to 25 percent). This seems to be primarily due to the greatly increased elite support for aid to the blacks in 1992 (56 percent approved in 1992 compared to 35 percent in 1988), while the loyal Democratic followers decreased in support (from 34 percent in 1988 to 18 percent in 1992).[34]

On balance, it appears there is greater congruence of the two parties today in the representation of mass opinions by their local elites than in earlier days. Republican followers are closer to their leaders, and the same is true for the Democratic intraparty picture. It is a picture of considerable ideological conflict between the two parties at both the mass and elite levels. Finally, one should remember that there is great *intraparty* factionalism also on policy; large minorities in each party dissent on certain key issue positions. We shall return to that subject later in this book.

THE IMPACT OF LOCAL ACTIVISM ON PUBLIC BEHAVIOR

This is *the* key question: Does the work of local party activists pay off; does it influence the public's views about politics and, particularly, its voting behavior? Long ago a student of Chicago politics in the twenties demonstrated in an experiment that if local party workers used special contacts and appeals to voters, they could increase voting turnout by about 10 percent.[35] We followed up this research after World War II with some experiments in Ann Arbor and demonstrated similar findings, including the finding that telephone contacts in campaigns can do as well as personal house-to-house contacts in getting out the vote.[36] At about the same time, we did a study of party precinct organizations in Detroit and found that strong grassroots party organizational efforts made a 10 percent difference in voting behavior in that city in the 1956 presidential election.[37]

Since these early studies, many others followed (over thirty special studies), in this country and in Europe, examining whether this local party effort continued to play a role in voting turnout and perhaps in changing party preferences. Almost every special study of local party effort has concluded that such effort has increased the vote, in local as well as national elections. Crotty tested this theory with a study of elections for state and county offices in the South (North Carolina) and concluded that party activity made a difference, often a substantial one.[38] At the same time, Kramer examined NES data from 1952–1964 and reported strong turnout effects but little preference effects.[39] More recently, a 1990 study of the electoral role of county party organizations concluded that if such organizations build up their strength, they will enhance their relevance by recruiting more able candidates for office, and if they do this, they will maximize their party's vote. "These findings indicate," they argue, "that [local] party organizations do matter."[40] Another study in the nineties of national voting behavior [using NES data 1952–1990] concluded that "party contact is clearly quite an important factor [in most elections] in predicting turnout. . . . aside from 1960 and 1986 party contact is statistically significant at the .05 level."[41] An interesting study of local party presidential activists in 1992 reported considerable effort by such workers on behalf of the presidential candidates and concluded that (1) "grass roots campaigns may have been more responsible for variations in local (electoral) outcomes" and (2) "the frequent Democratic advantage in local campaign effort may have contributed to its ultimate victory in 1992."[42]

One of the most interesting questions posed by scholars related to the contact efforts of local activists is to what extent is the decline in voting turnout in the United States since the 1960s attributable to the efforts of the local party? We know from all the research that party activism can have an effect and, secondly, that this impact varies over time and by locality with the strength of the local party organization and the level or extent of its mobilization efforts. The proposition that is suggested, then, is that the decline in national voter turnout is due to the decrease in the extensiveness and/or effectiveness of local party canvassing efforts. Is there evidence to support this? In their careful, systematic 1993 analysis, Rosenstone and Hansen have directly confronted that question and presented important research findings which provide a logical answer.[43] As evidence of decline in party contact, they use the responses to the regular NES question put to American adults in each national election year: "Did anyone from one of the political parties call you up or come around and talk to you about the campaign?" The percentage replying in the affirmative increased from 12 percent in 1952 to 29 percent in 1972 (and 28 percent in 1976), then declined to 24 percent in 1988. In 1992 there was a further decline in contact to 20 percent, but then an increase to 29 percent in 1996. So the percent of the public contacted has fluctuated greatly and today is as high as it was in the seventies. The decline in national voting turnout began after the 1968 election (and was very low in recent elections). So the coincidence of the curves for the decline of turnout and reported public contact is imperfect. Nevertheless, the earlier data on decline properly stimulated these scholars to see to what extent party contact was linked to the decline in the public's political behavior.

The elaborate analysis by Rosenstone and Hansen provides convincing evidence to solve "the puzzle of participation." We discuss the details of this study in Chapter 13 in our analysis of the factors related to voting turnout but will briefly summarize that study here. The researchers find, first, that those who were contacted were more likely to vote in presidential elections (a 7.8 percent difference) and in midterm congressional elections (a 10.4 percent difference). But of course, this is not convincing evidence by itself, since those who were contacted by the parties may have been the type of person who votes anyway. Further, there may have been other factors as important as party contact (age, extent of social involvement, feelings of efficacy, etc.). To be more certain, they then did a special simulation, using probit analysis, to calculate to what extent these factors contributed to the decline of electoral participation since the 1960s. They find that at least 50 percent of the decline in vote turnout is because of a decline in mobilization efforts.

Two points must be quickly underscored at this point. First, there were other factors which helped produce this deterioration in the American public's interest in voting; the "younging" of the electorate was one of these. Further, as Miller and Shanks point out, generational analysis suggests that the post–New Deal generation had a particularly low voter turnout compared to the pre–New Deal generation.[44] Second, the authors point out that it was not only parties that were responsible for declining mobilization, but also interest groups, social movements, and the efforts of candidates themselves. Above all, this analysis demonstrates that the contact efforts by parties are very important: if contact is high, there will be increased voting; if low, voting will decline. We know from other research that parties have been actively involved in getting out the vote and that these efforts have had an effect. We know also that for many years these party efforts have been "slack," at less than 50 percent efficiency. The important point is that if parties at the local level were not active at all in getting out the vote, the American electorate's voting turnout would be even more dismal.

Thus, party activism is clearly very relevant, even though we know that such efforts by themselves are not a complete answer to the "puzzle" of lower turnout in the United States. One should note also that the Rosenstone-Hanson study demonstrated that party mobilization efforts have other important effects: getting more people to work for the parties, getting more people to contribute money to campaigns, and encouraging more people to discuss politics with their neighbors and friends.[45] These are extra effects of party mobilization usually overlooked.

MINOR PARTY ACTIVISTS: CONTRASTS WITH THE TWO MAJOR PARTIES

The activists who were committed to Ross Perot in 1992 were in many respects quite different from the Democratic and Republican workers. Scholars obtained access to the lists of people (reported as about 500,000) who called Perot's telephone banks at Dallas volunteering to work for him. A sample of these has been studied and compared to those who attended party caucuses during the early nomination period of 1992 in two states, Virginia and Iowa. While a limited comparison, it provides some information about these "potential" Perot workers.[46] There were certain contrasts discovered in this analysis. One finding, as suspected, was that the Perot activists had very weak party identification: 17 percent were strong identifiers in contrast to over 60 percent strong party identification for the Democratic and Republican activists. Second, as suspected also, Perot activists had a very negative view of American government and of the two major parties. Thus, 41 percent of the Perot followers felt you could almost never trust the government to do what is right, compared to 7 percent and 24 percent of the Democratic and Republican activists, respectively, who held this view. As for their evaluations of the parties, 76 percent of the Perot followers said parties confused the issues, compared to 36 percent of the Democrats and Republicans who held that position. Third, the Perot activists differed also on certain issues (Table 8.10). They were much more conservative on affirmative action than both of the other parties and similar to the level of conservatism of the Republicans on health insurance and balancing the budget. But they did not differ extremely from the Democrats on the abortion question, environmental regulation, and a tax on social security. Their issue positions were not, therefore, so radical. In fact, they saw themselves ideologically as much less conservative than the Republicans.

The only other study of third-party activists is that of George Wallace's American Independent party in 1968, an intensive analysis in one Detroit district that was strongly pro-Wallace.[47] The author, Canfield, found that the Wallace campaigners were a rather special group of blue-collar workers (72 percent) who had stopped their education with high school (81 percent) but who had, for that time, respectable incomes (66 percent over $10,000). One-half were southern migrants, and they were predominantly Protestant (70 percent). Thus, in terms of occupation and education, Wallace's supporters were working-class, reflecting Wallace's appeals to the working class during his campaign. In addition, a fairly large proportion (25 percent) of Wallace's activists were young (under age thirty).

These social backgrounds were quite different, therefore, from those of the Perot third-party activists twenty-four years later. The Perot activists were well educated: 49 percent had a college education or more, and more than 40 percent reported incomes over $50,000. Thirty-eight percent were women. Only 8 percent were under age thirty. When asked about their practice of religion, over 40

TABLE 8.10 ▪ Comparison of Issue Positions of Perot and Regular Party Activists, 1992 (mean scores on a scale of –1 [liberal] to +1 [conservative])

Issue	*Perot Activists*	*Democratic Activists*	*Republican Activists*
Affirmative action	+.21	-.44	-.34
Abortion	-.04	-.06	+.33
National health insurance	-.40	-.81	-.43
Environmental regulation	-.52	-.79	+.19
Tax on social security	-.55	-.75	-.03
Balanced budget	+.52	+.04	+.66
Liberal-conservative self-placement index	+.22	-.54	+.81

SOURCE: Rapaport et al., "Activists and Voters in the 1992 Perot Campaign: A Preliminary Report (unpublished manuscript, 1994), 9-10.

percent said that they were not very religious. Thus, the Wallace and Perot activists represented quite sharply contrasted constituencies, persons from different social groups who at two points in time were dissatisfied with the two major parties and sought to replace one or both of them.

The Wallace activists in 1968 were preoccupied with the law-and-order issue (referred to by 55 percent) and with the "no-win" strategy of the American effort in the Vietnam War (mentioned by 20 percent). They were very concerned about the communist threat, 77 percent and 89 percent respectively claiming there was communist "influence" in both the Republican and Democratic parties, and 43 percent believing Martin Luther king was "Communist inspired." On civil rights, 69 percent opposed the integration of blacks. These were people quite strongly pro-Goldwater in the 1964 election (58 percent). On social-welfare issues the large majority (67 percent) believed that social security benefits and payments to the poor "are not much different than Socialism and Communism." These issue positions are strikingly different from those of the Perot activists in 1992 (Table 8.10).

While in some, limited, respects these two sets of third-party activists were similar (ideologically motivated, strongly critical of both the Democrats and Republicans, alienated from our system), they were very different in their social profiles, their issue positions and the level of their basic acceptance or rejection of the American system. Both mobilized large numbers of voters — 13.5 percent by Wallace and 18.9 percent by Perot. But neither form of extreme third-party movement could survive.

THE PARTY ACTIVISTS: SOME FINAL OBSERVATIONS

We have spent considerable space here dissecting and analyzing party activism at the local level. It was necessary to look carefully at the evidence in order to understand the activists and their role. The information in this chapter should help us comprehend better the nature and direction of U.S. parties, because it is our opinion that the local activists are the "hard core" of the organization. The activists are key party organizational actors and also key linkages between the public and their government.

The new empirical research on the activists has helped us discover certain types of knowledge. In some respects local activism has changed; in other respects it has not. Local party activism is alive and functional throughout much of the country, at the county, city, and precinct levels. Despite what the critics and debunkers say, party activism is a reality. Party activism exists (although we might like more of it), activists work hard in campaigns (although we might like more extensive and effective performance), and activists perform key functions in the party system (although we might like to see them do more). And these activists, with their diverse motivations and styles of activism, have a high level of satisfaction with the rewards they receive from such involvement. Over 50 percent tell us they wish to continue on in their work and aspire to more responsible party tasks. The morale of the majority is generally high, their commitment strong, their sense of satisfaction with party work impressive.

Our data suggest that over the years changes have been taking place in the nature of party activism. There is a "new inclusiveness" in the types of people involved: more women, more minorities in some areas, more young people, who are better educated and coming from different occupations. These local parties thus can be "opportunity structures." As this social renewal process takes place, the parties have demonstrated they have been adapting to the changing nature of American society and party life. Another evidence of change and adaptation is in the particular incentives which motivate activists. While this is still a complex and diverse picture, we tend to find somewhat less reference now to party loyalty and more engagement with issues and causes, while at the same time there is some decline in the need for social interaction and gratification as the basic impetus to activism. Hence, some motivational resocialization has occurred. Party activists are pragmatic but can also be ideological.

And then we come to activists' views about political issues. Here we found that as the public changes ideologically, so do the party activists. (One could also state this differently: as the activists change, so do their followers.) The process by which this has occurred is not precisely demonstrable. But there is an increasing issue and ideological distinctiveness or polarization between party activists, and perhaps the 1964 election and the civil rights acts of the sixties had something to do with this. In addition, party activists seem today to be closer to their own followers on issues of the day than previously; 1992 and 1996 are different from 1956. This intraparty representativeness in the relationship of local activists to their loyal supporters is a possibly significant development.

Finally, our analysis of several bodies of research on party impact suggests that party activism continues to "pay off," to be relevant, to get out the vote. The parties may go at this differently than they did before, employing newer technologies, more use of telephone canvassing, closer association with party mobilization strategies at higher levels, et cetera. However accomplished, almost every study reveals that strong party activist structures have increased the vote, depending on the state of strength of the opposition and the constraints of local political circumstances. Party activism "matters." We know that the basic equation we should keep in mind is: Every increment in local party mobilization can (under certain conditions) produce an increment in voter turnout. That blunt statement does not belittle the other forces influencing the vote, but rather states forthrightly a basic truth for American society that party effort by itself is very relevant. While in other systems voters may be mobilized by other means or may be the habitual product of other types of pressures, in our culture party activism in the form of personal contact with the voter is functional to turnout. Thus, who these activists are, how many there are, what they believe in, how they are trained and deployed, and what in the last analysis they do during campaigns and in between campaigns are of critical importance to our system.

NOTES

1. Steven J. Rosenstone and John M. Hansen, *Mobilization, Participation, and Democracy in America* (New York: Macmillan, 1993), 101.

2. Maurice Duverger, *Political Parties* (London: Methuen, 1951), 63-65.

3. William Chambers, ed., *The First Party System: Federalists and Republicans* (New York: John Wiley, 1972), 45-56.

4. Frank J. Sorauf, *Party Politics in America* (Boston: Little, Brown, 1968), 68-79. Earlier, in 1958, Key speculated that "the doorbell ringers have lost their function of mobilizing the vote to the public relations experts," *Political Parties and Pressure Groups,* 4th ed. (New York: Crowell), 375-76.

5. Beck et al., "Local Party Organizations and Presidential Politics," 1997 manuscript, 14.

6. See William Crotty, ed., *Political Parties in Local Areas* (Knoxville: University of Tennessee Press, 1986). The bibliography of such studies is excellent in this book, 235-44.

7. The California Club movement is described in James Q. Wilson, *The Amateur Democrat* (Chicago: University of Chicago Press, 1966), 110-25.

8. Dwaine Marvick, in Crotty, *Political Parties;* also Marvick, ed., *Political Decision Makers* (New York: Free Press, 1961); also Moshe Czudnowski, ed., *Political Elites and Social Change,* 1983.

9. Harold F. Gosnell, *Machine Politics: Chicago Model* (Chicago: University of Chicago Press, 1937), 64-66.

10. Crotty, *Political Parties,* 160.

11. Gibson et al., "Whither the Local Parties?" *American Journal of Political Science* 29 (1985): 142.

12. For earlier conventions see Warren Miller and Kent Jennings, *Parties in Transition* (New York: Russell Sage Foundation, 1986), 262; for 1996 see Larry Sabato, *Toward the Millennium: The Elections of 1996* (Boston: Allyn and Bacon, 1997), 108; for the 1992 conventions see the *New York Times,* July 13, 1992, and the summary of these data through 1992 in Paul Beck, *Party Politics in America* (New York: Longman, 1996), 237.

13. Sabato, *Toward the Millennium.*

14. James Q. Wilson, *Negro Politics: The Search for Leadership* (New York: Free Press, 1965).

15. Eldersveld, in Crotty, *Political Parties,* 94-99. See also a particularly good study of the link between demographic environment and party activist performance by Paul Beck, "Environment and Party: the Impact of Political and Demographic County Characteristics on Party Behavior," *American Political Science Review* 68 (1974): 1229-44.

16. Ronald B. Rapoport et al., "Activists and Voters in the 1992 Perot Campaign," a preliminary report, unpublished.

17. James H. Svara, *A Survey of America's City Councils,* research report of the National League of Cities, 1991, 12-14.

18. Joel D. Aberbach, Robert D. Putnam, and Bert A. Rockman, *Bureaucrats and Politicians in Western Democracies* (Cambridge: Harvard University Press, 1981), 60.

19. Samuel J. Eldersveld, *Political Parties: A Behavioral Analysis* (Chicago: Rand McNally, 1964), 124-26.

20. Peter Clark and James Q. Wilson, "Incentive Systems: A Theory of Organizations," *Administrative Science Quarterly* 6 (1961): 129-66.

21. See Denise L. Baer and David A. Bositis, *Elite Cadres and Party Coalitions* (New York: Greenwood Press, 1988), 32; also Jeanne Kirkpatrick, *The New Presidential Elite* (New York: Russell Sage Foundation, 1976), 114.

22. Warren E. Miller and M. Kent Jennings, *Parties in Transition* (New York: Russell Sage Foundation, 1986), 91-93.

23. See Samuel J. Eldersveld, "Motivations For Party Activism: Multi-National Uniformities and Differences," in Dwaine Marvick and Samuel J. Eldersveld, eds., *Party Activists in Comparative Perspective International Political Science Review* 4 (No. 1), 57-68. See also M. Margaret Conway and Frank B. Feigert, "Motivation, Incentive Systems, and the Political Party Organization," *American Political Science Review* 62 (1968): 1159-73; Philip Althoff and Samuel C. Patterson, "Political Activism in a Rural County," *Midwest Journal of Political Science* (1966): 10, 39-51.

24. Herbert McCloskey, Paul J. Hoffmann, and Rosemary O'Hara, "Issue Conflict and Consensus Among Party Leaders and Followers," *American Political Science Review* 54 (1960): 413.

25. William Crotty, John S. Jackson III, and Melissa Miller, "Political Activists over Time: 'Working Elites' in the Party System," unpublished manuscript, 1997.

26. Edward G. Carmines and James A. Stimson, 1989, *Issue Evolution* (Princeton: Princeton University Press, 1989): 102-3.

27. See also David Nexon, "Asymmetry in the Political System: Occasional Activists in Republican and Democratic Parties, 1956-1964," *American Political Science Review* 65 (1971): 716-30.

28. McCloskey, Hoffman, and O'Hara, "Issue Conflict."

29. Miller and Jennings, *Parties in Transition*, 181; Warren Miller, *Without Consent* (Lexington: University of Kentucky Press, 1988), 19, 32.

30. William Crotty, John S. Jackson III, and Melissa Miller, "Political Activists Over Time: 'Working Elites' in the Party System," unpublished paper, 1997.

31. This phenomenon was noted also by Miller in his study of the 1984 election, when he interpreted the 1980-1984 shifts, finding that "the net result was a clear leftward drift among Republican activists." Miller, *Without Consent*, 18-19. Similarly, in our Detroit study the "average liberalism" of Republicans increased by 15 percentage points 1984-1992 (to almost 60 percent), while the Democrats remained fairly steady at 85 percent. This is, of course, a major local variation from the national trend.

32. McCloskey, Hoffman, and O'Hara, "Issue Conflict," 422-24.

33. Miller, *Without Consent*, 35.

34. Crotty, Jackson, and Miller, "Political Activists."

35. Harold F. Gosnell, *Getting Out the Vote: An Experiment in the Stimulation of Voting* (Chicago: University of Chicago Press, 1927).

36. Samuel J. Eldersveld, "Experimental Propaganda Techniques and Voting Behavior," *American Political Science Review* 50 (1956): 154-65.

37. Daniel Katz and Samuel J. Eldersveld, "The Impact of Local Party Activity Upon the Electorate," *Public Opinion Quarterly* 25 (1961): 1-24. In a study in the same year, 1956, of the role of local party efforts and voting behavior in Gary, Indiana, a similar finding was reported. See Phillips Cutwright and Peter Rossi, "Grass Roots Politicians and the Vote," *American Sociological Review* 63, (1958): 171-74; also, Phillips Cutwright, "Measuring the Impact of Local Party Activity Upon the General Election Vote," *Public Opinion Quarterly* 27 (1963): 372-86.

38. William J. Crotty, "Party Effort and its Impact on the Vote," *American Political Science Review* 65, (1971): 439-50.

39. Gerald Kramer, "The Effects of Precinct-Level Canvassing on Voting Behavior," *Public Opinion Quarterly* 34 (1970): 560-72.

40. John P. Frendreis, James L. Gibson, and Laura L. Vertz, "The Electoral Relevance of Local Party Organizations," *American Political Science Review* 84 (1990): 225-35.

41. Peter W. Wielhouer and Brad Lockerbie, "Party Contacting and Political Participants, 1952-1990," paper presented at the annual convention of the American Political Science Association, Chicago, September 1992.

42. Paul A. Beck et al., "Party Effort at the Grass Roots: Local Presidential Campaigning in 1992," paper presented at the annual meeting of the Midwest Political Science Association, Chicago, April, 1994.

43. Rosenstone and Hansen, *Mobilization, Participation, and Democracy in America*, chap. 7. See also the discussion of this by Richard A. Brody, "The Puzzle of Political Participation in America," in Anthony King, ed., *The New American Political System* (Washington, DC: American Enterprise Institute, 1978), 287-324.

44. Warren E. Miller and J. Merrill Shanks, *The New American Voter* (Cambridge: Harvard University Press, 1996), 56-69.

45. Rosenstone and Hanson, *Mobilization, Participation, and Democracy in America*.

46. Ronald B. Rapoport et al., "Activists and Voters in the 1992 Perot Campaign: A Preliminary Report," unpublished paper. A further paper on this subject was presented at the annual meeting of the American Political Science Association in San Francisco, August, 1996.

47. James L. Canfield, *A Case of Third Party Activism* (New York: University Press of America, 1984).

CHAPTER 9

Parties and Leadership Recruitment

In 1996 there were 931 persons who somehow got their names on the ballots in the states as candidates for the 435 seats in the United States House of Representatives. In addition, 68 persons were granted candidacy for 34 seats in the United States Senate. In the race for president, there were at least twelve Republican candidates and a scattering of Democrats who considered a candidacy but none of whom seriously challenged President Clinton. There was, however, Ross Perot, who was a real, live presidential candidate, having won 18.9 percent of the 1992 vote. At the state and local levels, probably 20,000 persons somehow got their names on the ballot for legislative, executive, and judicial positions. Of the millions of Americans who could have tried, or could have been instigated, to be political candidates, only a minuscule percentage made it. Why and how did these 20,000+ candidates make it? How did they get by those "gatekeepers" who made these candidate-selection decisions? How do we go about doing this screening process in the United States? And is our recruitment process satisfactorily done?

The American system of nominating candidates for public office is unique in the world. It is the only system that gives ordinary citizens a direct vote on who should be their candidates for offices, from the presidency down to the county sheriff. In 1996, almost 25 million Americans voting in presidential primaries told the Republican and Democratic parties who they preferred for their party's candidate for president. In 1992 almost 33 million did; in 1988 over 35 million did. In addition, these same millions in direct primary elections throughout the country expressed their party preferences for Congress, as well as for state and local public offices. Ours is, thus, a massive, complex, wide-open, populist system of candidate selection. In other countries this selection task is performed either by national party committees or conventions, or by local party, or constituency, committees. In these other systems, only loyal party leaders and members, usually dues-paying members, are allowed to participate in the selection process. In the United States anyone who is eighteen years of age and older, who meets residency requirements, can vote in the candidate-selection process. Parties may hope these voters are loyal Democrats and Republicans, but in fact we rarely have a meaningful test to make sure that only party loyalists participate. In the United States this means that the contest for candidacy takes place before the entire interested public, and with the public's participation, during the campaign and *before the primary election. That* is when would-be candidates compete for office — by having their credentials for public office debated and evaluated by the entire public, or the segment of it that is interested. In other countries, this debate evaluation and vote on credentials is rarely performed by the mass electorate; activists and leaders perform this screening of candidates for the public.

It is necessary to keep in mind what we mean by leadership "recruitment." What is involved in this process, a universal phenomenon which occurs periodically in all societies, authoritarian and democratic? By way of definition we can say that the recruitment process involves the discovery, screening, grooming, promoting, and finally certifying of leaders for governmental positions. It is during this process that the credentials (or eligibility criteria) for elite status are determined, particular mechanisms for arriving at agreement about preferred candidates are utilized, political support groups for particular candidates are formed, and presumably the relationships between elites and their support groups are cemented. Much goes on, therefore, in the process of recruiting persons for governing positions in the society, and theoretically the circumstances and conditions of what goes on have a considerable effect on how these leaders behave when they get into office. To understand the behavior and performance of the elite in any society, it is extremely important to understand the process by which it arrived at elite status.

In modern democratic societies a set of procedures, channels, and pathways have developed for the recruitment of leadership which is *in theory at least* more open, inclusive, and egalitarian than in nondemocratic societies. However, it is a process which, in democratic societies, usually does not *directly* involve the public — neither in Britain nor in continental European systems. In the United States, large numbers of citizens may be directly involved.

In traditional and authoritarian societies, the recruitment of leadership is the prerogative of a family, or a small ruling group, which determines the credentials for leadership, does the screening of eligibles, and makes the final decisions. It is a system that is closed, exclusive, and nonegalitarian. The accessibility of the elite and its permeability by non-elites is one of the major differentiating characteristics of a democratic society. At least the opportunity for being a candidate for office is not hierarchically controlled. This is not to say that in a democratic society the elite group is completely open, nor that its composition is completely determined by elections. Any observer of American political leadership is aware of the difficulty for lower-class and less-educated individuals in seeking public office or in getting administrative posts. The same is true for women, blacks, and minorities of certain nationality. Furthermore, it is very difficult to get rid of certain incumbents in elite positions — the system seems often to favor those who are entrenched in power. Nevertheless, in a democratic society, in contrast to a closed authoritarian or traditional society, the channels for achieving a leadership position are more diverse, the opportunities for someone who seeks political elite status are much greater and exist particularly for those who are non-elites, for those out of power.

The complex process of leadership recruitment in democratic societies involves political parties, to a greater or lesser extent. One reason political parties came into existence in such societies was the need for specialized structures to bring people together who would evaluate, screen, and sponsor potential leaders and mobilize support on their behalf in relationship to particular group goals. Parties are modern democratic structures developed for this purpose. They are not the only type of structure to engage in recruitment, but they are often the major one, and if they did not exist in a true democracy, they would soon be created. It is natural and inevitable in a democratic community for individuals to band together to promote their political interests, and in the process they will do the following: (1) *identify* individuals, from the large pool of possibles, who have the qualifications for public office and who would promote the group's interests; (2) *test the electability* of such individuals and begin to build a preliminary political coalition in support of such individuals; and (3) *sponsor* these individuals and give them the group's endorsement, so that the voting public will be able to evaluate and identify with these individuals in terms of the group interests represented. There may be "self-starters," of course, particularly in the United States, but they usually seek the support of a group.

The recruitment of leaders should be seen as a *search process,* which can be broken down into certain stages (see Figure 9.1) Part of this process is going on in groups, part of it in the public more generally. Of course, at the same time that groups (parties) are screening and exploring the qualifications and support of particular would-be candidates, individuals (self-starters) may also be considering the possibility of candidacy without particular group support. In most such cases, however, eventually the individual seeks group support and ideally will get group sponsorship. The entire recruitment process has two critical ending points — first, when the group formally designates and sponsors a candidate, and second when the public decides among a variety of candidates, endorsed or not endorsed. The first of these critical decision points is often called the nomination and the second the election.

Democratic societies differ in their elite recruitment patterns, particularly in the role that parties play in this process. First, countries differ in *eligibility requirements,* or in the qualifications necessary to be considered for public office. There are, of course, formal requirements: citizenship, age limits, and in certain systems

FIGURE 9.1 ▪ Party and Group Recruitment as a Search Process in Democracies

	Stage 1 Discovery	Stage 2 Negotiation	Stage 3 Sponsorship	Stage 4 Campaign	Stage 5 Election
Available Pool of Possible Candidates	The Group Identifies Persons It May Support	Group Tests Electability: Assesses Extent of Support Inside and Outside the Group	Group Formally Decides to Endorse the Individual	Mobilization of Public Support	Voters Finally Decide
	Processes within the Group or by the Group			Public Process	

residence in the area. More important are the informal expectations, which are difficult to prove as *requirements* but which are assumed, and those who do not meet them may seek office but will rarely be selected. Educational achievement of a certain level is expected for elite status, and white males in our society have a much better record of acceptance than blacks, other ethnic minorities, and women. Particular skills are often considered advantageous, as attested by the high proportion of lawyers in the bureaucracies and legislative bodies of certain countries. It is sometimes argued that the ideological conformists have a much better chance to be selected than the unorthodox.

Second, the *channels* through which persons move in the recruitment process may vary considerably. Civic groups, labor unions, business societies, the church — these and other channels may be important. Political parties are certainly a major channel. In Great Britain, after education at Oxford or Cambridge, promotion through the party to the parliament is the major route. In Western European democracies, sometimes education at prestigious universities is also very helpful. In the United States, sometimes nonparty groups are as important as parties or work through parties. In some communities, indeed, such groups take over the recruitment functions. Finally, there is always the self-starter who does not follow any prescribed channel but decides to mobilize by himself the political coalition he or she needs in order to seek high office. This is a phenomenon found much more in the United States than in Great Britain or in Western European countries.

The way in which leaders are recruited in modern societies is critical for the performance of the system. This process of recruitment is critical because in it decisions are made concerning the types of leaders needed — the skills, personal qualities, group characteristics, and orientations toward the political process which are necessary for effective government. If the elite recruitment process is done poorly, not only will inept leadership emerge but the policy output will be inadequate or defective, and the public will lose confidence in the system. Therefore, recruitment is very important for the *political system,* just as it is very important to the *party as a group* with particular goals that hopes to compete successfully for political power. For the party to remain credible and stay in political business, recruitment of able leaders is critical.

The importance of the leadership nomination process is underscored by the practice in nondemocratic societies of the state controlling the way in which candidates are finally designated for public office. In a society such as China, by way of illustration, the Communist party decides who the candidates will be. No other group has a role, nor does the public, except recently in some village elections.

Who makes a party nomination? and *how*? are the basic questions. Is it to be made by a small claque of unrepresentative leaders, by a larger but indirectly representative body, by the loyal party supporters, by the public generally, or should all levels of the party be involved? And if there are stages in the process, whose decision is to be final? Further, is it to be an open process of true deliberation, or closed, with deliberation a facade? These are the questions that have plagued the American people since 1789.

MAJOR CHARACTERISTICS OF POLITICAL ELITE RECRUITMENT IN THE UNITED STATES

The political leadership in the United States emerges as the result of a continuous and highly varied set of recruitment circumstances and conditions. It is important, first, to remember that the system of government influences this recruitment. We have a federal system and a presidential system and therefore must select leaders from a nationwide constituency for president, as well as, at the other extreme, officials who run the fifty states, the local mayors, city councils, and county boards. And because of the belief in the separation of powers between the legislative, executive, and judicial branches, as well as in the public's right to participate, there is a complex set of opportunities for public involvement. This includes the recruitment of different types of leaders: judges, governors and other state administrative and executive personnel, and, at the local level, sheriffs and other county administrators, and mayors, as well as legislators at the national, state, county, city, and rural-township levels of governments. Thus, thousands of elite positions at all levels are screened and certified each year on the basis of considerations that may be primarily local.

Second, this nation is a relatively decentralized and autonomous system, so that every state has the right to adopt its own procedures for the recruitment of officials at the state and local levels of government; there is no uniform system. This variation in procedures is a result of political culture differences in the states — some state parties wish to be (or are required by state law to be) much more disciplined and exclusive than parties in other states.

Third, the American elite recruitment system provides for multiple opportunities for candidacy. Lester Seligman has described five types of party candidates in the United States:

1. Promotion — the selection of a person who has worked in the organization, seeks the support of the party, and is awarded the nomination on this basis
2. Conscription — the persuasion of someone in the party organization to be a candidate, on the basis of his/her work in the party, even though the person has no desire to run for office
3. Co-optation — the persuasion of someone *outside* the party to become the party's candidate
4. Agency — the selection of a candidate who is the representative of a major nonparty group (such as a labor union leader, business leader, black leader, and so forth)
5. Self-starter — a candidate who did not come up through the party ranks or other groups and who decides on his/her own initiative to run for office, despite the initial absence of group support[1]

All of these types of candidates are possible in the United States, particularly the last type, the self-starter. This contrasts with, for example, the British and Continental systems, where without party endorsement the likelihood of a successful bid for elite status is rare. Under our system, the maverick and the insurgent have an opportunity to oppose the establishment candidates by declaring their intention to appeal to the public for support in the primary election of the party. Similarly, other groups may decide to present their representatives as candidates for office. On occasion union leaders, for example, run for office in the party primary.

Fourth, as a consequence of these possibilities, it is a recruitment system in which the party organization often does not have exclusive responsibility for decisions on the certification of candidates. Indeed, this is one of the major changes which has taken place in the American system — the parties have lost their central role in the recruitment process. The organization may seek out possible recruits, may encourage people to seek public office, may actually work to mobilize a coalition of support behind possible candidates, but the decision at the final certification stage may be made by a voting public, many of whom may not be strongly involved with the organization. This again is in striking contrast to European and British systems, where the final decision on candidacy is made by the organization, and one which the organization will stand behind. The crucial element of party organization responsibility is often missing in the American system. And, consequently, the voters may blame the party for leadership behavior, but in actuality the party as an organization may not have approved the candidacy selection or may not really have been involved in the final decision.

Fifth, American recruitment decisions often are not majority decisions — by either the party or the public. Rather, they may be decisions made by a plurality of participants in a primary election in which there is a multiplicity of candidates. In certain parts of the United States, notably the South, there is an effort to create a majority decision by the use of run-off primaries, in which the two highest winners in the first primary confront each other. But in actuality in the South and in other states, there is a strong possibility that a person will be certified as a candidate who secures the backing of only one-fourth or one-third of the party supporters. This raises serious questions concerning the legitimacy of the decision and the extent to which it will enlist the wholehearted support of the party rank and file. Plurality nominations are often not genuinely authoritative and thus they weaken the role of the party as well as result in nominations of persons who are not well qualified.

The American political culture, as it has developed over the years, is linked to these special characteristics of our elite recruitment process. Disciplined party systems are inclined to attempt to control the procedures more tightly than we do. Unitary (non-federal) systems will not reveal such a great variety of mechanisms and techniques for doing the job. Where interest-group politics is associated more closely with party politics, elite recruitment may be much more simply the certification of interest-group leaders by the parties. Where conflict is frowned upon, there may be less opportunity for self-starters and challenges to the established leadership. Political systems differ in their cultural norms, and these in turn are reflected in how the system goes about the elite recruitment task.

One of the most important questions for our analysis in this book is, to what extent do parties still play an important role in this recruitment process? In the last ninety years, we have gradually increased the role of the public in the selection of candidates, via the direct primary. Have the parties adapted to these changes in political values and procedures, and thus have they retained considerable influence over the recruitment process? Some of the material in the preceding chapters suggests that they have. We will continue to explore that question in this chapter, first by examining state nomination procedures and then by looking at the presidential nominating system.

LEADERSHIP RECRUITMENT AT STATE AND LOCAL LEVELS

In what some have called the "good old days" of politics in the nineteenth century, local parties, whether Federalists or Republicans, Democrats or Whigs, would hold private meetings or "caucuses" and by show of hands or by quickly prepared ballots decide who their candidates would be for many local offices. Sometimes attendance at these meetings was restricted, and the meeting places (often in the back of a saloon bar) were secret. On the other hand, it is reported that sometimes any person who happened to be interested could wander into such a meeting and participate. The local party was a law unto itself; it was a "private association." The period from 1832 to 1888, as Leon Epstein has so carefully described it, was a time of unregulated political recruitment. Not only did parties informally select their candidates as they saw fit, they also prepared their own election ballots and distributed them to whomever they wished.[2] This early era of the American party experience reflected an interest in decentralized politics, local control, autonomous or private parties, and, to some extent, popular participation in leadership selection. By 1832 we had replaced the earlier legislative caucus system, by which a small group of elites in legislatures selected candidates for office, and established the delegate convention method. This was a new institutional form in which delegates from local party organizations met in convention to make decisions on who should be the party's candidates for president, legislator, and a host of other national, state, and county officials. Beneath this convention system, at the very base of the political system, were "the local parties" and their "caucuses." Their behavior was unregulated, irregular, unorganized, often manipulated, and "open" (in a loose sense) to the mass public. These local caucuses were thus already sometimes called "primary elections."[3]

This "private" and unregulated state of affairs began to change in 1888–90 when we adopted the "Australian ballot system." This system had many features, one of which was the requirement of secret voting. But a most critical aspect of this new ballot system was that the government, *not the parties,* took over the preparation of election ballots and the printing of them, at government expense. Further, it was provided that government officials, local and state, would place the names of candidates for office on the ballot and that these officials would secure these names of candidates from the duly authorized and functioning leaders of the Democratic and Republican parties. These parties secured the right to have their names on the ballot because of their vote in preceding elections. These new regulations meant that the parties now had legal status, became "public" organizations (not just private associations), and were then subject to state laws as to *how* they should nominate their candidates in an authoritative way for placement on the final governmental election ballot. States began requiring parties to make their candidate selections according to certain rules. And courts in the states began to uphold the constitutionality of such government rules. We thus had come through three historical periods in which we changed our candidate selection process:

1790–1832: selection by legislative caucuses, and then by "mixed caucuses" (of legislators and other leaders)

1832–1888: selection by delegate conventions, with local parties free to act as private groups in the selection of delegates and of candidates for local offices

1888–1903: state regulation of parties and the way they selected delegates to conventions and their local candidates.

In 1903 our system entered a fourth stage when we adopted the direct primary system. Beginning with Wisconsin, we rather quickly moved to what are called "mandatory," state-regulated primaries in which the public could directly express preferences for party candidates for public offices in state-run nominating elections. Such state laws described specifically how such direct primaries were to be conducted: under state and local government supervision, at public expense, at particular times of the year, spelling out what type of primary should be held, specifying who in the public would be eligible for participation. By 1912, a majority of states had adopted such direct primary laws, all but four states by 1917, and today all states have such direct primaries for at least some state and local offices. Presidential primaries, which came to full use much later, will be described later. It should be noted that there was considerable "pressure from below," from the public, to popularize the candidate selection process. However, one must remember that it was party leaders, the elites, who in the last analysis made the decisions to change our system, first to the delegate convention and then to the direct primary. While leaders probably preferred the "backroom" negotiations at conventions, and the control that provided to them and to the party as such, these party elites were the decision makers who changed this system. In effect they voted to dilute and weaken their (the party's) authority and the role of the party as an institution.

In the history of the United States, therefore, we have experimented with different nomination procedures, changing from legislative caucus to delegate convention to direct primary. In each case it has been a change in the answer to the basic question of *who* should make the nomination decision. The decision which evolved over the years to decentralize and popularize the nomination function — to take it out of the hands of party leaders and put it in the hands of the public in a separate election — is unique. It stands in contrast to the British system of nominations in which the decision on candidacy remains to this day in the hands of the party, whose leaders and supporters at both the local and central levels make the nomination decision. The American emphasis is on mass involvement in the nomination decision by persons not necessarily loyal to, and active within, the party organization. Today there is one *general* answer to that question with the fairly universal adoption of the direct primary at the state and local level. But each state has its own *specific* answer to that question.

MAJOR TYPES OF STATE AND LOCAL NOMINATION PROCEDURES

All nomination procedures can be classified into nonpartisan and partisan procedures, each containing various subtypes.[4]

Nonpartisan Procedures

Independent Filing. In many states a person can get his or her name on the general election ballot by filing a petition with a specified number of signatures (usually very large and sometimes requiring geographical diversity). He or she may create a new group for this purpose.

Nonpartisan Primary. In at least two-thirds of the cities, all with nonpartisan elections, the designation of candidates takes place in a special primary election in which party labels do not appear on the ballot. Theoretically parties are not involved in leadership recruitment and designation, but often they are active behind the scenes. If, for example, ten persons are to be finally elected to the city council, the primary election will narrow the field of would-be candidates (as many as one hundred perhaps) to twenty, who then compete in the general election.

Partisan Procedures

The Caucus. In rural areas, often a township meeting of partisans is held (such as the Democrats of Superior Township or the Republicans of Superior Township) to directly choose their candidates for township clerk, supervisor, treasurer, et cetera. This direct democracy approach is still used in certain areas every year.

The Delegate Convention. Representatives (delegates) from county or other local party organizations assemble in a plenary body called a convention to select the nominees. It was the widely used procedure at both the state and county levels of government until 1900, when it began to give way to the primary. In some states it still is used, however, as a nominating device, although supplemented by the primary. In Utah and Connecticut, for example, party conventions are still required to nominate candidates, but in Connecticut a challenger who receives 15 percent of the convention vote can force a primary. In Alabama and Virginia, the party has the option to hold a convention rather than a primary. In several states the parties meet in state conventions before the primaries and ballot on nominees. There are other states which hold conventions to make extralegal endorsements before the primary. Finally, in some states, such as Michigan, certain state offices are still filled by convention nomination (lieutenant governor, secretary of state, attorney general, state treasurer, superintendent of public instruction, regents of the state universities, and justices of the state supreme court).

The Direct Primary. This is a partisan nomination method with direct mass participation and decision making. In a separate election, usually at public expense, the loyal party supporters (but in some states any citizen regardless of party loyalty) go to the polls to select the party candidate from among two or more aspirants who were placed on the ballot by convention decision, by the petition method, or by filing a fee.

The primary has become the most widely used method at the state and local level and is used in all fifty states in some form. What these primaries have in common is *direct voting* (rather than indirect representation through a convention), *mandatory* character (since they are required in almost all states for both parties), and their *public* operation (administered by state and local officials, financed by public funds). There are several types of primaries.

- *Open Primaries.* Anyone can usually participate in these, there being no tests of party loyalty. Party preferences usually do not have to be declared in order to secure a ballot; however, the voter cannot vote in the primary of more than one party. Over twenty states have such primaries, although scholars differ as to how "open" they really are. In the wide-open, or blanket, primary, the voter may vote in the primary of more than one party but may not indicate two preferences for one office. (Thus the voter may help the Republicans pick a candidate for sheriff, the Democrats a candidate for governor, and the Socialist-Labor party a candidate for Congress.) Three states, Alaska, Washington, and Louisiana, use this method.
- *Closed Primaries. Theoretically* only those citizens who meet some test of party loyalty or commitment are permitted to vote. They are usually required to register for a party before the election, although some states permit a declaration of party affiliation at the time of voting as sufficient. There are three types of tests: *previous allegiance* or support, *future intention* to support the nominees of the party, and *present affiliation,* or current declaration of support. The voter under this system may be asked to declare and make public a party preference before being permitted to vote. Actually, the applications of these tests may be so loose that often it is possible to participate in a party's primary without too much evidence of prior commitment. Certain states, however, have deadlines for shifting from one party to the next. Up to twenty-seven states have closed primaries.

There is one more basic distinction to be made among state primaries — whether or not they are majority oriented. Nine southern states plus South Dakota use the runoff primary, that is, a second primary election if no one in the first primary secures 50 percent of the vote. In the runoff, the two top contenders oppose each other. Similarly, in Iowa a postprimary convention makes the nomination if no one gets a majority in the primary. This is to assure a majority winner (no matter how superficial the majority decision really is) in states where one party is so dominant (historically, but no longer today, the Democrats in the South, and the Republicans in South Dakota and Iowa) that the person who wins in the primary is the probable winner in the final election. The runoff primary and the postprimary convention, then, are devices to legitimize the party decision.

A survey of this bewildering array of nominating procedures seems to indicate that each state has developed its own special approach to the task of leadership designation. In actuality, there are two basic distinctions in these procedures and, thus, in how state nominations can be classified: (1) the accessibility of the nomination decisions to the general public — that is, the extent to which the citizenry generally has the opportunity to participate, irrespective of party loyalty or activity; and (2) the encouragement and tolerance of party organizational attempts to influence if not control the nomination decision.

Although other distinctions can also be made (for instance, majority-plurality), the two principles of accessibility and encouragement help to differentiate among these systems. On the one hand is the question, Does the procedure adopted by a state assume that any citizen has the right to help decide what candidate the party should nominate, or should this be the exclusive right of a more limited set of party leaders and party faithful? On the other hand, how involved should the party organization be and does it have the prerogative of *structuring* the nomination decision, if not actually making it? These two basic questions may appear to be the obverse of each other, but in actuality they are not, because it is conceivable a state could have a direct primary which is accessible to all but which also permits (or even requires) the party organization to make a preliminary designation before the primary or a postprimary decision if the primary results are inconclusive. In a sense, this is exactly what some of the states have done. The twofold issues involve *organizational responsibility* and *citizen involvement.* Both are desiderata of those designing a nomination system, although one might seem more important than the other.

If the state systems are classified by these dimensions, we can see how both principles are in fact manifest in the procedures adopted. Thus, in approximately one-fourth of the states the primary is used in conjunction with a requirement for, or tolerance of, the party organization's formal involvement in the process. In about one-half of the states the primary is theoretically closed to party supporters, but the party organization's role is not specified. In only a minority of the states does it appear that the nomination is thrown completely open to the public with apparently minimal party organizational influence. From this it appears that the states have given quite different answers to the question, Who should make the nomination decision? And often the answer is ambiguous.

Since 1900 the United States has experimented with various approaches to the nomination problem at the state level. The legislative caucus system was denounced by Andrew Jackson and his followers in the early nineteenth century and rejected by convention. The delegate convention method was viciously attacked in the latter part of the nineteenth century for being unrepresentative, corrupt, and manipulative. The charges were often made that the delegates were bought and that the will of the party rank and file was ignored by the bosses controlling the conventions. That was the basis for the attack on the convention in Wisconsin by the Progressive wing of the Republican party under Robert M. La

Follette, which led to the adoption of a mandatory, statewide primary there in 1903. It was the result of a protest against establishment forces within the party and the way in which the leadership manipulated and controlled conventions. Thus, the attack on the conventions by those who were left out — by the have-not activists in the parties — convinced people that a change was necessary. Had there been an effective, acceptable, democratic convention method of nomination by 1900, or some acceptable intraparty procedures which were considered democratic, such as the British had, the primary might not have been necessary. But people were angry at the abuses of the system. They demanded regulation. They got the primary, which quickly swept most of the states. The American political culture, with its emphasis on populism and mass participation, its concern about bosses, its extreme belief in the right of protest and the protection of insurgency, particularly in areas of one party dominance — these elements aided and supported the reform movement. Forced to accept a reform of the system, the parties have had to adapt to the new conditions of leadership recruitment which these reforms imposed. The basic idea of the direct primary caught on, but its specific implementation varied from state to state. But what is its impact on the system, on public involvement, and on party organization?

THE EFFECTS AND CONSEQUENCES OF THE DIRECT PRIMARY

The controversy over the primary has been heated. A variety of strong arguments was made for the primary in the era of its widespread adoption, and since that time an equally articulate set of arguments has been made against its adoption. Many of these arguments are summarized in Table 9.1. Essentially, of course, the primary adherents emphasize the virtues of mass involvement and the need for protest and competition. The opponents at heart feel that if we want responsible parties, nominations should be left to the organization, and that interlopers should not interfere with the organization's performance. As one scholar has put it, "You wouldn't let General Motors stockholders in to vote at a Ford stockholders meeting merely because there is some possibility of their purchasing a Ford!" Similarly, it is argued that Republicans should not be allowed to enter a Democratic primary to help the Democrats pick their standard bearers, because then the Democrats as an organization could not be held responsible. Nor should independents, who are not committed to any party, be permitted to tell the parties whom they should nominate.

Open primary advocates say, however, that since the candidates who emerge from the party designation procedures may be elected and affect their lives profoundly, there should really be considerable opportunity for everyone to be involved. Closed primary advocates say, "Let's limit participation to bona fide partisans (although let's not be too rigid in defining who these are and let's allow them to make up their minds shortly before primary election day or on that day), but let's really give these people a guaranteed voice in the decision." Anti-primary enthusiasts

TABLE 9.1 ▪ A Summary of Arguments in the Direct Primary Controversy

Pro-Primary	*Anti-Primary*
1. It permits direct popular participation in nomination decisions.	1. Primary participants are unrepresentative of the party supporters — in commitment to the party, in attitudes and beliefs, in social and political characteristics.
2. It provides opportunity for protest against the established party leadership controlling the party organization.	2. The conflict produced by primaries pluralizes the party, leaving scars that are not easily healed.
3. It may contribute to the development of a membership idea through its tests of party affiliation.	3. It takes the nominating function out of the party organization, thus rendering the organization less relevant and making participation and membership in it less meaningful.
4. It enhances the competitiveness of the political process by giving many more would-be candidates an arena to compete for offices.	4. The competition that develops in the primary is not where the party needs it the most — in areas where the party is strongest.
5. It provides a pretest of the election strength and popularity of candidates.	5. The primary makes party success more difficult because it does not permit the party to put together a slate of candidates representing different geographical, ethnic, and ideological factions in the organization.
6. In one-party dominant areas it assures that leadership selection will be a diverse and competitive process in that party.	6. The party's ability to integrate candidate selection with its program is impossible under the primary, since the platform is adopted by the convention but the candidates run by themselves and can ignore the platform.
7. It may encourage independents to become involved in party politics.	7. Cross-voting in the primary can be either irresponsible or mischievous: Supporters of party A may vote in the primary of party B and may never expect to support party B's candidate in the final election; indeed, they may try to nominate B's weakest candidate.
8. It results in different types of candidates from those of the convention method, and therefore it will lead to different types of politics emanating from this elected leadership.	8. Primaries often produce minority nominations that are not considered truly legitimate or authoritative and will not be backed by the entire corps of party supporters.

say, "We already do that through the representative democratic procedures by which delegates are elected to conventions." And so the debate rages, with very little evidence that the U.S. public is ready to backtrack from its increasing support for the direct primary system.

Direct primaries have been in existence for over ninety years and have become the major institution for leadership recruitment at all levels of our political system. Much research has been done on the big questions, and much continues to be done and needs to be done. Systematic research has not been done on all arguments. There are certain key concerns which scholars have addressed in this research. We shall attempt to summarize here the evidence produced, which can help us to evaluate the performance of our system under the direct primary system.

The Primaries and Voter Participation

The first major question concerns the extent and nature of citizen participation in the direct primaries. In the early studies of state primaries, for the years 1926–1972, it was found that in non-Southern states, 27 percent to 35 percent of the eligible voting population on the average voted in primaries.[5] In the South the average vote was higher, close to 40 percent in these early days. The higher southern vote was understandable, since in the one-party states the primary contest was often more important than the general election. More recent studies (up to 1986) indicate that the turnout percentage has not increased in statewide primaries, remaining perhaps somewhat lower at about 30 percent.[6] Thus, from the beginning a fairly large minority of American voters took advantage of the opportunity to participate in candidate selection — 30 percent of the eligible voting population is a large number of voters. As we shall see in our analysis of turnout in presidential primaries, in the 1992–1996 period, from 24 million to 33 million voters participated. There is no question that the primaries involve many citizens, and always have.

There is considerable variation by state; over time, in primary election turnout, scholars have found that the type of election, the type of primary, and the competitiveness of the leadership contest are linked to the size of the primary vote. Local primaries draw a smaller percentage of voters than do state level primaries. "Tightly closed" primaries, which restrict (or try to restrict) voting to those who are declared party supporters, have had lower participation than open primaries (a 15 percent difference).[7] Another relevant factor is the number of candidates contesting the primaries from 1946 to 1976, those states with a median number of five to seven candidates tended to draw a somewhat larger turnout than primaries with fewer candidates. But the difference is not great, an average of five to six percentage points.[8] Finally, Jewell and Olson found that those states with higher turnouts in general elections also tended to have the highest turnouts in primary elections. Obviously there are exceptions to these patterns. But certain political cultural conditions do seem to exist, and to influence voter interest in primaries. We shall see evidence of this again when we analyze the presidential primaries.

The Primaries and Political Competition

One of the laments of students of American politics is that genuine competition for public office is often very limited. This limitation is due to many factors, including the fact that challengers are discouraged from running because of the "costs" of candidacy, the limited success of challengers in the past, and incumbents' tight hold on offices. Congressional representatives are a case in point: over 90 percent of incumbents regularly win; there are usually few "newcomers" in a new Congress. In the 1996 House elections, for example, there were only 53 of 435 (12 percent) seats which were "open" (i.e., not contested by an incumbent). Only 21 percent of the incumbents who ran, or 5.5 percent, were defeated. And only two of these were defeated in the primary. Further, this decline in successful competition worsened in the eighties. Jacobson summarizes the trends: "competition for House seats that were held by incumbents fell to distinctly lower levels in the mid-1980s, extending the historical trend. . . . Because potential candidates and contributors behave strategically, uncompetitiveness feeds on itself. The fewer the incumbents who are defeated or even threatened with defeat, the more reluctant good candidates are to run and contributors are to support challengers."[9] Challengers did much better in House elections until 1970, and then the decline set in.

The role of direct primaries in opening up the candidate-recruitment process, one of the early arguments for the primary, can be argued in different ways. First, it is clear that the primary has produced more visible candidates for many state offices than before. Studies of gubernatorial campaigns reveal that in states with strong two-party competition, over half of the primaries will be contested by three or more candidates. These are not just candidates considered as "possible" candidates in smoke-filled rooms at conventions, but candidates who run active campaigns for public support. The decline in challengers for incumbent congressional seats is somewhat a similar phenomenon for other offices. But one must remember that under the primary system there are still many multicandidacy contests. The primary enhances that opportunity, at least for some types of offices. The second question, however, is whether the primary enhances the *possibility of defeating* incumbents. The record of the primary is spotty or uneven in this respect. One recent election in which primary election candidates were relatively successful was the congressional elections of 1992. In that House election, 24 incumbents (16 Democrats and 8 Republicans) were defeated, 20 of them in the primary. While the percentage of all incumbents (348) defeated was still small, the important point is that the defeats occurred in the primary. The election of 1992 was a deviation, no doubt (in 1988 there were only two primary defeats, and in 1996 only three), yet the potential for mounting a successful campaign in the primary was clearly demonstrated. But the probability of success, previous research has indicated, is greater if no incumbent is running.[10]

A third question raised by scholars earlier, perhaps more important, is to what extent do primaries exist in states and districts where the party is strong and progressively more successful? Can one argue that the primary may be a competitive antidote to one partyism in an area? V. O. Key, in his early classic work, claimed

Recruitment of Candidates at the Local Level: The Trends toward Diversity

In order for parties to remain viable and competitive, they must recruit candidates for public office carefully. And they must diversify candidate selection over time to include representatives from different social sectors in their populations. An example of how this was done is provided by the case study of Ann Arbor, Michigan. A study was done of all candidates for local office (council and mayor), of the Republican and Democratic parties, from 1946 to 1992. The results reveal how the parties differed in candidate selection, and how the parties changed over this 46-year period. Here are the findings:

Candidates for Council and Mayor, Ann Arbor, 1946–1992 (in percent)

	Democrats		Republicans	
Personal Background	*1940–55*	*1980–92*	*1942–55*	*1980–92*
1. Women	11	38	3	26
2. Blacks	3	14	4	6
3. Under age 40	42	64	17	44
4. Occupation: business	23	7	40	50

It is interesting to see how the parties attracted different types of people to run for mayor and council, as well as how both parties changed over this period, in a sense adapting to social change as well as competing with each other from certain types of candidates. Parties need to appeal to a broad social spectrum while also maintaining the support of their most loyal constituents.

SOURCE: Samuel J. Eldersveld, *Party Conflict and Community Development* (Ann Arbor: University of Michigan Press, 1995), 153–58.

that his research on U.S. Senate elections demonstrated a "tendency for the frequency of primary competition to increase with the general election strength of the party." His study of the Senate elections from 1920 to 1960 provided the evidence for this: where Democratic strength was low, the percentage of close primaries was low (11 percent); where the Democrats were strong, the percentage of close primaries was much higher (51 percent).[11] Since Key's research, studies of state primaries for other offices suggest sometimes only marginal confirmation of his finding. But the basic point is still true — there often *are* open primary contests in areas of party dominance, as well as equal party competition. The frequency of these contests varies greatly. And 50 percent of "safe" seats may not be really contested. But on the other hand, 50 percent of safe seats *are* contested, often in close elections, and sometimes turn the incumbent out of office.

THE ROLE OF PARTIES IN THE RECRUITMENT PROCESS

The relevance of the party organizations for this central function of the political system, namely leadership selection, has been questioned for some time. For some scholars of parties this is the main issue, the main deficiency in the performance of our party system. And they would say further that this is the real evidence of party decline. In the early days of the convention and caucus, it is argued, the party organization was dominant. With the onset of the primary, the conclusion was that the party organization lost much of its influence, with a consequent loss in the relevance of the organization in putting together a candidate ticket which was representative of the party's interests and programs and supporters. The argument was, thus, that the organization lost its competitive and integrative roles. The masses of party loyalists, and even independents, presumably took over this function from the party.

On the other hand, there are those who argue that our party organizations, after declining in the immediate postwar period, have been strengthened since the seventies, at the national, state, and local levels, and that as a result they are playing a more important role again, including a role in candidate selection. In previous chapters we reviewed the evidence of the revitalization of our parties and concluded that, indeed, parties are stronger today. However, we took the position then, and would state again here, that our party organizations have not merely recovered much of their earlier capacities but have demonstrated considerable adaptive skills in responding to the "new politics" in America of the eighties and nineties. And that is the basic concept to keep in mind — the concept of the adaptive American party organization, in leadership recruitment also.

In support of the argument that parties still play an important recruitment role, the first point to be made is that so far as the *formal* institutional aspects of the direct primary are concerned, the party politicians have retained some control. There are still states which use conventions for a variety of purposes, including nomination of certain candidates. In other states endorsement by the party organization is permitted, or even required, before the primary.[12] And in a few states this endorsement may close out other candidates. Further, of course, it was the party politicians in state legislatures who, when confronted with the pressure to accept direct primaries, adopted "closed primaries" (in approximately twenty-seven states), in order to try to restrict voting in primaries to committed partisans. These efforts to influence the formal process and retain some sort of organizational role in the early days have certainly not allowed the party organization to assert control, but they have maintained in many states an "organizational presence." Informally, of course, the party often seeks to play a selection or endorsement role and may discourage candidacies or assist candidates with a commitment of organization resources.

One interesting development at the national presidential level beginning in 1984 was the action by the Democratic National Committee to adopt a rule that 14 percent of the delegates to the national convention would be party officials and representatives who did not run as delegates in the presidential primaries. Thus, many members of congress, state and city officeholders, national committee members, and other party officeholders secured convention seats as uncommitted delegates. The proportion of such party regulars with automatic delegate status at the Democratic convention was increased in 1996 to 20 percent. Thus, the party organization continues to attempt to adapt to the new rules of the game but also seeks to modify them in certain directions to maintain a party organizational role.

With the resurgence of state and local party organizations, many candidates realize that it is to their advantage to maintain a close relationship with the local parties. This was always true to some extent, despite the adoption of primaries. Early studies of legislators revealed that often the district, county, or city organizations promoted or instigated candidates to run for these positions, put organizational personnel and registers of voters at the service of such candidates, assisted them with propaganda activities and mobilized the vote on election day. If the party organization was well developed, state legislators and congresspersons relied on them for much assistance. Further, most of these candidates came up through the party organization, and their activity often led them to work with that organization when they became candidates.[13] When incumbents are running for reelection, if the local party is strong the candidate may be strongly endorsed and supported, and potential opponents may be discouraged. When there is no incumbent, the party may play a role in searching for a viable candidate and providing advice, research, money, personnel, and much work. When the local party thinks an incumbent (of the opposite party) is vulnerable, the organization may also work hard to secure and support an able challenger. There also have been constituencies in which the party is much more important than the primary per se, where the local organization is so dominant that it selects the candidate and controls the primary vote overwhelmingly, if there is a primary contest at all. In the early days of machine politics that was indeed the situation; the party machine controlled and the primary was virtually irrelevant. For one classic study of such a situation, one should read the work of Leo Snowiss.[14] On a smaller scale, this is still a possibility in certain communities.

Today, with the closer relationship between national, state, and local party organizations that we discussed in Chapters 6 and 7 there has certainly been a revival of party organizational involvement in leadership recruitment and candidate selection. The funds available in 1996 that we described earlier are significant: the Democratic National Committee raised $99 million and gave 56 percent of this to state and local organizations; the Republicans raised $110 million and gave 43 percent to state and local organizations. Much of this money was given to help candidates in their campaigns, thus linking the candidate more closely to the organization.

It is interesting that in recent years the courts have defended this right of state party organizations to challenge the primary laws in their states. In fact, the courts have held that if a state party wanted to open up its primaries, contrary to state law, it could do so. This is what happened recently in Connecticut, North Carolina, and West Virginia.[15] In another case, Justice Thurgood Marshall said: "a state can't justify regulating a party's internal affairs without showing that such regulation is necessary to ensure an election that is orderly and fair."[16] How far parties could go in asserting their rights to determine how candidates should be selected has not been determined, however.

CONCLUDING OBSERVATIONS

Historically we have, since 1800 in the United States, seen an inexorable populist drive in our politics, including the pressure to decentralize parties and thus the nominating process, and, beginning in 1903, to turn the nominating function over to the public. The direct primary has been the ultimate institutional mechanism by which to do that. It has swept the country, state by state, and overtaken the presidential selection process also, as we shall see in Chapter 10. The parties have had to adapt to this reality. Indeed, it has been the party leaders in state legislatures who have actually passed the laws taking the nominating function from the party organizations and giving it to the people: the party elites gave the power of the elites over to the masses! We have discussed the way in which this unique American system works and the consequences which it seems to have had. While our research is not completely adequate to demonstrate conclusively all these consequences, certain key impacts or effects are apparent. There is impressive evidence that direct democratic participation in the primaries has occurred. Millions of citizens have taken advantage of the opportunity. While one may wish for a higher turnout than 25 to 30 percent, one cannot ignore the magnitude of the turnouts. On the other hand, it is clear that the types of control which the party organizations used to have does not exist. But the parties have adjusted and have developed new ways to influence primaries and the nominating process. While they no longer directly select their candidates in an organizational meeting, they do have much influence over that selection process. Thus we have a mixed system, partially organizational, partially primary; partially elite controlled, partially mass determined.

Whether this has harmed our party system and political system generally is hard to say. If one desires more organizational responsibility, we must admit there is less today. If one prefers more citizen involvement, we must admit there is much more. The extent to which that involvement is deliberative and representative is another matter. The system is no less competitive than it was before, and probably a little more. The candidates who have merged under the primary system are probably as able as, but not more able or better qualified than, they were in the preprimary system period. The evidence on this is actually nonexistent. A major concern of scholars is that the party organization has been weakened. However, there is less evidence of that today than previously. Ours is a system in which significant change has taken place in how we select our leaders. The present system seems to work reasonably well. And the parties are adapting to it. As one scholar put it, perhaps with some resignation, the primaries are "now part of a well established system, [which is] a distinctively American development."[17] This is certainly true, but we should add that even if it is "well established" it will continue to be criticized, defended, and continuously changed.

The United States seems to be torn between the two principles of wide-open accessibility of nominations to the public generally and some role for the organization in assuming responsibility for leadership selection. If we were to specify what type of nomination system we should have *ideally,* most people would agree that we want nominations arrived at through a process that is

- *participative,* providing for meaningful citizen involvement;
- *competitive,* so that real choices are available to those who participate in the selection process;
- *representative,* in the sense of reflecting accurately the desires and opinions of the people who are most attentive to politics and concerned about the nature of our political leadership;
- *structured,* in the sense that those organizations under whose label and program candidates can be held responsible for the leadership recruited;
- *majoritarian,* so that the decision in the selection of leadership is not the act of a factional minority to which the majority dissents.

It is clear that our system does not optimally meet these requirements. But one may well ask, what system in what country does meet them? And, can all of these requirements be met simultaneously by the same system? Probably not. Since we have been moving away from the pro-party organization model to the populist model of nomination, we have obviously been opting for more participation and open competition, and less structured party control. There may be significant consequences and "side effects" which can result from this. The heavy public support for the dominant party in a safe area, while not a monopoly, could lead to atrophy of the opposition. There is also some indication that the divisiveness of hotly contested primaries may lead to bitterness and the withdrawal from politics of those activists on the losing side in the primary. In two-party competitive areas, the party with a hotly contested primary may hurt itself so badly that its chances of success in the general election are impaired. Finally, strong arguments can be made that the divisive nature of the nominating process may have an impact on the subsequent process of policy making at the state level. Governors elected in bitterly contested primaries may find it difficult to work with opponents in the legislature. These tendencies leave advocates of strong, cohesive, responsible parties critical of the primaries. They emphasize two valuable aspects of the old nomination system: the restriction of the nominating function to those loyal and active in the organization, and conducting the deliberations in a forum which will permit the development of a consensus among the many factions in the party support structure. This is in the British and European tradition of responsible parties. For the primary advocates this is unacceptable, particularly because it is contrary to their conceptions of what most enhances democracy. After much historical experimentation, American culture, with its emphasis on populism, its paranoia about party leadership, and its antipathy for organizational discipline, seems to have led to the direct primary. But it is still an open question as to whether its costs for the system, and particularly for a responsible party system, are not greater than its contributions to the democratic spirit.

NOTES

1. Lester Seligman, et al., *Patterns of Recruitment:A State Chooses Its Law-Makers* (Chicago: Rand McNally, 1974), 191.

2. Leon Epstein, *Political Parties in the American Mold* (Madison: University of Wisconsin Press, 1986), 158-62.

3. Ibid., 160.

4. See Malcom E. Jewell and David M. Olson, *Political Parties and Elections in American States* (Chicago: Dorsey Press, 1988); L. Sandy Maisel et al., "The Naming of Candidates: Recruitment or Emergence?" in Maisel, ed., *The Parties Respond* (Boulder: Westview Press, 1990), ch. 7.

5. Austin Ranney in Herbert Jacob and Kenneth N. Vines, eds., *Politics in the American States* (Boston: Little, Brown, 1976), 72; V. O. Key, Jr., *Politics, Parties and Pressure Groups* (New York: Crowell, 1964), 378.

6. Jewell and Olson, *Political Parties and Elections in American States,* 110.

7. Ibid., 144.

8. Ibid.

9. Gary C. Jacobson, "Congress: A Singular Continuity," in Michael Nelson, ed., *The Elections of 1988* (Washington, DC: Congressional Quarterly Press, 1989), 132-33.

10. See Jewell and Olson, *American States,* 1978, 132.

11. Key, *Politics,* 438.

12. See Sandy Maisel, ed., *The Parties Respond* (Boulder: Westview Press, 1990), 147-49.

13. Among these early studies, see particularly David M. Olson, "The Electoral Relationship Between Congressmen and their District Parties," paper presented at annual meeting of the American Political Science Association, Washington, DC, September 1977; Barbara Greenberg, "New York Congressmen and Local Party Organization," Ph.D. dissertation, University of Michigan, 1972.

14. Leo Snowiss, "Congressional Recruitment and Representation," *American Political Science Review* 60 (1966): 627-39.

15. Leon Epstein discusses these cases in his book, *Political Parties,* 260-70.

16. In *Eu v. San Francisco County Central Committee,* cited in Beck, *Political Parties in America* (New York: Longman, 1997), 69.

17. Epstein, *Political Parties,* 108.

CHAPTER 10

Presidential Nominations: The New Model

Nothing is more important in a democracy than the process by which the contenders for the top executive office emerge from the many aspirants and pretenders to that office. This nomination process tells us not only what type of leadership is preferred but also the dominant interests and forces behind this leadership and the type of governance the society will now have. An examination of this nominating process tells us also something about the quality of the leadership selection system, how genuinely competitive it is, how rational, how responsible, and even how democratic. By late March in 1996 the American parties had selected the three top candidates for the presidency: Robert Dole had won enough primaries to control a majority of delegates to the Republican convention, Bill Clinton long before had been given the Democratic nomination, and Ross Perot had erected his Reform party and had set in motion a process assuring him the designation as its candidate for the presidency. What was the system and process by which these selection decisions were made? How did our nominating system screen out all the other would-be candidates? And was it an effective process?

The manner in which we Americans select our presidential candidates is unique. Scholars in other countries are usually bewildered by it and critical of it. In Western democracies there is no desire to copy it, although in one new Asian democracy, Taiwan, it is considerably admired and is being imitated. In parliamentary systems usually, however, the process by which a party selects its candidates for the top executive position is much different. The party caucus in the legislature selects its candidate for prime minister, usually from among its members and usually from its members in the lower house of parliament. The leaders of the parties in the parliament thus become the candidates eligible for the prime ministership, dependent on the vote the party receives in the election. There is no direct public role in voting for party candidates for the prime minister position. The selection of candidates for chief executive in a parliamentary democracy, then, is a party organizational decision, not a decision in which the public has a formal role.

Over the years the United States has gradually evolved its own system for presidential nominations.[1] From 1790 to 1832, congressional leaders in their party caucuses played a major role in the selection of men like Washington, Jefferson, Adams, Madison, Monroe, and Jackson. The demand for more representation in these decisions led to the adoption of the delegate convention system in 1832, the conventions consisting of party representatives picked by state and local party caucuses or committees. In 1908, because of criticism of the behavior of these conventions, pressure developed to decentralize the process further by combining so-called "preferential primaries" with the national convention system. Woodrow Wilson was a great advocate of this decentralization and pushed for universal adoption of presidential primaries. Actually, by 1916 twenty-six states had adopted some form of presidential primary. Over half of the national convention delegates were picked by the public directly, and in many of these states also there was an opportunity for voters to express their preference for a specific presidential candidate for the party. But the enthusiasm for these primaries declined, several states repealed their primary laws, and the presidential nomination up to 1968 was still really left to the national convention, although some delegates were selected in primary elections. So the period from 1908 through 1968 was one in which we were still adhering to the national convention system but experimenting, in a sense, with the presidential primary idea. This hybrid system worked fairly well, in the judgment of many observers, although there were serious convention deadlocks (such as for the Democrats in 1924 and the Republicans in 1920 and 1952). Conventions were hard pressed sometimes to make decisions on nominees who could rally the party faithful. The Republicans had a split convention in 1912, the Democrats in 1948. Sometimes outsiders were able to win the party nomination, as Wendell Wilkie did for the Republicans at the 1940 convention and Eisenhower did in 1952. Generally, however, the conventions picked well-known and experienced party leaders after a long-drawn-out process of local and state delegate selection, and a tense and complex coalition-building process before and at the convention. Not everyone went home from these conventions satisfied, but usually the convention decision was accepted and in fact resulted in hard campaigning by the party faithful.

Nevertheless, the national convention continued in the sixties to be criticized as unrepresentative of the party's supporters, as well as denounced as an elitist institution. The demands for reform became more insistent. As a result, two types of action took place: (1) the parties revised the credentials for representation, opening the conventions to minorities and women; and (2) state legislatures adopted laws for the popular election of delegates in presidential primaries in which candidates for delegate to the convention declared their presidential preferences. The number of states with such presidential primary systems increased from seventeen in 1968 to twenty-nine in 1976 to thirty-five in 1988 to forty-three (Republicans) and thirty-six (Democrats) in 1996. In 1996 over 88 percent of the Republican and 63 percent of the Democratic delegates were selected in such primaries.

THE NEW POPULISM MODEL OF PRESIDENTIAL NOMINATIONS

Since the late 1960s we have had a new system, process, or model for the selection of presidential candidates. There are certain key features of this new American system that should be kept in mind:

1. Voters participate directly in candidate selection and can declare their candidate preferences by selecting delegates to the national convention who are committed to a particular candidate. While a few states still use a "caucus system" rather than a primary, these caucuses of the local or district party organization have also become more open to large numbers of party supporters who can vote for candidates committed to presidential aspirants.
2. The national convention is retained and meets in July or August every four years. But its presidential nomination function is reduced to ratifying the decision of the presidential primaries (and of the caucuses in a few states).
3. The presidential primaries are public elections conducted by states at public expense on dates determined by state legislatures. The state prepares the primary ballot and administers the primary election.
4. The costs of candidate campaigning are partially paid for by the national government. Every serious candidate must register with the Federal Election Commission and can opt for public funding. For each presidential election year, there is a ceiling on the amount a candidate can spend in the preconvention period. The federal government will match half of the funds collected by the candidate (of a certain type). Thus, in 1996 the Clinton and Dole campaigns could spend in the preconvention period a maximum of $45 million, $30 million collected by their party and a matching $15 million from the federal government. Candidates can opt to not use public funding and then can spend as much of their personal funds as they wish.
5. States differ in the specific types of presidential primaries they adopt: "closed" primaries (limited to loyal partisans) or "open" primaries (in which anyone can vote); "proportional representation" (number of delegates allocated strictly according to the proportion of the total vote) or "winner-take-all" systems; selection and allocation of delegates to each candidate at the local/district level or at the state level. The variety of primaries is considerable.
6. It is the responsibility of each candidate for the presidency to get on the ballot in each state, by meeting the specific state requirements for this. There is some discretion by state election officials in deciding this. Candidates can petition, parties can also inform state officials, and generally candidates registered with the Federal Election Commission are accepted.

In a more general sense, one must understand the basic characteristics of the American nomination practice. Ours is unique in the sense that there is no established leadership succession system. In parliamentary systems the legislative career route leads to the top. Not in the United States necessarily. Ours is a fairly open and accessible system. There is ample encouragement for self-starters, outsiders, those not necessarily connected with the "Washington Beltway." Governors can compete as well as senators and Republicans. A businessman like Ross Perot or Steve Forbes as well as a Rev. Jesse Jackson or a Colin Powell or a Shirley Chisholm. As Theodore White wrote, "Nowhere can men [sic] gather together on their own initiative and self-selection . . . and then rush the bridge of state with greater chance of success."[2] Ours is also an elaborate, multilevel, diffuse process. Above all, it is a populist, decentralized process. Candidates are faced with a formidable burden of appealing to many levels of leadership: the city politicians, the state party leaders, interest groups, those controlling the mass media, and the public. Putting together an organization that is an expanding coalition of supporters at all levels and communicating effectively with all these people, facing different primary deadlines in different states, each with their own special political conditions — these are daunting tasks. And, above all, expensive tasks. The fundraising is an endless obligation if one is to mount a credible campaign. Long ago Bryce said that candidates for the United States presidency had to be "wise and prudent athletes." Today that is more true than ever.

This new model stands in sharp contrast to the "old, classic, brokered convention" model which existed from 1832 to 1968. This earlier model had several important elements. Its basic purpose was to nominate for the presidency a person who it was hoped would achieve both of these goals: contest the presidential election very well, if not win it, and unify the party nationally, if only temporarily, during the period of the campaign. In the elaboration of this dual purpose there developed a set of operational practices and norms which, given the conditions of American politics, were crucial for party success. One of the basic elements of the classic convention model was the acceptance by all leaders that a *single* nomination would be made by a *majority*, hopefully a strong majority, of the relevant party personnel. Plural nominations spelled disaster for the party (as, for example, 1912 proved for the Republicans with the nomination of two candidates, Roosevelt and Taft). Narrow majority decisions, while acceptable, were not optimal (hence the two-thirds majority rule of the Democrats until 1936, the unit rule, and resolutions to make the nomination unanimous at all conventions despite the division and conflict which actually ensued). Second, the preconvention activities of candidates in appealing to the public and to the state and local elites were important aspects of the process, but they were indeed preliminaries. The maneuvers and intrigues of the convention might be influenced by public opinion, but the convention remained the final decisionmaker.

Third, the convention under this old, classic model arrived at its final decision through power brokerage and political bargaining techniques by which consensus was built. This was the essential character of the convention. Denis Sullivan and his colleagues, in their study on convention decision making, have specified certain assumptions: Hierarchical party leaders with authority do the bargaining, they are uncommitted and thus can bargain from strong positions, and conflicting elements in the party put aside their differences finally and work together for party victory.[3] Ostrogorski recognized early this basic character of the convention around 1900. He said the presidential nomination was "left to the professional politicians" who placed a "high commercial value" on contributing votes to the future president. "In reality, it is all a matter of bargaining: they calculate, they appraise, they buy, they sell, but the bargain is rarely stated in definite terms."[4] In the process of bargaining, however, there is a very canny analysis of, and much debate about, how the party can win while appealing to those interests in the electorate who have to unite behind the candidate. The task of the party convention was to pick "the man most likely to win."[5] And this results in different calculations and different theories, with the majority party, supposedly, having a different nominating task from the minority party.

Finally, the classic model implies that not only was the presidential nomination the consequence of power brokerage, but the total product of the convention, when it was handled properly in all its classical elegance, was a bargain. The convention does other things than nominate someone for the presidency. It also picks a vice-presidential candidate, adopts a platform, selects a national chairman (handpicked by the presidential nominee), and possibly adopts proposals concerning the party organization. The major elements of the party have to be given some recognition of their contribution and role and an incentive for their involvement in the campaign. Thus, in Los Angeles in 1960 when Kennedy barely won the Democratic nomination, the East got the presidency, the South the vice presidency (Lyndon Johnson), the liberals got the platform (chaired by Chester Bowles), and the Western moderates got the national chairmanship (Senator Henry "Scoop" Jackson).

The classic model placed much reliance on the party organization's role and function, meeting in convention, to make the crucial decisions for the party. As Theodore White colorfully expressed it, in terms bordering perhaps on caricature at times,

> As recently as 1960 one could write of a Democratic convention as it used to be — a universe in itself, a nucleus of 30 or 40 tough-minded power brokers, making decisions behind closed doors, while outside the thousands who swelled into convention city made carnival. Tough, surly or corrupt as they might be, the power brokers understood what gave the Democratic party its unique power — its ability to absorb new groups, whether Irish, Italian, Jewish, black, ethnic or labor leaders. The national convention was the anteroom to national executive power.[6]

Much of this classic model is now obviously out of date. The convention no longer is the tumultuous bargaining arena of the past. A hierarchy of party leaders controlling delegations and delivering blocs of votes for presidential nominees no longer exists. The conventions' functions have changed, as we will discuss later. The winning candidate is usually known well before the convention. The most prominent party leaders are still there, as are the party activists from the states and localities, and their interaction is important, but the context and meaning of their actions are fundamentally changed. Thus, the old model has given way to the new populist model. This leads, then, to our description of the different stages and factors in the new presidential nominating process as it operates today.

PREPRIMARY PERIOD: THE EARLY BLOOMING AND DEMISE OF POTENTIAL CANDIDATES

Of the millions of Americans who are interested in politics and perhaps the hundred thousand who have moved up to some party or governmental elite position, and the one thousand or fewer who hold (or have recently held) office as governor, U.S. senator, or representative — perhaps fifteen or so give some thought to being a candidate for the presidency. And they do this within a year after the November national election. At least within the "out party," which does not control the presidency, there are several aspirants who become visible very early, either because they tantalize the press with this possibility or because supporters are publicly encouraging them to consider the race. There follows then a two-year period, stage one of the process, leading up to the presidential primaries (which begin in February of the election year), during which the country goes through a special series of actions, events, and developments presumably helpful to decision making by the public and to a realistic evaluation of the "presidential quality" of would-be candidates by the politicos. Open to question is whether this two-year period is necessary and useful for evaluating these candidates, persuading some of them to drop out, demonstrating the particular appeals and capabilities of others, and gradually leading the American public to a rational decision. There can be no doubt that this pre-primary period begins early and is perceived as critical for presidential candidates. By October 1997, Steve Forbes already had begun his preparations for the campaign of the year 2000, with a new campaign organization having 57,000 dues-paying members, active fax machines, and many radio ads.

The timing of the candidate's announcement is a critical decision, of course. One strategy is to seek to establish early credibility and announce early, hoping to head off rivals through good organization and indications of strong support (Phil Gramm, who announced in February 1995, is an example). Alternatively, the strategy may be to indicate availability without announcing in order to evaluate the extent of support while appraising the success of others (Colin Powell in 1995). And then there used to be the wait-and-see strategy (Gerald Ford in 1979), with no plan to compete in the primaries and hoping for success in a pluralized convention. Under the populist model, this strategy today is unlikely to lead to the nomination.

The big question in this early period is: which leaders can demonstrate that they are serious, viable candidates? The aspiring candidates have to make some crucial decisions, staking out the political territory they feel is theirs and the political approaches that will win them the nomination. They have to decide, first, what groups or interests within the party, or even outside the party, they wish to appeal to (or they may wish to generalize their appeal) while, second, deciding on what style that appeal will take. They must decide what image they wish their candidacy to communicate publicly. In addition, of course, they have to decide on the timing of their declaration of candidacy, delaying or expediting it on the basis of certain calculations of advantage. Recently it appears that early declarations are encouraged by the intensity of the competition and the necessity of collecting enough

William J. Clinton as a Native-Son Presidential Candidate

In order to win the White House after the passage of the 1964 Civil Rights Bill, the Democratic party nominated southerners five times and the party was victorious four out of the five efforts. Johnson won once (1964), Carter won once (1979), and Clinton won twice (1992 and 1996). In fact, these years plus the Kennedy victory in 1960 are the only times in the last half of the twentieth century that the Democratic party captured the White House. This success of southern Democratic presidents has become a special phenomenon of American politics. It could repeat itself in the 2000 election if vice president Gore of Tennessee captures the party's nomination and the general election in November.

From the disputed presidential election of 1876 to 1948, the South (defined here as the eleven states of the old Confederacy) voted solidly Democratic in every presidential election except 1928. In all of these election, the South has given almost all of its electoral votes to the Democrats, and at the state level the Republican party was nearly nonexistent as a competing political entity.

In the 1948 to 1960 period, however, the South began to desert the Democratic party in the presidential elections, four states going to the States Rights party in 1948, and then some states switched to the Republican party in 1952 (four states), 1956 (five states), but in 1960 only two (Florida and Tennessee). The major turning point came in 1964 when Goldwater announced his opposition to the Civil Rights Bill and his support for states rights and racial conservatism. That year five of the eleven states switched to the Republicans. The southern Democratic bloc started to crumble. The Republican party's "southern strategy," developed by Goldwater and Nixon, furthered by Reagan and Bush, assisted this shift. And some southern leaders like Democratic senator Strom Thurmond switched parties. Eventually the southern bloc became Republican. Only Texas among the eleven states of the Solid South voted Democratic in 1968. Five states cast their Electoral College votes for George Wallace's American Independent party. Native-son candidate Johnson halted the Republican resurgence temporarily in 1964. Carter halted it in 1976 and Clinton did so in 1992 and 1996. The Democratic vote increase in these southern states from 1992 to 1996 ranged from 0.5 to 9.0 (the mean was 3.8 percent). However, on the eve of the 2000 election the Republican party may counter the Democratic winning strategy with a native-son of their own from this region.

SOURCE: Hanes Walton, Jr., *Reelection: William J. Clinton as a Native-Son Presidential Candidate* (New York: Columbia University Press, 1999).

funds to establish the credibility of the candidacy to qualify with the Federal Election Commission for public funds. Early campaign personnel decisions also have to be made, as well as decisions on organizational strategy and publicity.

The key decision, however, for the aspirants in the pluralized "out party" remains that of image — how to picture oneself before the party and the public. This, of course, depends basically on (1) what has been the image of the candidate up to this point; (2) how the other candidates are likely to picture themselves; and

(3) what is needed to win. Recent research suggests that four basic styles develop, depending on whether one runs as a "partisan" (appealing to the party faithful) or "nonpartisan" and whether one is an "ideologue" (espousing a particular philosophy) or a "coalitional" candidate (seeking to mobilize a wide variety of supporters).[7] For the Republicans in 1996, Buchanan was perhaps the partisan ideologue, while Dole was more the coalitional candidate; Forbes might be considered the somewhat nonpartisan ideologue, and, if he had run, Powell perhaps would have been the nonpartisan coalitional candidate.

While candidates are attempting to establish themselves in the public eye during the preconvention period and to demonstrate their attractiveness to party leaders at all levels, the public presumably is trying to sift through the available candidates and begin eliminating those who have no appeal. The credentials on which these decisions are made are not altogether clear. Thus Stephen Hess, in his discussion of the "routes to the presidency," suggests that the process has become more open than previously. It used to be accepted that the presidential candidate should come from a large state with a large electoral college vote; from those who have had executive experience as governor in such states; and not from among the minority groups of American society — particularly Jews, blacks, perhaps Catholics. But these assumptions have proven invalid in recent years. Small states like Arizona, Arkansas, South Dakota, and Georgia have supplied presidential candidates. Governors are now challenged by U.S. senatorial and congressional leaders. And minority groups are considered in many circles very eligible for the presidency, particularly after 1960. Thus the process has become more open and the pool of eligibles has been enlarged. Hess argues that the qualities of intelligence, political expertise, a "transcending honesty," a "feel of the nation," a "sense of history," a capacity to communicate and persuade, and tremendous physical stamina are some of the qualities which the public and political leaders use to evaluate the many contenders during the pre-primary period.[8]

During this early period, before the presidential primaries begin, many candidates, announced or possible, withdraw. The preprimary period serves a "screening" function. For the 1996 election there were at least twenty would-be-Republican candidates. Twelve of them pulled out of the race by the end of 1995. These included Jack Kemp, Dan Quayle, Colin Powell, Pete Wilson, Arlen Specter, and Newt Gingrich. Various factors were responsible for these withdrawals: the prospect of raising the \$25–\$30 million necessary to wage a successful campaign, poor standing in the polls, health considerations, family considerations (particularly reluctant wives), and inability to develop a distinctive appeal that would provide a credible status in the competition in the primaries. In 1992, similarly, for the Democrats there were twelve prominent candidates initially. Only five got through the preprimary period: Clinton, Tsongas, Harkin, Kerrey, and Brown. Others such as Gephardt, Bradley, Cuomo, Wilder, and Jackson withdrew earlier. They were deterred by the same types of considerations as the Republicans, although each case had its particular explanation.[9]

For those candidates who persist during the pre-primary stage, their campaign tasks and deliberations become increasingly complex and demanding. Bob

Dole's case in 1995 illustrates this well.[10] First was his indecision to run, forcing him to overcome his misgivings while being pressured by friends and family to commit himself. Even when preparing his papers to file for candidacy with the Federal Election Commission, Dole wrote a note that he had not yet made up his mind, which upset his wife and campaign team much. The early polls proved to be reassuring, however; one by the *Washington Post* gave him a 62 percent approval rating, compared to 35 percent for Gingrich and 45 percent for Clinton. Further, a poll of Republicans in March revealed 42 percent support for Dole (although only 15 percent "definitely"). Second, Dole had to make an early decision on who was to be his campaign manager, picking Scott Reed, formerly deputy chair of the Republican National Committee. Reed's decision was contingent on the development of an effective working relationship with Dole's Majority Leader staff (who some felt were too "liberal"). Next, there was continuous pressure on Dole to develop a "message," a problem with which he and his team struggled for many months (some complained this took most of 1995).

Having assembled his own campaign team, consisting of staff from his Senate office, a press secretary who had worked for Reagan, a strategist who had been a long-time advisor and political director at the Reagan White House, and a top organizer who had helped him defeat Bush in Iowa in 1988, as well as his wife Elizabeth and also the wife of the columnist George Will, his team began to develop other key aspects of his campaign. They had to plan his advertising, the fundraising activities to support these ads, the identification of key leaders who were Dole supporters and who would run the Dole operation in each of the states, and his own issue strategy for the campaign. Dole met with other political leaders and candidates throughout 1995, including a very important session with Colin Powell in which Powell said he was "keeping his options open." In the meantime, he was disturbed by Phil Gramm's campaign, the entry of Steve Forbes (with his 17 percent flat tax), and Buchanan's announcement of candidacy in March 1995. The August straw poll in Iowa that put him in a tie with Gramm (25 percent each) was upsetting, as was the New Hampshire poll which put Forbes ahead 31 percent to 22 percent. Dole made early visits to both these states, which were not as reassuring as he had hoped. (Eventually Dole won Iowa with 26 percent, far below his 1988 victory, and lost New Hampshire to Buchanan, 28 percent to 27 percent, with Lamar Alexander a close third at 23 percent).

During the entire period, of course, Dole's team was preoccupied with how to raise and spend the $45 million allowed (up to the convention) and how to win the majority (996) of the delegates needed for nomination. The withdrawal of other possible opponents was certainly gratifying (Wilson in September 1995, Powell and Specter and Gingrich in November). This left him with Gramm, Buchanan, Forbes, and Alexander as major contenders by the end of the year. Meanwhile the direct confrontation between Dole-Gingrich and President Clinton over the balanced budget and collateral agreements on Medicare, Medicaid, education, and environment led to a government shutdown. Dole was exasperated and finally on January 2, 1996, pushed through a Senate resolution to open the government again. The preprimary period ended in January 1996 for Dole with a variety of misgivings — about his issue positions, his opposition, his image, his organization and polling (he insisted on a major shakeup of his campaign team in February), and his prospects for the Iowa caucus (February 12) and the New Hampshire primary (February 20).

The above pre-primary narrative of one major candidate's experiences is duplicated every four years in our presidential sweepstakes, sometimes only in the "out party," sometimes in both parties. What occurs is that a multiplicity of would-be candidates test the extent of their appeal to the public, the media, their fellow party leaders, and those with big money. They need to know the strength of their support and the probability that they can win that magic number of delegates which will give them the majority for the nomination. (That number was 996 for the Republicans in 1996 and 2,150 for the Democrats.) In this effort they engage in a variety of activities. They seek to build a significant coalition of supporters within the party, or demonstrate that they already have such a coalition. They seek commitments from fellow party leaders, in Washington, in the states, in the counties and cities. They seek to build a campaign organization with respected specialists in the media, in fundraising, in polling, in strategic decision making. They attempt to mobilize by February of election year the millions of dollars they know it takes to have a viable candidacy. They work on their issue positions hoping to discover and develop a message that is distinctive and catching. They appeal throughout the country in direct public appeals in speeches, ads, and rallies. They then watch the public opinion polls carefully. In the process of all this, we engage in a major nationwide leadership screening function, before the American public, or that segment of the American public attentive to politics. The public's role is significant, the media's role is significant, and the role of money is critical. By the end of this pre-primary stage, at least half of the potential candidates capitulate. Those candidates who have persuaded themselves that the signs are hopeful remain.

What factors decide whether a candidate survives this early stage? They may differ for particular candidates. Access to big money is relevant for all, but more worrisome for certain candidates than others. The results of early polls may dissuade some (as Gramm). A candidate needs good enough polling figures to warrant some optimism and a supportive wife and family (which it appears Powell did not have). A major deterrent to many is the tenacious scrutiny by the media. And then there is the matter of good health (a factor for Quayle and a concern of Dole's supporters). Finally, of course, is the ideological positioning of a candidate in relation to other candidates — not too extreme (although this certainly did not apply to Buchanan), and capable of mobilizing a large body of partisan supporters. Finally, there has to be a burning desire to run. In Powell's statement of withdrawal, he said he did not have "a commitment and passion to run the race."[11] One may well ask, Is this, then, a rational, responsible, and genuinely open system? Above all, is it a system that in this early stage encourages and assists in the evaluation of candidates on their leadership skills and potential? We will set those questions in front of us also as we now describe the caucuses and primaries, which constitute the second stage of this process.

THE PRESIDENTIAL PRIMARIES AND CAUCUSES

This is stage two of the nominating process. It really begins when Iowa holds its caucus and New Hampshire its primary in February of election year. (Actually, in 1996 Louisiana "stole a march" by holding its caucuses on February 6, but this was unique, since Iowa and New Hampshire have always been the earliest.) While it is the primaries that people pay most attention to, one must remember that not all delegates are selected by that procedure. In 1996 one-third of the Democratic delegates were selected by caucuses and state or district conventions, as were 14 percent of the Republicans. The caucus thus remains a critical element in the nominating system. The important aspect of this stage is that people can now finally participate actively in the process by casting votes for their preferred presidential contenders. And many do vote, more in primaries than caucuses, but many do take advantage of this opportunity. In 1996 almost 25 million voted in the primaries, and hundreds of thousands in caucuses. From February to June, the nominating decision is indeed in the hands of citizens who wish to be involved.

As we indicated earlier, the present system with its heavy emphasis on direct presidential primaries developed after 1968. Up to that time, the primary existed but was not yet the instrument for popular control of the final nominating decision. Nominations were still in the hands of the national conventions. True, the primaries had their role in the process, from time to time. In 1952, for example, Dwight Eisenhower did score impressively with write-in votes in primaries, establishing his appeal over Senator Robert Taft, but the final decision was left to the bargaining process of the Republican convention. Estes Kefauver won in the Democratic primaries, but the convention selected Adlai Stevenson. Similarly, John Kennedy in 1960 established his vote-getting ability, despite his Catholicism, in certain primaries (particularly Wisconsin and West Virginia), but his majority was not assured until the convention met in Los Angeles. And even in 1964 the primaries were not decisive. Republican Barry Goldwater did not do well in the primaries, losing to Henry Cabot Lodge in New Hampshire, New Jersey, and Massachusetts, to William Scranton in Pennsylvania, and to Nelson Rockefeller in Oregon. His victory with 51.5 percent of the vote in California was important, but his victories in the selection of delegates in the conventions of nonprimary states were more decisive. From 1968 on, however, the primary became the central channel in the road to the presidency. Candidates ignored the primary at great risk to their candidacy.

The adoption of the presidential primary by a majority of states after 1968 is obviously the first major aspect of this development. The proportion of delegates selected by primaries jumped strikingly in the seventies (Table 10.1). While only one-third of the delegates were thus selected in 1968, today over 70 percent are. The Republicans have increased steadily, to 88 percent in 1996, while the Democrats have fluctuated more over the years. Yet it is clear that once the populist reform movement set in during the seventies, state after state jumped on the bandwagon. Large numbers of voters responded to the opportunity to directly participate, particularly in the years of sharp party competition: 26 million in 1976, 32

TABLE 10.1 ▪ The Increasing Adoption of Presidential Primaries, 1968–1996

	1968	*1972*	*1976*	*1980*	*1984*	*1988*	*1992*	*1996*
1. Number of states with primaries								
Democrats	17	23	29	33	24	37	39	32
Republicans	16	22	28	32	30	36	38	40
2. Proportion of delegate votes determined by primaries (in percent)								
Democrats	38	61	73	72	62	67	70	63
Republicans	34	53	68	76	71	77	79	88

SOURCE: William Crotty and John S. Jackson III, *Presidential Primaries and Nominations* (Washington D.C., Congressional Quarterly Press, 1985), 63; *Congressional Quarterly Reports*; Austin Ranney, *Participation in American Presidential Nominations 1976* (American Enterprise Institute for Public Policy Research, 1977), 6; Gerald Pomper et al, *The Election of 1996* (Chatham, NJ: Chatham House, 1997), 46–47, 60; Ibid., *The Election of 1992* (Chatham, NJ: Chatham House, 1993), 49–50.

million in 1980, 35 million in 1988, 33 million in 1992. In 1984 when no one opposed Reagan, there actually were no Republican primaries. And in 1996 when Clinton was clearly the front runner, only 10.7 million Democrats participated (but 14 million Republicans).

To get a measure of the extent of turnout in primaries, we can use the primary total vote as a percent of the general election total vote (Table 10.2). The turnout rate is unexpectedly high, particularly for Democratic voters (except in 1984 and 1996). States vary considerably in the extent to which citizens avail themselves of the opportunity to vote in these primaries. Note the illustrations of states with different "norms" in state party voting (Table 10.3). Thus, there are party differences in turnout, variations in turnout by the competitiveness of the election, and state political cultural differences over time which are reflected in voter interest in participation in primaries.

TABLE 10.2 ▪ Presidential Primary Vote as a Percentage of the General Election Vote, by Party (for comparable states each year: only those states where a primary was held for both parties)

	1976 (11)	*1980 (30)*	*1984 (24)*	*1988 (36)*	*1992 (36)*	*1996 (32)*
Democrats	65	71	45	58	51	33
Republicans	45	39	*	26	39	42

*In 1984 there were no Republican primaries.

Note: The number of states used in the calculations each year is in parentheses.

TABLE 10.3 ■ Variations by States in Presidential Primary Turnout, Primary Vote as a Percentage of General Election Vote (averages for elections 1976–96 in selected states)

State	*Democratic Vote*	*Republican Vote*
California	73	57
New Hampshire	64	62
Florida	74	38
New Mexico	76	30
Montana	76	52
Oregon	61	59
Tennessee	44	31
Massachusetts	56	30
New Jersey	44	18
Range	*49 to 76*	*18 to 62*

CHANGES SINCE 1968 IN THE OPERATION OF THE PRESIDENTIAL PRIMARY AND CAUCUS SYSTEMS

The presidential nominating process is regulated by two legal entities, the national committees of the two parties and the state legislatures. Both have been interested in developing a method of nomination that is focused on two questions: What type of primaries (and caucuses) should we have, and when, in the election year, should they be held? The reform of our system began after the raucous Chicago Democratic convention in 1968. The Democratic party set up the McGovern-Fraser Commission that worked from 1969 to 1972 and proposed a series of reforms.[12] It was followed by the Mikulski Commission (1972–1973). They made the following proposals:

1. All delegate selections should be conducted publicly with timely public notice of meetings and elections, procedures to be open to all party supporters and held in convenient public places. *The Public Notice Principle*
2. Voting in primaries and caucuses should be restricted to Democrats only. *The "Closed Primary" Principle*
3. Candidates for the delegate position to the national convention should state their preferences for the presidential candidate they prefer, or indicate clearly that they are not committed. These preferences must be publicized. *The Preferential Commitment Principle*
4. There shall be a proportionate distribution of delegates in each state for each presidential candidate, based on the total vote cast in the primary or the caucus. *The Proportional Representation Principle*
5. Each state must adopt an affirmative action plan for the selection of delegates to include blacks, women, and youth in proportion to their presence in the state's population (later changed to "as indicated by their presence in the Democratic electorate"). *The Modified "quotas" Principle*

These party reform commissions argued that all states, and Democratic parties in these states, should be forced to implement these rules. The Democratic National Committee concurred in this and set up a Compliance Review Commission which would monitor each state party's actions to see if they complied with these rules. At first there were many objections, but by 1972, forty state parties met the requirements completely and the rest of the states met them in succeeding years. However, efforts were made later through other party commissions to overturn these reforms. Specific new reforms on timing of primaries, filing procedures, and at-large delegates were added, but the basic principles decided on by the two early commissions were not repealed. There was one rule that in the process became more strongly enforced, that delegates are "bound" to the presidential candidate they committed to and must honor that commitment until released by their preferred candidate. This was a principle much debated at the 1980 Democratic national convention and was the focus of the Kennedy-Carter conflict at that time.

These are the major principles laid down by the Democratic national party, then, as to the character and functioning of the primary and caucus process. The Republican party also set up a commission in 1968 to study reform and did adopt certain reform proposals at its national convention in 1972. But the Republicans left to the states for the most part the decisions on types of primaries. The problem, of course, was, and is, that state legislatures are the bodies which decide what type of system will be used in each state. That is, the party leaders in each legislature decide whether there will be a primary or caucus, when it will be held, and what type of primary or caucus system will actually operate in the state. This sometimes results in a conflict between national party rules and state legislation, which the courts have to adjudicate. In recent years the courts have usually ruled in favor of the political party committees.

There are many ways in which the states vary in their type of primary systems. Most states have some type of closed primary which seeks to limit participation to bona fide partisans, so that Democrats vote in Democratic primaries and Republicans in Republican primaries. Wisconsin is one state which has always had open primaries, but was required in the eighties by the Democratic National Committee to change to a closed primary. This was true also of other open primary states such as Michigan (which responded by electing delegates in closed party caucuses at one time). Recently the conflict over this matter has been resolved by the DNC's adoption of a more tolerant policy permitting the few open primary states to continue as they were.

Another controversy emerged over the failure of states to strictly follow the proportional representation rule. In 1988, for example, Jesse Jackson complained bitterly that he was not being allocated the proper number of delegate votes in some states, based on his total public vote in the primary. Related to this is the question of what the voting "threshold" should be for a candidate to be given *any* delegates — in the early days, securing a minimum of 15 percent of the vote was suggested (by the Mikulski Commission), but this was later changed to 20 percent by the DNC. The idea was to provide a fair distribution of delegates to votes and to discourage frivolous candidacies. Further, the original idea was to get away from the "winner-take-all" primaries of the past which gave all the state votes to the candidate whose slate of delegates received the largest primary vote. We have largely distanced ourselves from "winner-take-all" schemes while not entirely embracing precise proportional representation, which varies somewhat by states.

Caucus systems vary by states also. In some states there is an elaborate "ladder system" of caucuses, with local meetings selecting representatives to district caucuses, which in turn select delegates to the state party convention, which actually decides who the delegates will be and what proportion of delegates will be allocated to each of the viable presidential candidates who competed in the caucuses. In a few states the delegates are actually elected locally based on the vote in party caucuses for each presidential candidate.

The rules of the Republican party have been less strict and defer to the state Republican organizations, as well as to state legislation. However, the Republican convention and National Committee did lay down certain guidelines similar to those of the Democrats. Already in 1980 the Republican delegate selection rules stated that only voters "deemed to be Republican" could participate in a Republican primary, caucus, mass meeting or mass convention (except where state law otherwise mandated). They also had a broad anti-discrimination rule, an equal participation of men and women advisory, and a requirement that precinct, ward, township, and county meetings would be open. Thus, the Republicans began to accept some of the same basic principles laid down by the Democrats in the 1969–1972 period.

"FRONT LOADING": A MAJOR CURRENT PROBLEM

One of the most controversial aspects of the new presidential primary system today is the question of timing. State legislatures (consulting with the parties) must decide when each state should hold its primaries. From the beginning, New Hampshire was recognized as the first state to hold its primary, just as Iowa was permitted to be the first to hold its caucuses. These would be held in February or March. Then would follow all the other states at well-spaced intervals, a few states at a time, until June when states such as California, Ohio, New Jersey, Tennessee, and Oregon, would complete the primary season. As a result, in 1968, for example, twelve weeks would pass before 50 percent of the delegates were selected in the primaries, with 47 percent of the delegates picked in the primaries of the last three weeks. This type of primary season presumably provided time for voters to assess candidates rather deliberately on the basis of their performance in a series of primary contests, and theoretically kept alive candidate chances for the nomination, while also placing great demands on them to campaign throughout the country. It also gave the later states a sense of being a meaningful part of the nomination process. In some presidential primaries, the candidates had to compete to the end.

This system of spaced-out primaries changed in the seventies and eighties, with a radical change occurring in the 1988–1996 period. The basic change was the rush by states to hold their primaries earlier. In 1996, twenty-nine states set their Republican primaries between February 20 and March 26 (and twenty states held Democratic primaries in this period), resulting in the selection of 65 percent of the total number of Republican delegates (and 59 percent of the Democratic delegates) by the fifth week (compared to 10 percent to 15 percent in that period in 1964).[13] There were extreme examples of state legislatures changing their primary dates in order to get into the nomination process early. California jumped from June to March 26, Ohio from June to March 19, Oregon from June to March 12. As a result, in 1996 barely 20 percent of the delegates were picked in primaries held after March. A majority of states wanted to be front-runners.

This phenomenon of "front loading" was the product of the trends in previous years in the character of the process. It had become clear that candidates who did well in early primaries achieved a certain momentum, most of the media attention, and easier access to financial support. Thus, performance in the early primaries could be determinative of success. Early primaries also could terminate or certainly dim the presidential aspirations of candidates, as presidential nomination history has revealed. Even President Johnson in 1968 decided not to run for reelection after Eugene McCarthy's strong vote in the New Hampshire primary. In previous years it was clear that the "winnowing" of candidates occurred early in the primary season. Thus, if voters of a state were to be assured of any influence in the presidential nominating process, having an early primary was vital. The calculations of party strategists also might lead to the desire for front loading. Crotty reports that such was the strategy of the Carter team in preparation for the 1980 election.[14] Another illustration of this was the decision in 1988 to concentrate southern and border-state primaries on one date, March 8 (actually, sixteen states held their primaries on that date). The hope of southern leaders was that this would enhance regional clout in the process. Unfortunately for the South, Dukakis and Jackson emerged as the victors. For a variety of reasons, thus, it seemed to make sense for politicians and states to shift their primaries to February or March if they were not to be irrelevant to the nomination process.

The effects of front loading and of states switching their primaries from June to March is not easy to document. Early withdrawals from the race occurred in the early years as well as more recently. In 1984, for example, after March 13 (an early version of Super Tuesday), only Mondale and Hart were left to contest the Democratic nomination. In 1988, after March 8 only Dukakis won in the remaining primaries (except for Jackson in the District of Columbia and Senator Simon in his home state of Illinois). In 1992, after March 10 (Super Tuesday) only two Democratic candidates besides Clinton survived (Tsongas and Brown), and although they continued to compete, they won no primaries. In 1996, Gramm withdrew (February 14), then Alexander and Lugar (March 6) and Steve Forbes (March 14), leaving the field to Dole and Buchanan (who won no primaries except in New Hampshire, but refused to retire, securing finally 21.6 percent of the primary vote). Thus, front loading may expedite withdrawals, although this early winnowing is noticeable also in previous years.

An interesting question is whether states that switched from a late primary date to an early date induced a higher primary voter turnout. One might argue that voters would turn out in larger numbers if they sensed that their state's primary was now possibly more influential. But there is no strong and clear evidence to support this expectation. California is a case in point. It shifted its primary in 1996 to March from June. In certain earlier years in Democratic primaries, it had strikingly high turnouts (measured as the party primary vote's proportion of the general election vote for the party). In 1976 and 1980, it had over 90 percent turnout (in 1980 the Carter candidacy was challenged by Kennedy, who won the late California primary). In succeeding years the California turnout was close to 70 percent, but declined to 49 percent in 1996. Since Clinton was not closely challenged in late primaries in 1996, this is not a good test. Perhaps the year 2000 will provide a better test. Other states that also shifted to early primaries and did not have higher turnouts were Ohio, Tennessee, and Georgia. Oregon is an exception on the Republican side. When it shifted its primary to March 12 in 1996, its turnout increased to 76 percent, much higher than in previous years (its vote in 1996 was split: 51 percent for Dole and the remainder divided among the other three candidates). Perhaps it is too soon, and the evidence too incomplete, to see turnout effects of front loading. As of now, other factors, including differential state "turnout cultures" seem to be more important than the time of the primary.

THE PRIMARY VOTERS: WHO ARE THEY AND WHAT FACTORS ARE LINKED TO THEIR VOTE?

We know that millions of Americans participate in the primaries, 33 million in the very competitive primaries of 1992, 25 million in 1996. These voters total over 40 percent of the general election vote cast in these same (thirty-two to thirty-six) primary election states. Many questions beg answers about these voters: how representative are they of the general electorate? Are these primaries confined to party supporters, or are there also independents and party crossovers among them? How divided are these voters within their own party on issues and ideology? These three questions provide a beginning focus for our analysis. Fortunately, both in 1992 and 1996 the *New York Times* published results of surveys of voters leaving the polls in twenty-nine Democratic primary states in 1992 and in twenty-eight Republican primary states in 1996. These surveys provide us with some basis for responding to the above questions (Table 10.4).[15]

There has been much argument in the past over the representativeness of the primary voting electorate, with different scholars using varying definitions of representation: representative of the public, or of the regular party supporters, or of the party activists. The data from these surveys for the two parties suggest that although the demographics of primary voters are relatively close to those of the loyal partisans in the voting public, there are some differences. If we compare primary voters for each of the parties with the party supporters in the general election, we find that young voters are not closely represented, particularly in the Democratic primaries. The same is somewhat true also of those with low family incomes. On the other hand, blacks are proportionately well represented in the

Democratic primary, as are education groups. On balance, then, the primary voters seem fairly well represented *demographically*. The same is also true of their *ideological* and *partisan identification* proportions. The contrasts between the parties in these respects are reflected well in the comparison of the two samples of primary voters and general election voters (Table 10.4).

One should note particularly that the primaries in both parties attract large numbers of independents (over one-fifth of all primary voters), which of course is anathema to party regulars who want the primaries completely closed. Indeed, a few self-declared Republicans and Democrats report voting in the other party's primary. Hence, here is evidence of the relative openness of the primaries. Finally, one should note the implicit division, if not polarization, in each party on ideology. The primary voters are indeed a diverse group. The 1996 study documents this particularly for the Republicans for certain key issues.[16] For example, 38 percent of Republican primary voters favored a constitutional ban on abortion, 57 percent were opposed. On trade policy, 43 percent felt it creates more jobs, 39 percent believed it costs more jobs. When asked whether Buchanan's policy positions were too extreme, 56 percent agreed, 41 percent disagreed. Sharp divisions are found in both parties, and perhaps the primaries expose them and give people an opportunity to express them.

TABLE 10.4 ▪ Characteristics of Party Voters in Primaries and the General Election (percentage of each group of the total party vote)

	1992		1996	
Characteristic	*Democratic Primary Voters*	*Democrats Voting for Clinton in the General Election*	*Republican Primary Voters*	*Republicans Voting for Dole in the General Election*
1. Age: 18–29	12	21	10	14
2. Race: black	14	15	*	3
3. Education: high school or less	35	32	23	25
4. Family income: under $15,000	15	19	*	8
5. Party identification:				
Democratic	67	68	4	10
Independent	20	20	21	22
Republican	4	12	75	68
6. Ideology:				
Liberal	35	33	9	5
Moderate	45	54	33	38
Conservative	20	13	58	57

* No data available.

SOURCES: Adapted from Gerald Pomper, *The Election of 1992* (Chatham, NJ: Chatham House, 1993), 61; and Pomper, *The Election of 1996* (Chatham, NJ: Chatham House, 1997), 54–55. Also, *New York Times Reports*, July 12, 1992, 18; March 31, 1996, 24; November 10, 1996, 16.

One of the most interesting questions investigated by scholars of the presidential primaries is how rational the voting decision is in these primaries. Given the multiplicity of candidates, the hectic campaigns, the deep and persistent media coverage, and the conditions under which the primaries are held in each state, one might well expect the voters to be bewildered, conflicted, and confused. Rather than being critical of the primaries, however, recent research tends to emphasize the logic and reason behind the primaries and the capacity of the voter to sort things out and make sophisticated, even "rational," decisions. Bartels' excellent study based on the 1980 and 1984 primaries demonstrates how individual vote choices are formed.[17] Candidates achieve early momentum in the race by early, and sustained, electoral success, while the media provide selective coverage. The voters respond, despite only a minimal level of information initially, to the ongoing primary events with increasing certainty about their vote choice. Their political predispositions are brought into play and, linked to their evaluations of the candidates, lead to a perception of the most viable, winnable candidate. The process is a dynamic one, therefore, for candidates and voters.

A subsequent study of voting in the 1988 primaries carefully analyzed the basis for the support for Bush and Dole in the Republican primaries and Dukakis and Gephardt in the Democratic primaries. The authors concluded that the majority of respondents were "sophisticated" voters. They distinguished between those who voted their preference feelings for a candidate ("sincere voting") and those who decided on the basis of which candidate had the best chance of winning ("strategic voting"). They discovered a high level of strategic voting. Here is a summary of their findings:

Presidential Primary Voters, 1988 (in percent)

Basis for the Vote	*Republicans*		*Democrats*	
1. Sophisticated vote decisions				
a. Voted for most viable candidate even if against personal preference	53	67	44	57
b. Voted for viable candidate who was also most preferred	14		13	
2. Sincere voting Voted for preferred candidate although candidate ranked low in rating on viability	10		14	
3. Irrational voting Clearly voted for candidate the voter knew was neither viable nor highly preferred	10		12	
4. Ambiguous basis for voting; unclear as to respondent's knowledge or feelings about candidates	11		17	

The conclusion from this study was that, while some voters may have made sincere but foolish choices, few were explicitly irrational. The authors interpret their findings thus: "citizens decide how to cast their presidential primary votes in sensible, even rational, ways. . . . Amid a nomination process often referred to as chaotic and capricious, our evidence nonetheless suggests that primary voters are not simply fickle."[18]

THE PRESIDENTIAL PRIMARIES: EVALUATIONS

There can be no doubt that the primaries have radically altered the presidential nomination process. We have thrown the process open to the public, and a very large minority of voters have accepted the opportunity and challenge. Further, the primaries have become popular, and therefore the prospects for returning to the classic national convention method are slim. Rather, the questions scholars and columnists debate are how well the system is performing and how we should attempt to reform the way the primary system works while keeping the basic system intact.

What are the key bases or criteria to use in evaluating the presidential primary system? The first concern is the level and nature of citizen involvement; more American citizens are participating today than ever before. Yet they are a minority, about 30 percent to 40 percent of the general election voters, varying in turnout greatly by state and year. From available recent data, they also appear to be, with some exceptions, representative of general election voters. As for the quality of voter participation, research suggests that the voting decisions of most of those who do participate are for the most part "sophisticated" and even rational. The major question on involvement may be that the primary system does not provide all citizens an equal opportunity to influence the presidential candidate selection. There may be a fairness problem, in two respects: (1) not all candidates are on the ballot or compete before the electorates in all states (candidates may see certain states as not critical or may not have the resources to campaign in all states); (2) the voters in states holding primaries late in the season, sometimes when the decision is already made, may be irrelevant to the final decision. In 1996, for example, when Dole had wrapped up the nomination in April by securing the 996 delegates voters he needed, the vote in the fifteen remaining states was strategically unimportant. The voters in these states had to acquiesce in the decisions made in the earlier states. One could also argue a third point (the other side of this coin) that those citizens voting in early states like New Hampshire, Arizona, and South Carolina (particularly in 1996) have the chance of providing the early momentum for a candidate that affects media attention and access to funding. Thus, while the voting decisions may be "rational," the institutional process may not be fair (and in that sense not "rational").

The second question to consider in evaluating the primary system is whether it is a careful, high-quality screening or winnowing process. Leadership recruitment systems are supposed to systematically evaluate the candidates in terms of their intellectual, personal, and political skills and attributes and certify only the most capable. It is presumably a truly competitive process that political leaders with experience can properly aspire to participate in. Ours is a long-drawn-out ordeal covering at least two years (and really longer) with no clear indicators as to what the expected leadership succession is to be. It does seem to screen out "the lunatic fringe" (as Harold Lasswell observed), but it may not really give all the competent contenders a fair chance. A U.S. senator called it a "glorified crap game." While this label may be extreme, it is true that the screening is often more haphazard than systematic, less welcoming and encouraging than forbidding, and less competitive and open than we might prefer. It is quite clear that to be a truly viable contender, a candidate needs $25–$30 million in the primary season or that candidate must drop out. To mobilize this money, he or she must spend a tremendous amount or allocate from personal resources (as Forbes did in 1996). Yet money is not by itself enough. Even Forbes' money was not enough to keep him in the race, and he retired from it on March 14, along with Alexander and Lugar. In 1992 the major opponent of Clinton after the early primaries was Tsongas, but even though he had won six eastern primaries and three caucuses he said he had to retire on March 19 because "the resources are not there." It is difficult to speculate what would happen if money did not play such a major role in our presidential primaries and if people like Tsongas or Forbes could compete on an equal level of financial solvency with the major contenders to the end of the primary season. In short, one wonders whether the present system provides a genuinely fair opportunity for all candidates to demonstrate their viability.

The third question is what impact the primaries have on the party system. Does this new "populist model" have a deleterious effect on the party organizations? It is obvious that the party organization per se, acting through the national convention, no longer exercises the final decisional authority on the selection of the candidate, nor is there now at the convention the "behind the scenes" deliberation and bargaining that characterized the process in the past. To some this means that the party organizations have been bypassed, and subverted, by the primaries, leading to party decline in the United States. On the other hand, it is true the party organizations at all levels are deeply involved in the primary process. One must remember first that the delegates usually have come from the party organizations and therefore have important organizational ties. The party organizations do run the caucuses in the ten or more states that do not have primaries. They participate with the presidential candidates in the selection of delegates to district and state conventions as well as the delegates which are bound to presidential candidates at the national convention. The state party organizations, in alliance with the national party, also do a great deal of fundraising in connection with the primaries, directly for an incumbent president and indirectly assisting the fundraising of other candidates. Further, under the recent Democratic party rules, the state organizations participate in the selection of "at-large," so-called "super delegates" (totaling up to 20 percent of each state's delegation) to the national convention. However, we must realize that, in the last analysis, delegates are not really responsible to the party organizations but to the presidential candidates to whom they are pledged.

It is this diminution of the party organizational role, this lack of responsibility in the system, which some scholars still decry. In addition, the sequencing of the primaries, the phenomenon of the "sudden-death primary," and "front loading," with their implications of unfairness to able candidates, gives further pause for thought. Finally, if electoral democracy means giving each citizen an equal opportunity to participate, in a context of similar vote-choice options and with a feeling of equal vote relevance, then the primary system we have is imperfect. It is not maximally democratic, responsible, or institutionally rational.

Ever since the presidential primaries gained momentum, after 1968, and the early Democratic reform commissions spelled out the rules for these primaries (the McGovern-Fraser and Mikulski Commissions, 1969–1973), there have been efforts by the so-called "regulars" to overturn the work of the "reformers." The Winograd Commission (1995–80) and the Hunt Commission (1980) set up by the Democrats made serious efforts to impose limitations on the primaries. They particularly pushed for the "super delegate" rule by which the party regulars (congresspersons, senators, Democratic National Committee members, governors, and state party officials) could be seated at the national convention without running as delegates in the primaries. But aside from that proposal, which was implemented, other attempts have been thwarted. At the 1980 Democratic convention, for example, the Kennedy forces opposing President Carter sought to have the "bound delegate" rule repealed (requiring delegates to support their presidential candidate who they preferred at the time of their election as delegate). This effort to repeal was defeated. Another reform was completed for the 1992 Democratic convention: the requirement that all delegates should be allocated to candidates by proportional representation (which still in 1988 had not been accomplished completely). Candidates, however, do have to secure 15 percent of the vote to be awarded any delegates. Yet today there remains some interest in modifying the system further, particularly in changing the scheduling of primaries and doing something about "front loading." How this will play out in the deliberations of the national party committees and the state legislatures is not clear. What scheme for spacing primaries would be acceptable to the politicians, if they expect to change the present system at all, is not known. Extreme solutions have been proposed, such as a lottery to decided which states each year can have early primaries; or four regional primaries (or primaries grouped on some other basis) on the first Tuesdays of March, April, May, and June; or a "one-shot" national primary for each party on one day. This latter proposal raised difficult questions: when to hold such a national primary, who should be permitted to vote, what to do about plurality nominations, et cetera. For the time being, we are faced with accepting the present primary-caucus system, with all its liabilities and virtues.

THE NATIONAL CONVENTION: CHANGED ROLE AND FUNCTIONS

Although it has lost its special role of selecting the party's presidential candidate, the national convention is still a major forum of party action. It remains the plenary body, which can adopt new rules, create new structures, decide on the party platform, and bring together once every four years leaders and activists from all over the country to share experiences, interests, and strategies for the party campaign. It is a large gathering — over 4,000 elected delegates to the Democratic convention in 1996, and almost 2,000 to the Republican. Meeting in August (or sometimes in July) as they do, they set the stage for the campaign. It is still a major media event, although different from the way it used to be in the days before the presidential primaries settled candidate selection decisively, as is the case today. In the past it was often a spectacle focused on the battle among the major contenders: Roosevelt versus Smith in 1932, Eisenhower versus Taft in 1952, Kennedy versus Johnson in 1960, prime examples of the convention as the final political arena where the nomination was decided. In those days this was high drama. "There is something about a national convention," said H. L. Mencken, " that makes it as fascinating as a revival or a hanging." The scene, he said, could be "exhilarating" and "gorgeous," or "preposterous" and "obscene."[19] We have come a long way from those days of political theater. Now the convention performs its roles, without such excitement.

The convention is still organized much as in the past. It consists of 50 state delegations varying in size from 19 (Wyoming) to 424 (California) for the Democrats and 12 (Vermont) to 165 (California) for the Republicans.[20] States are apportioned the number of delegates by the national committees based on different formulas: the Republicans tend to emphasize population size, the Democrats place more emphasis on the size of party vote in previous elections. Deliberations do still take place in these state delegations, but they are less important as voting units, and less controlled, than they used to be. Since the delegates are bound to different presidential candidates, these loyalties still have some meaning at the convention. However, the candidate who has won a majority of the delegates by the time of the convention (Dole and Clinton in 1996) essentially takes over the major decisions on procedure at the convention, their staffs and campaign teams working closely with the national committees to try to orchestrate an effective and peaceful and unifying convention extravaganza. There are basic convention committees still having important functions: the Platform (or Resolutions) Committee is a major one, as are the committees on rules (for the convention procedures), permanent organization, and credentials. This latter committee used to be the battleground for fights over contested delegations from the states, which recently rarely occur. Each state has representatives on each of these four committees. On occasion, there can be sharp contests today between rival sets of party leaders. In the 1992 Republican convention, for example, there was a confrontation between the Bush loyalists and the Christian Coalition, which secured control of the Platform Committee, over the language to be used in the platform on the abortion question, homosexual marriage, and the use of public funds to support certain types of art. In 1996, however, the Dole campaign team prevented a platform fight over these and other issues. The platform remains today an important document laying out party ideology, goals, and policy commitments.

Essentially the national convention's major function now is to formally ratify the nomination of the presidential candidate and to launch the campaign, with all the glamour and fireworks the party can muster. The vice-presidential candidate, picked by the presidential candidate, is also nominated. The character of each party's convention is usually different. The manner in which the parties present themselves to the American public differs, and this can set the tone for the last weeks of the campaign. In 1992 Clinton was presented as a moderate candidate, in contrast to Bush's appeal to the Republican conservatives. At the Republican convention, Pat Buchanan delivered a withering attack on Clinton and the Democrats, and the conservatives saw that much of their platform was adopted. The religious "right" was in its element. And this emphasis repelled Republican leaders such as Senator Lugar, whose retort to all this was "you don't build majorities [this way]."[21] In 1996, on the other hand, Senator Dole's forces controlled the Democratic convention, and the theme was different. General Colin Powell gave a major speech supporting Dole; Buchanan was not allowed to speak. Elizabeth Dole and Hillary Clinton gave major speeches. Events such as these, including Dole's announcement that Jack Kemp would be on the ticket as the vice-presidential candidate, and, of course, the acceptance speech by the presidential candidate, were attempts to inject enthusiasm into the convention. The media attempted to cover both conventions with interesting stories and commentary, providing roughly equal time to the two conventions. However, in 1996 it was reported that viewership was down 20 percent compared to 1992.[22] It was difficult for the media to generate the dramatic stories to maintain public interest. The convention under certain circumstances may have an effect on the popularity of candidates. Poll data for the 1992 campaign suggests that Clinton's support jumped considerably at the time of the convention, while Bush's support did not.[23] The primaries have certainly diminished the central dramatic role of the convention today, but by no means has it become irrelevant.

In the past, the convention was the arena of party bargaining, of much discussion of campaign strategy and candidate electability, within the context of selecting the candidate. It was also a time for uniting each of the very diverse Republican and Democratic parties as national structures of belief and action, behind their major leaders. These functions may now be lost, a serious matter to some observers and scholars. To them, the resolution of internal party conflict and the integration of the party organization behind a presidential standard-bearer is very important. But the convention must still be seen as a conclave of the faithful who have at the convention a unique opportunity to interact with leaders and activists from all parts of the country. As Pendleton Herring noted long ago, "the value of the convention lies in its permitting the rank and file of the party to participate physically and emotionally in a common enterprise. Here are the men [sic] who must carry the brunt of the campaign."[24] In the past the convention had many values, many of which still exist today. It can provide new leaders an opportunity to be heard. It can provide activists a real chance to interact and share experiences and ideas. It can help mobilize public support. It can define the party's ideology and policy positions more explicitly. It can give interest groups a chance to be heard. It can enhance a party's image or detract from that image. Thus, while it no

longer selects the candidate, the convention is still a key element in the party system, even today in its own way legitimating presidential candidate selections and launching the presidential campaign.

NOTES

1. For a good description, see Leon Epstein, *Political Parties in the American Mold* (Madison: University of Wisconsin Press, 1986), 88-108.
2. Theodore H. White, *The Making of the President 1960* (New York: Atheneum, 1980), 328.
3. Denis G. Sullivan, Jeffrey L. Pressman, F. Christopher Arterton, *Explorations in Convention Decision-Making: The Democratic Party in the 1970s* (San Francisco: W. H. Freeman, 1976), 122-24.
4. M. Ostrogorski, *Democracy and the Organization of Political Parties* (New York: Doubleday, 1964), 133-37.
5. James Bryce, *The American Commonwealth* (New York: Macmillan, 1916), 187.
6. Theodore H. White, *The Making of the President 1972* (New York: Atheneum, 1973), 209.
7. Gary R. Orren, in Seymour M. Lipset, ed., *Emerging Coalitions in American Politics* (San Francisco: Institute for Contemporary Politics, 1978), 130-34.
8. Stephen Hess, *The Presidential Campaign* (Washington, DC: Brookings Institution, 1978), 27-38.
9. Gerald M. Pomper, ed., *The Election of 1996* (Chatham, NJ: Chatham House, 1997), 26-36; *The Election of 1992* (Chatham, NJ: Chatham House, 1993), 42-44.
10. Bob Woodward's study of the 1996 campaign, *The Choice: How Clinton Won* (New York: Simon and Schuster, 1997), describes in detail the day-to-day developments in Dole's and Clinton's campaigns. The material for this paragraph draws on that narrative.
11. Ibid., 309.
12. An excellent review of the reform commissions is found in William Crotty, *Party Reform* (New York: Longman, 1983).
13. William G. Moyer, in Gerald Pomper ed., 1997, *1996 Election*, 22-26.
14. Crotty, *Party Reform*, 77-78, 83.
15. See the reports of these polls in Pomper, *1992 Election*, 61, and Pomper, *1996 Election*, 54-55. The *New York Times* July 12, 1992, 18; and March 31, 1996, 24, published these results originally.
16. Pomper, *1992 Election*, 138.
17. Larry M. Bartels, *Presidential Primaries and the Dynamics of Public Choice* (Princeton: Princeton University Press, 1988). See also John H. Aldrich, "A Dynamic Model of Presidential Nomination Campaigns," *American Political Science Review* 74 (1980): 651-69.
18. Paul R. Abramson et al., "'Sophisticated' voting in the 1988 Presidential Primaries," *American Political Science Review* 86 (1992): 55-69.
19. Quoted by Theodore H. White, *The Making of the President 1964* (New York: Atheneum, 1965), 231-32.
20. Delegates are also seated from the District of Columbia, Puerto Rico, Guam, Virgin Islands, American Samoa, and, for the Democrats, "Democrats Abroad."
21. See Ross K. Baker's discussion of this in Pomper, *1992 Election*, 63-68.
22. Pomper, *1996 Election*, 87.
23. Pomper, *1992 Election*, 115.
24. Pendleton Herring, *The Politics of Democracy* (New York: Norton, 1940), 238.

CHAPTER 11

The Campaign Process

The American election campaign, as in other democratic societies, is a significant episode in the political life of the society. It covers a long time period, beginning immediately after the last election. It is often hectic, confrontational, very combative. While much time is given to the presentation of substantive policy positions, there is also much indulgence in personal attacks. Negative advertising is as common as positive advertising. Media coverage is extensive. Media advertising by candidates and parties, as well as by other interest groups, blankets the country. And the campaigns are very expensive. In 1996 all campaigns in the United States — national, state, and local — cost an estimated $4 billion.[1] The cost of electing the president alone in that year was $700 million. The campaign goes through a series of stages. There is the early period of the announcement of candidacy, during which would-be candidates "reality-test" their probabilities. This period merges with the pre-primary period during which candidates stake out their positions, organize their teams, and target their partisan clienteles. We then come to the primary period itself (four months for the presidential aspirants) and then the post-primary battles when the candidates for the two parties face each other (June to November in presidential campaigns). This is the time of perfection of campaign techniques, gloves-off adversarial combat, convention hoopla, the advertising blitz, the public debates between candidates, vote mobilization at the "grass roots," and the careful scrutiny of the many surveys and polls. Often the campaign goes down to the wire, but even when the polls suggest a clear winner, as in 1996, there is always the last-gasp, desperate, exhausting marathon campaign, in the hope for a reversal of fortunes by the "late-deciders," which indeed has happened frequently in our experience.

Our campaigns, in many respects, are quite different from those of other democracies. The political system environment is different, and the norms of campaign behavior are different. Parliamentary democracies usually have much shorter campaigns. For example, the formal campaign period is twenty-one days in Britain, from the call for a new election at the dissolution of Parliament to election day. While campaign preparations may begin much earlier, the length of time involved is nothing like in the United States. The multilevel campaigns in the United States also are a special feature. Of course, the role of the party abroad in selecting candidates shortens the campaign and also changes the character of campaigning. Then, too, the role of the media is more circumscribed in Europe and elsewhere, with limits on the amount that can be spent and on the time allocated to parties and their candidates. The role of interest groups is also limited, since most allowed spending is channeled through, and controlled by, the parties. Proportional representation systems on the Continent result in somewhat more campaigning at the national level and less at the local level. The single-member-district plurality system in the United States means legislative campaigns are focused more on candidates at the state and district levels. Political party (and candidate) budgets for campaigns are very small in Europe compared to the United States, and the proportion of such budgets spent for television programs is minuscule compared to the United States (for example, in Germany only 3 percent to 5 percent is spent on television). The type of campaigning in Europe is different: more reliance on newspapers (although television use is increasing in some countries); different types of efforts to mobilize the voters (very little house-to-house canvassing on the Continent, but quite a bit of such canvassing in Britain); more emphasis on party and ideology, less perhaps on candidate personalities and backgrounds. Hence, the institutional and the normative contexts for campaigning differ. In the United States our campaigns are long and expensive because our system both requires and permits such campaigns, and our norms tolerate and expect it.

With all this prolonged and expensive American campaigning, one might well expect campaigns to be of critical importance in determining political outcomes. Yet there are those who have argued that other factors, particularly the state of the economy, are more determinative of election results. Scholars have attempted to predict the result of elections long before the campaign begins, using models that give little relevance to the campaign itself. Generally, however, most observers would return to the classic doctrine that "campaigns matter," although campaign techniques that matter today may be quite different from those in the past. One type of evidence comes from studies of campaigns tracing the ups and downs of candidate fortunes during campaigns, as revealed by the polls. These studies suggest strongly that there are "causes" (actions by candidates and their organizations) which produce "effects" (the nature of the fluctuation in candidate's rankings in level of public support). Thus, in 1988 Dukakis was leading Bush in July 50 percent to 33 percent, but after the conventions in August, Dukakis fell to 40 percent and never recovered. In 1992 Bush held a 10-percentage-point margin over Clinton in the early months (in the two-party division of poll preferences), but all this changed in July and August, with a surge of voter interest, apparently generated by the campaign, especially the conventions and debates, leading to a sharp swing to Clinton by September, which he never relinquished. On the other hand, long before the 1996 election, in fact already a year earlier, Clinton held a lead of 48 percent to 35 percent over Dole, and this relative advantage in the polls did not change much over the next year (although the final percentage difference was 8 percent).[2] In all three cases, one can well argue that actions during the campaign played a significant role, in the change in the relative popular support for the candidates in 1988 and 1992 and also in the maintenance of the strong lead for Clinton in 1996.

All of this is very speculative, of course, and forces us to try to specify more precisely what are the objectives and possible effects of election campaigns. If we focus on these two parameters, we can structure the discussion of campaigns better and analyze our research findings on the meaning and relevance of campaigns. We can divide our discussion of campaign objectives and effects into four categories:

A. *Candidate Objectives and Effects*

Generally: The candidate's goal is to design and carry out a strategy to identify a target set of voter groups that must be mobilized in order to assure victory, and to develop a message and technology to persuade these voter groups to support the candidate.

Specific requisites: The candidate must have a clear "message" (ideology, issue positions, image, theme); a team of committed campaign managers; a media strategy; an adequate fundraising plan; polls and surveys to help plan the campaign and to monitor performance; an adaptation capability, so that the campaign can respond over time to new conditions and events and to counter the opposition's appeals.

B. *Party Objectives and Effects*

Generally: As Downs said long ago, "a political party is a coalition . . . seeking to control the governing apparatus"; to do this, "they formulate policies in order to win elections." The party in a two-party system is preoccupied with securing or maintaining majority control, and election campaigns are conceptualized as the critical events in an electoral democracy relevant to that objective.

Specific requisites: The party sees its tasks as helping select viable candidates, probable winners; working with and advising campaign teams to strengthen a candidate's strategy; channeling available (minimal) resources to those candidates with the greatest probability of success; and developing a national party image in the country in a variety of ways to establish "the national party" as a visible institution with an important political role to which party followers can relate. Although many would argue that this party role is diminished of late, others would point to its resurgence.

C. *The Media's Objectives and Effects*

Generally: The media's primary goal in the campaign is to provide adequate coverage of campaign events and personalities in order to satisfy mass demands for information and news. A secondary goal is to influence the way the citizen sees the campaign, by imposing its own conception of what is important, interesting, and exciting about the campaign. It thus is not merely a servant of candidates and parties but an independent actor in the campaign process.[3]

Specific aspects: The media place much more emphasis on the "horse race" aspects of the campaign than on the issue and pragmatic arguments of the candidates. The media selectively provide coverage to candidates who win early in the primaries, on the judgment that these are most newsworthy. Third-party candidates rarely are given much coverage over the long haul of the campaign, unless their candidates have particular personality appeal. The media feel that they must provide the public with as much information as possible on the private lives of candidates and that they have the right and responsibility to influence election outcomes by endorsing candidates.

D. *Interest Group Objectives and Effects*

Generally: Unlike in European countries, interest groups in the United States have explicit objectives and considerable resources, independent of the parties and candidates, in connection with campaigns. They seek to realize group aims by influencing the selection of candidates, their policy positions, and their electoral success. They seek to influence the parties' platforms as well as their leadership and strategies.

Specific aspects: Millions of dollars are spent by interest groups through their political action committees, either in direct contributions to candidates or party committees, or on media campaigns to influence voters in support of particular candidates. Many groups, from labor unions to the Christian right organizations, have large memberships whom they can influence to support presidential candidates or senatorial and congressional candidates in states and district elections. In 1992, the amount raised for U.S. House and Senate races was $490 million; 40 percent of this came from "PACs and other committee contributions."[4] In addition, of course, interest groups spent much on behalf of the presidential candidates.

The fundamental question in all this discussion of effects is: What is the campaign's relevance for, and effects on, the *public*? This is the most critical viewpoint from which to evaluate campaigns. For many citizens the campaign is the one time they relate to the political process, the one time they become attentive to what goes on in their political world. They may become involved as activists or as interested spectators. Their image of the political process may be influenced by what they are exposed to and how they interpret their contacts with the campaign. An important question is how the public perceives the candidates, the parties, the media, and the interest groups as they play out their roles in the campaign. Are ordinary citizens informed, involved, perceptive, deliberative about the vote decision as a result of the campaign, or disaffected, cynical, and alienated? Is their reaction to the campaign a major factor in their low level of voting participation? William Riker stated very well this focus for the study of campaigns:

> Election campaigns are a distinguishing feature of representative government. They force voters to think about government, and they force officials to think about voters — and that is really the heart of representative government. According to democratic theory, rulers are supposed to be responsible to the ruled, and it seems to me that campaigns are the nexus of that responsibility.[5]

Do campaigns meet these sets of objectives — for candidates, parties, the media, interest groups, and most important, the individual citizen and the public? Do they have an impact and make a difference, do they have significant consequences for American politics? We think they do, despite the constraints of the American political system. They obviously are important for persuading voters to support the candidates. They can help parties gain political power or maintain their dominance, under certain conditions. And they are important for the role of the media and interest groups in our system. They may have *positive* or *negative* effects on the public's perception of our politics and the public's willingness to participate in politics. The evidence is not always clear on these points, as we shall see in the ensuing discussion. But the campaign must be seen as an important opportunity for citizens and their leaders, and their parties and interest groups, to know each other better and to communicate with each other, and to play an important part in the ongoing democratic process.

It is claimed that American campaigns have changed greatly in the past thirty years. One can argue that today they use more advanced technology, in polling, in computer simulations, in predictions of voter behavior in response to differential campaign techniques, in new approaches to mobilizing voters, in the experimentation and development of media messages and advertisements, et cetera. In Chapter 15 we discuss these technologies in detail. Campaigns are also much more expensive than ever. Further, it is claimed that the role of parties and interest groups has changed, with some arguing that parties are less important today. For a time, as Herrnson says, party machines were dominant, then parties became "peripheral," and now again the party organization is playing an important intermediary role.[6] It is also contended that there is more "mudslinging" and negative advertising by politicians today. Some would even say, on the other hand, that campaigns are more "scientific" today. All this may be true to some extent. Indeed, the campaign process is constantly changing, as is true also for our political system generally. As the society changes (technologically, socially, even culturally), those who run campaigns adapt to such change if it is relevant for the achievement of their objectives.

Yet we come to the realization that the basic character of the campaign remains the same. As Riker put it, "the very heart of the activity of a campaign" is today, as previously, "how politicians persuade voters to support them."[7] How this overriding objective is achieved still challenges us: How can a message be put together and communicated, by the candidate and party (or interest group), through what most effective means of rhetoric and communication, so that the target populations necessary for victory can be activated and mobilized, while the opposition forces can be reduced to defeat? That is still the heart of what campaigns are all about. The means and techniques and tactics may vary over time and become more refined, even more scientific, but the essence of the campaign goal and challenge persists.

THEORETICAL MODELS OF THE POLITICAL CAMPAIGN

Many scholars have worked with models of campaigns and American voting behavior. Perhaps the most famous is "the funnel of causality" model proposed in 1960 in *The American Voter*.[8] This conception combined long-range factors (such as citizen political predispositions and preferences) and short-term, more proximate, factors (such as perceptions of candidates and issue positions) as explanations of voter decisions. The same model with elaborations is used recently in an update by Miller and Shanks of that earlier study.[9] Other versions of this model emphasize similar or different factors or "causes." In the study of campaigns, we need to keep in mind two basic types of models, one explaining voter decisions, as in *The American Voter* and other studies of the vote decision, and a second model that focuses on how the candidate and his or her party or team conceptualize the strategy and factors of the battle plan for the campaign.

Every campaign has a basic thrust to it, however subdued and inarticulate it may be. True, for some campaign managers the idea of a fixed strategy is abhorrent, the argument being that the campaign operation must be flexible and adaptive (sometimes called incremental) so that the candidate and his team can respond easily to new issues and developments. Despite different degrees of rigor in specification of a strategy and its implementation, a basic concept for the campaign does usually exist.

Political campaigns vary in specifics, but one notes that they have four common characteristics. They utilize the principles of a communication model and operate with certain basic ideas of how best to communicate with the voters. They employ a coalition-building model, wherein victory is conceived in terms of appealing to certain interest sectors of the public who can be mobilized to produce a majority. They are based on a model (or theory) of attitude influence, which assumes that citizens' political perceptions and attitudes can be appealed to, or changed, in such a way as to influence their voting behavior. They implement a model of ideological competition, appealing to the public with certain ideological orientations, however general and obscure, in such a way as most efficiently to attract votes on the basis of issue positions while also counteracting the ideological appeal of the political opposition.

First it is necessary to see the campaign as a communication process (see Figure 11.1). Each party with its candidate and strategy team is confronted with the basic question of deciding how to communicate most effectively in order to maximize chances for victory. The public has to be visualized as not just an aggregate of voters, but as subgroups of voters with differing interest orientations (defined in terms of partisan predispositions, ideological perspectives), social-class status, and demographic characteristics such as age, race/ethnicity, religion, and sex. The party thus must make certain communication decisions, such as what interest sectors should be appealed to, with what types of messages (differentiated for these interest sectors), using what types of media, in order to achieve the effects of penetration and to minimize the possible effects of boomerang and bypass, and how to do this to counteract most effectively the appeals of the opposition.

According to this model, the essence of campaigning is the "rational" planning of the entire operation to adapt the messages and appeals to selected groups in the population. Clearly, in this ideal model one must have considerable intelligence about the behavior of the voters in each targeted interest sector as well as

FIGURE 11.1 ▪ A Communication Model of Political Campaigning

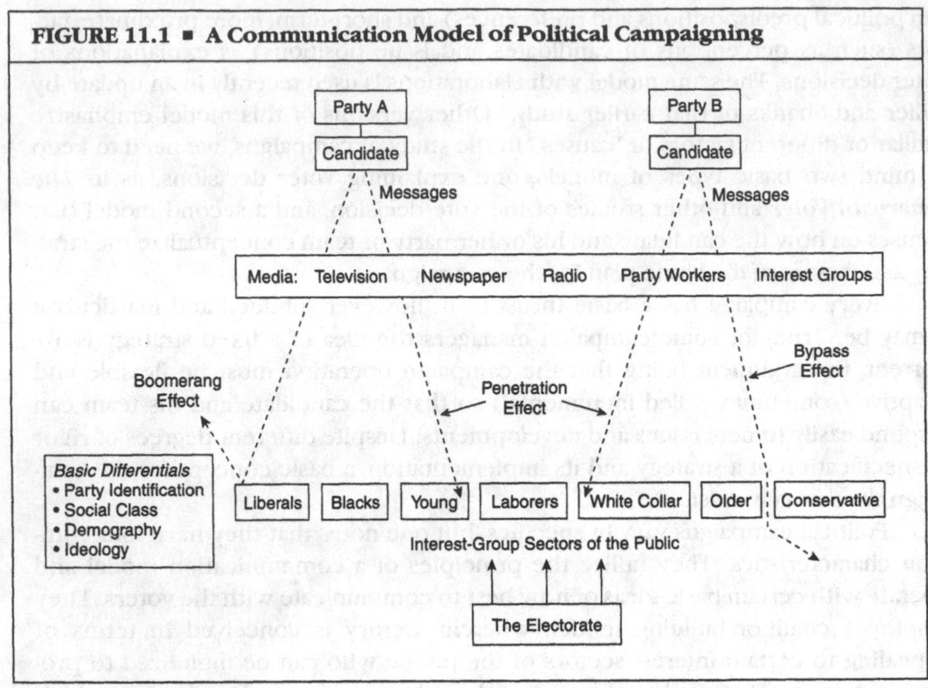

great expertise in the use of media. The problem with such a rational model, of course, is that it makes assumptions that are not tenable and which, therefore, require constant revisions of strategy.

Thus, as L. Froman has pointed out, this type of rational-choice model has to face up to, for example, the realization that there are groups that are not politically neutral at the outset of a campaign but have a history of support for a particular party; voters with partisan predispositions that limit their availability to persuasion, including their selective exposure to campaign appeals; and candidates with biases that also limit the credibility of their appeals to certain types of voters.[10] These and other considerations have to be continuously reviewed in the implementation of campaign decisions. Yet each campaign has its own communication assumptions and goals.

A second model follows. A major element in campaign strategy is the need to put together a coalition of interest groups which, if appealed to properly, will constitute the winning margin. The particular interest groups which are targeted will, of course, vary greatly by community. As part of the strategic problem, there is always the possibility that an appeal to one group will offend another group of supporters. It is then necessary to estimate the probability of this occurring, the extent of such damage, and ways of minimizing losses or offsetting such losses with other appeals.

The difficulty in devising a campaign strategy that is a design for victory in interest-coalitional terms is that individual voters are complex human beings. To identify their dominant interest orientation and appeal to that interest is necessary, but such calculations are not always easy. Further, the absence of political solidarity in interest-group sectors of the American public, as well as the lack of homogeneous party loyalty, makes it difficult to mobilize a strong majority in such groups. Elmer Schattschneider called this the "law of imperfect political mobilization of interests."[11]

Third, another way of thinking about campaigning is to see it as an ideological competition. In most campaigns, after all, two or more candidates face each other, each with his or her own issue positions on current problems facing the community. What they say (or do not say) about each of these issues in the context of political combat presumably plays a role in providing greater vote support for one candidate and less for the other. In a sense, this is an application of Anthony Downs' theory, discussed earlier. Since the majority of citizens theoretically will refuse to vote for parties (and candidates) that take positions too extreme for them, and will support that party whose issue positions are closest to theirs, the party must adapt to this reality. The party and its candidate must devise a strategy that is multidimensional, appealing to groups on differential fronts. A party may take a liberal position on welfare to attract poor voters and a conservative position on government spending and taxation to attract more affluent voters. This suggests a campaign strategy flexible enough to tailor appeals to different groups using different issues, which requires determining which issue is most relevant for each group and collectively developing and adopting a set of issue positions with the greatest chance of mobilizing support.

Finally, any campaign process model must incorporate a theory of ***individual*** behavior that is the consequence, in part at least, of attitudes and perceptions of the political environment brought into play by these campaign stimuli. As the campaign unfolds, the individual's basic orientations to politics are activated, involved, and utilized in a perception of the candidates, the parties, and their appeals. Usually the campaign does not change attitudes, although it may well influence immediate perceptions of political events and actors. It is not attitudinal change that normally occurs but the development of preferences (for candidates, policies, and parties) linked to basic attitudes and colored by immediate perceptions. These voting-preference decisions, often arrived at early in the campaign, are then reality tested during the course of the campaign by discussion with friends, associates, and family members and by exposure to the media's analysis of politics, leading ultimately to a voting-booth decision on election day. The campaign strategy, therefore, must face the necessity of developing messages that (1) influence the voter's perceptions of the candidate-actors; (2) engage basic attitudes in such a way as to lead to particular voting preferences; and (3) throughout the campaign reinforce these initial voting-preference decisions. The process can be visualized, then, in the type of model suggested by Figure 11.2.

These four approaches, or models, of campaign strategy tend to fit together. They suggest that campaigns have to be planned and yet flexible; they must have a basic strategic concept that is also adaptable; they must be aimed at particular

FIGURE 11.2 ▪ A Stimulus-Response Model of the Campaign

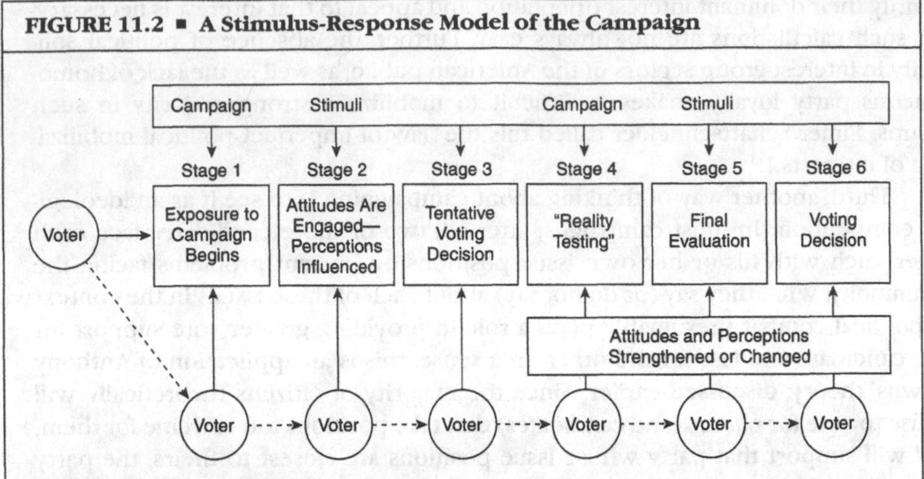

SOURCE: Bernard Berenson, Paul Lazarfeld, and William McPhee, *Voting* (Chicago: University of Chicago Press, 1954), 278. The diagram has been modified and supplemented for use here. Earlier discussions with Daniel Katz were very useful in developing these revisions.

groups and yet be aware that there are cross-pressures; they must calculate how to develop messages that penetrate while accepting the probability that many appeals will bypass. Above all, the campaign must have a communication theory, an ideological self-consciousness, a coalition-building theory. Finally, there is "the law of the unearned increment of politics"— that parties get votes that surprise them, for which they did not appeal. And such votes increase as the campaign gathers momentum.[12]

Planning and managing a campaign is obviously very demanding, comprehending a complex set of tasks. We often distinguish between the "strategy" of the campaign and the "tactics." "Strategy" is the basic conception of the campaign's message, focus, constituency, target populations, and theme. "Tactics" include the operational techniques and measures for realizing strategic goals. Often campaigns are less than successful and make significant mistakes in either strategy or tactics, or both. There are many conditions complicating campaign management: lack of information on the state of public opinion, particularly on what will persuade voters to maintain or change voting preferences; lack of knowledge of which communication and media contact will pay off and which will be ineffectual; the diversity of group interests within the target constituencies and uncertainty as to how appeals need to be differentiated in order to be effective; the multiplicity of issues that might be relevant for the vote and uncertainty as to which of these are salient for the public and which will determine the vote; the need to constantly monitor the opposition's efforts and develop effective counter strategies; disagreements as to how to best market the candidate's image, how to best call attention to the candidate's strengths and play down his or her weaknesses; lack of clarity as to how best to appeal to the critical interest groups so as to use their strength in support of the candidate. In the United States, of course, media costs, limited finances, and campaign length exacerbate the problems of campaign management. American history is replete with examples of successful campaigns: FDR in 1932, Truman in 1948, Reagan in 1980, Clinton in 1992. And of course there must have been also many unsuccessful campaigns in these same years: Hoover, Dewey, Carter, and Bush. Detailed studies of these campaigns and the knowledge we have gathered from the surveys of voting behavior since the 1950s, linked to these successful and unsuccessful campaigns, can provide us with some basis for discussing the effects of campaigns and the factors and forces that explain campaign success or failure.

CAMPAIGN EFFECTS

To demonstrate that what goes on in a campaign *itself* was effectual *by itself* has always been difficult. Does it matter what parties and candidates and their teams of workers do in a campaign? Or, to put it more bluntly, can you sell a president? There have always been skeptics of the relevance of the campaign. Yet today the research on campaigns and on voting decisions suggests more strongly than ever that campaigns do matter. The early classics in the field, one should note, argue that there are effects. Thus, the 1948 study by Lazarsfeld and his colleagues at Columbia, *The People's Choice,* reported the following types of campaign effects:[13]

Possible Effects of the Campaign	*Percent of Sample*
Reinforcement	53
Activation	14
Conversion (from one party to another)	8
Partial conversion (from a party to independent)	6
Reconversion (from one party to another and then back to the original party)	3
No effect	16

The authors of *The American Voter* in 1960 in a sense confirmed the observation that campaigns have an effect: "only about one person in twenty said the campaign had failed to reach him through any of the principal media of communication."[14] The earlier Michigan study of 1954, *The Voter Decides,* had also clearly demonstrated, by an analysis of the time of voter decision in 1948 and 1952, that a fairly large number of citizens were "switchers" (particularly in 1948 when Truman defeated Dewey) who were influenced by the course of the campaign.[15] While such early studies emphasized that campaigns do have an effect on the voting decision (more reinforcement and activation than conversion), they did not answer the basic question of *how* campaigns have an effect, what strategies, events, candidate actions, or media ads and presentations seem most relevant and where the campaign fits into the overall model of how voters decide. The position of some scholars is that certain specific salient actions can "deliver the vote." Others see campaigns as part of a long-range voter decisional process. Miller and Shanks, speaking particularly about the 1992 election, argue that "the effects of the campaign should have been to activate the long-term predispositions," particularly the partisan and policy-oriented predispositions, in terms of which the voter evaluates the issue positions of candidates and compares their personal qualities and records.[16] *Which* campaign considerations are important and *how* they interact with predispositions are still not very clearly delineated, however. We will summarize the state of our knowledge about campaign effects, keeping in mind these alternative ways of viewing the campaign.

THE MASS MEDIA ROLE

Because of the great emphasis on the mass media in discussing our campaigns, it is necessary to start with the differing interpretations of the media's role. One must distinguish two aspects of this matter: (1) the influence of the media per se, as an independent force or determinant of effects; and (2) the role of the media as it is used by parties and candidates. First we deal with the media as an independent actor.

American election campaigns are distinctive because of their heavy reliance on television, which appears much more important today than newspapers and radio (Table 11.1). In the early postwar period that was not the case; radio and newspapers were the more important media for political information. Actually, some recent studies reveal a small decline in the public's reliance on television, as our 1996 data indicate. Yet, if the public is asked directly to name their *primary* source of information about the campaign, television (November 1996) ranks far above the other media: television 56 percent, newspapers 20 percent, radio 10 percent.[17]

TABLE 11.1 ▪ Public Reliance on the Mass Media: Changing Roles (in percent)

Followed the Campaign Through:	*1952*	*1956*	*1976*	*1980*	*1996*
Television	51	74	89	86	72
Radio	70	45	45	47	19
Newspapers	79	69	73	71	60

Note: The 1996 question was: "How did you get most of your news about the presidential campaign?" In other years the question was: "From which news medium would you say you got the most of your information about the campaign?"

SOURCE: National Election Studies, 1952–1980; Roper Center for Public Opinion Research, *America at the Polls 1996* (Storrs, CT: Roper Center, 1997), 174.

Thus, the dissemination of information about both candidates and campaign events is clearly one very important role for television and, to a lesser extent, for radio and newspapers. The presidential debates are one particular type of television program that reaches millions of voters, and up to 40 percent of these viewers feel the debates are "helpful" and that they "learned" something from them.[18]

As the campaign proceeds, it is clear that voters feel they are gradually accumulating more information. They were asked by the Roper survey at various time points, January to November 1996, whether they had "received enough information about the presidential candidates to make an informed decision about whom to vote for?" In January only 31 percent said yes, by September it was 69 percent, and by election time 87 percent felt adequately informed.[19] One should be careful, however, in assuming public satisfaction with all aspects of the media's coverage, because in 1996 68 percent also felt that the coverage of the candidate was "confusing and unclear." Similarly, 60 percent would prefer to hear directly from the candidate than to hear or read what the media say about the candidate.[20] This pattern of response suggests considerable ambiguity in the actual level of public satisfaction with the informational role of the media.

There is further reason for concern about what the media does in a campaign. In our discussion of the presidential primaries, we have already noted the *selective coverage* of the media during the primary period. The media obviously make their own decisions after the early New Hampshire primary (and Iowa caucus) as to who should be given the most space and time, which candidate(s) have the "momentum" and which have no chance. This is what Zaller calls "the rule of anticipated importance," which of course applies also to many events, issues, and activities in the general election.[21] In quantitative allocation of space given the two major candidates after they have been nominated, the networks did provide fairly equal coverage during their two conventions in August 1996, but then gave Clinton the advantage until late October, when the coverage converged again. In the meantime, however, Ross Perot's coverage by the networks was minimal.[22]

A considerable amount of research has been done on the "priming" that the media engage in, the way the media influence citizens' evaluations of presidential candidate performance by selectively presenting news stories or programs that will portray the candidate positively or negatively. Subsequent ratings of the candidate reflect immediate viewer responses to these news stories. In one example, it was demonstrated that citizen attitudes and subsequent ratings of Reagan were "primed" by a program on the Iran-Contra Central America scandals. Thus the substantive focus of the discourse in the media may have a strong effect on the public.[23]

This type of research can lead to large theories about the manipulative role of the mass media. Zaller's model of media politics assumes a "deep seated tension between journalists and politicians over control of news content." Journalists want to have autonomy in deciding what they report, politicians want to "get their story out." So journalists and politicians compete and the public has to accept what it gets.[24] In the early days of media research, it was concluded that the media had

TABLE 11.2 ▪ Public Views of the Media's Role in the Presidential Campaign of 1996 (in percent)

Public View	*Agree*	*Disagree*
1. "The media lead candidates to avoid issues and perform for the cameras."	83	17
2. "The media have too much control in defining issues."	77	23
3. "The media give undue advantage to front-runners."	76	24
4. "The media stories about the campaign generally provide unbiased accounts of what is happening in the campaign."	32	65
5. "Television coverage seeks to influence the election outcome."	69	26

SOURCE: October 4, 1996 Surveys, reprinted in the Roper Center for Public Opinion Research, *America at the Polls 1996* (Storrs, CT: Roper Center, 1997), 176–79.

"minimal effects" on election campaigns. Today, much scholarship emphasizes the significant role the media have in "mediating" candidate and party messages. Message revisions or deletions or de-emphases may help or hinder a candidate.[25]

The public's perception of the mass media is often derogatory (Table 11.2). The negativism and even cynicism in the public's response patterns is considerable. People do feel they get "too much" on the "horse race" aspect of the campaign (57 percent, compared to 2 percent "too little") and "too little" on the issues (45 percent, compared to 8 percent "too much").[26] These are disturbing findings. Yet one must remember that a majority of the public (58 percent) also reported in 1996 that "the media helped me make up my mind about who to vote for." To decide from such mixed responses *what* the effect of the media is and how it is accomplished, is questionable. To decide that the public's evaluation of the quality of the media's performance is rather low is not difficult. As another Roper study demonstrated in 1997, the overall level of confidence in the media has been declining in the past twenty-five years. These data are presented in Table 11.3[27]

The decline in confidence is clear. In other data in the above surveys, the percentage reporting "hardly any" confidence increased for the press from 15 percent to 39 percent over these years and for television from 22 percent to 42 percent.

TABLE 11.3 ▪ Public Confidence in the Media
"Would you say you have a great deal of confidence, only some confidence, or hardly any confidence in [these institutions]? (answers in percent)

"A Great Deal of Confidence"	*The Press*	*Television*	*The Scientific Community*
1973	23	19	37
1980	22	16	42
1996	11	10	39

CANDIDATE STRATEGIES AND DIFFERENT TYPES OF CAMPAIGNS

Campaigns should be studied not only because of the way the media performs but more importantly because of what messages the candidates and their parties use in order to persuade voters. The strategy of appeals in one sense is the crux of the campaign. Campaigns differ greatly in terms of their basic character, their basic strategic appeal focus. Among the many ways in which campaigns differ strategically, we can immediately think of three:

1. The extent to which they are very confrontational, i.e., use of negative attacks on the opposition as a major way to get votes: are they extreme and bitter conflicts, or moderate and benign?
2. The extent to which one or two dominant issues constitute the focus of the campaign, in contrast to campaigns where a plethora of smaller issues are utilized: is the focus on a central substantive question or a multiplicity of secondary issues?
3. The extent to which the personal attributes, careers, and character traits of candidates become a large issue in the campaign, in contrast to highly depersonalized campaigns: is much time or very little time devoted to attacking and defending a candidate's person?

Other dimensions by which to distinguish campaigns obviously can be used. These three may provide at least a beginning in our thinking about campaign strategy. A brief look at recent presidential campaigns may be instructive.

The 1996 campaign is perhaps illustrative of a relatively low-key campaign with limited drama in which candidates utilized a variety of issue strategies rather than a single dominant issue.[28] It was a campaign which the press found rather dull, and studies reveal that network coverage was low, with considerably less time given to it than in 1992 or 1988.[29] The context for the campaign might well have suggested a bitter battle. The Republicans had just taken over Congress in the off-year 1994 election and were riding high in Washington, promising a confrontation with President Bill Clinton. His approval rating in late 1994 had slipped to 42 percent. With Republican control of the House and Senate, Clinton was morosely pondering his presidential relevance. With the advice of Dick Morris and other advisors, a plan was evolved, based on the famous "Triangulation Strategy" of moving to the political center, taking moderate positions on a variety of questions, even co-opting Republican issues if necessary, and thus combining his tried-and-true liberal Democratic vote on the left with an appeal to a variety of other voters: the elderly, previous defectors from the Democratic party, independent voters, even suburbanites and fiscal conservatives. Clinton argued that the economy was strong, the budget deficit was down, and that he had worked hard on such matters as family leave, tax breaks for college tuition, and portable health care, all of these issues emphasizing his concerns for the family. On the other hand, in his contacts with the Democratic "left," he was strongly in support of affirmative action and environmental regulation, and he expressed concern for what the Republicans would do to Medicare. Eventually he would join the Republicans in a balanced-budget action as well as sign a conservative welfare-reform bill. His basic attack, which was relatively subdued, was on Dole's proposal to cut taxes (arguing that Dole's previous record could only inspire skepticism on this matter). Further, he neatly argued that the Republicans were at fault for the partial shutdowns of the national government during the budget crises in early 1996. Putting the Republicans back into the presidential office was, Clinton argued, too great a risk. Thus, using a multiple-issue strategy, Clinton fought back into ascendancy in the campaign. His approval jumped from 42 percent in 1994 to 60 percent by November 1996.

Dole's strategy was difficult to decipher. In fact, there were many news stories in 1995 and 1996 asking what his "message" really was. The context for Dole and his team was forbidding by early 1996. People were turning against the Republicans and some of their leaders (particularly Gingrich) for failure to deliver on promises. There was division in the Republican ranks. Dole had five challengers in the primaries. Buchanan, who had strong support on the "right" (with Gramm), did well in the early primaries. And then Forbes appeared, with his flat-tax proposal, as well as Governor Alexander. In addition, the economy was healthy. Dole's personal liabilities were obvious — his age, his ties to the Washington establishment, his less-than-exciting speaking style. The Dole team's strategy in a sense paralleled Clinton's "triangulation" approach, but with a distinctive Republican emphsis. The Republicans also wanted to control "the center" and to that end even supported issues like the minimum-wage increase and portability of health insurance. But they also worked to shore up their support on "the right" with Dole's opposition to third-term abortion and to the ban on assault weapons. Desperately seeking to establish a message, he came out in favor of the 15 percent tax rate and the end to the Internal Revenue Service. His credibility on this issue was not high with the public (*Newsweek* reported that the majority of voters by a 70–19 percent margin thought Dole had proposed this only to win voters). Above all, polls reported that a minority of Americans (35 percent) felt Dole had "a clear plan for solving the country's problems."[30] Dole did not really attack Clinton on the "character issue" until late in the campaign, and these attempts then seemed to be ineffectual. As a result, the gap between the two candidates widened in approval ratings, so that by election time there was actually a twenty-point difference. This was a campaign, then, in which a very determined president and his campaign staff, with a carefully crafted strategy, explored a variety of issues to reverse his approval ratings and decisively defeat an opponent who certainly had a chance at the campaign's beginning but never put forward, nor implemented, an articulated message capable of wresting the presidency from the Democrats.

If we look at other campaigns, we will see differences along the lines of the dimensions we utilized earlier. We obviously cannot go into detail here on the many interesting campaigns of recent years. We can say something by way of summary about the 1992 and 1988 campaigns that will suggest how different they were from each other and from the 1996 campaign.

In 1992 the presidential campaign came close to being one in which a single issue was dominant: the state of the economy.[31] We recall the Democratic strategists' cry, "It's the economy, stupid!" The economy was alleged to be in a decline; growth was down ("there are three million more Americans out of work today than when Bush took office"). Further, it was claimed, there was decline in the American family's standard of living. Another economic worry was the decline in our competitiveness with Japan and Germany. Clinton, the challenger, exploited this issue mercilessly, maybe even exaggerating it. Bush's strategy was to play it down and argue that the economic decline was not as bad as claimed. But he failed to establish this position. Bush did not present an economic plan until rather late in the campaign (in September, in Detroit), and it was such a general agenda that the news coverage was limited. Meanwhile, Bush was under attack within his party for violating his 1988 campaign pledge of "no new taxes." He tried to divert the campaign to other issues, especially to Clinton's imperfections of character (avoiding the draft) and his poor job as governor of Arkansas. On the "integrity" issue there seemed to be some temporary impact on voters, but no lasting effect. Polls reported that the views of Clinton's integrity fluctuated during the Republican attacks in October, but by November 69 percent of the public said Clinton had the honesty and integrity to serve as president.[32] In the last analysis, the economy as the central issue could not be avoided. Clinton worked this issue effectively, appealing to a broad range of voters "center" and "left of center," while holding on to the core groups in his party's "theoretical" coalition: blacks, ethnics, labor, women, and urban voters. Various facets of the economy issue could be used differentially with these groups. In the meantime, Ross Perot assisted in keeping the economy issue salient. It was astounding, as one reviews the course of this campaign, how Bush's approval rating fell — from 88 percent in March 1991 to 47 percent by January 1992 and to 35 percent by the November election. Clearly the economy issue with all its ramifications, as it was articulated and communicated to the American public by the Clinton team, seems to have been a major factor in the fall of George Bush. We say "seems," because studies of the 1992 vote question to some extent the exclusive relevance of "the economy" as the decisive issue.[33] Miller and Shanks point out that, although 19 percent of those voting for Clinton and Bush mentioned unemployment as the "most important problem" and another 22 percent mentioned "the economy," other issues were perhaps of equal or greater importance. How much "credit" should be given to economic issues such as aid to the disadvantaged, compared to social questions, tax policy, or, for example, homosexuals in the military, is difficult to estimate. They conclude that one must take a "more global assessment of the role of policy-related conflicts in elections."[34]

The 1988 presidential campaign is an excellent example of the negative "attack" strategy, of the extreme personalization of political appeals and their success.[35] The context for the campaign suggested a clear route to victory for Bush: a strong economy, fairly low unemployment, no real inflation, strong Republican performance with Reagan in the previous two elections. Also, the Republicans had apparent control of key issues: family, national defense, tough stand on crime, patriotism, et cetera. Yet the stock market plunged by five hundred points in October, the big federal deficit and trade imbalance causing some concern. In addition, Bush had considerable opposition in the early months from opponents in his own party.

And he was late in developing an effective strategy. By May of 1988, Dukakis was leading by ten percentage points, a lead lasting into August. The image of Dukakis was that of a moderate Democrat with a good record of balancing the budget in Massachusetts. It was from May on that the "attack" on Dukakis by the Bush campaign team was conceived and implemented as the only way to elect Bush, given his relative disadvantage in the polls. The heart of the attack strategy was to portray Dukakis as (1) soft on crime, particularly in granting early furloughs to Willie Horton in Boston; (2) weak on patriotism, particularly Dukakis' position on the Pledge of Allegiance, his veto of a bill requiring teachers to lead students in the pledge; and (3) his label as a "liberal" exemplified by his boasting a card-carrying membership in the ACLU. The "liberal" label had different meanings: "economics," supported high taxes and big government; "social," emphasized the rights of criminals; "defense," opposed new weapons and was too willing to trust the Soviets. While there were many other issues, such as Bush's promise of "no new taxes" and the Democrats' capitulation to special interest groups, as well as Dukakis' attack on Bush's Iran–Contra role, the "liberalism" attack on Dukakis was the dominant theme. And Dukakis refused at first to honor it with a reply. By August the polls began to change in Bush's favor, and Dukakis never recovered. His debate performance on the symbolic patriotism issue was not effective, and when asked point blank in the second debate whether, if his wife were raped and murdered, he would favor an irrevocable death penalty for the killer, Dukakis calmly said no, and many felt this response was inadequate. In the last days of the campaign, Dukakis finally denounced the "garbage" and "lies" of the campaign, but the damage had been done. He lost, 46 percent to 54 percent. Evidence that the attack strategy may have been successful is found in the analysis of polling data, before and after:[36]

The 1988 Campaign: Poll Results

	May	*September*	*November*
A. Percentage of electorate who said Dukakis was a "liberal"	27	36	56
B. Percentage of those seeing Dukakis as a "liberal" who preferred:			
Dukakis	39	19	30
Bush	50	72	61

Of those who did not label Dukakis a "liberal," a majority supported him. Dukakis recouped some support in the late stages, but the attack strategy seems to have achieved its objectives.

Do negative ads and an "attack" strategy pay off in votes? There is no convincing, extensive study evaluating effects of such campaigns. In 1988 Louis Harris reported, from his polling results, that 71 percent of the voters were disturbed by the negative campaign (in fact, 39 percent were "fed up" with it), *but* that "the simple story of the election is that the Bush commercials have worked, and the Dukakis commercials have not."[37] However, clearly establishing a cause-effect rela-

Presidential Campaign Strategies and the South: Playing the Race Card

To capture the South, the major party nominees developed different campaign strategies for playing the race card in the region. President Johnson set off this matter when he embraced, supported, and institutionalized the civil rights movement with the passage of the 1964, 1965, and 1968 civil rights legislation. From 1964 on, race became a cleavage and wedge factor in the presidential elections in the region. But presidents have played the race card differently.

Unlike Johnson, who appealed to the African American voter, Nixon appealed to the southern white voter. Carter didn't play the race card in the South but in the North, in order to capture the African American vote in the big industrial cities of that area. Andrew Young and the King family went directly into the northern churches to rally the electorate over the heads of the African American elected officials, who were quite skeptical of Carter, the born-again southerner. It worked.

Reagan, like Nixon, wooed the South and its leaders. As early as 1964 and continually until he was elected in 1980, Reagan traversed the region and spoke to small groups at house parties and social gatherings, telling his audience that he was one of them. Then in 1980, he launched his presidential campaign in Philadelphia, Mississippi, the place where three civil rights students had been brutally murdered. At that place, he embraced states rights and offered his opposition to busing and affirmative action. It worked. He carried the entire region except Georgia in 1980 and the entire region in 1984.

Clinton, like Carter, played the race card in reverse: not in the South but in the North. In the critical New York primary, he dispatched several busloads of civil rights leaders from his state of Arkansas, including the African American mayor of Little Rock, to churches in the African American community where they praised him and urged mobilization in his behalf. But during the primary season, he distanced himself from Jesse Jackson and criticized some of his supporters, and in the general election never mentioned race or offered any policy to deal with the race problem. Before he ended his presidency, however, he began a national conversation on race and accepted a report from his Advisory Commission on the subject in 1998. Overall, Clinton was quite moderate and flexible in his use of the race card.

As these campaign strategies dealing with race reveal, the campaigns in the South and to some extent in the North involved the development of an approach to the racial issues, to either neutralize or exploit it, with different degrees of success.

tionship based on voter perceptions or opinions is difficult. There is a tendency for voters to perceive others as more likely influenced than they themselves were. There are many factors that influence voting decisions and one must be cautious about jumping to simplistic explanations, such as the power of negative ads. Further, scholars have found that the public rates negative ads lower in influence than debates, news, and non-negative ads.[38] It is probably true that many Americans are "fed up" with such campaigns. In 1992 we saw advertising questioning

Clinton's character, as well as ads criticizing Bush's promises of millions of new jobs, his record on environmental regulation, and his proposals to cut college aid. But there was nothing like the blatant, even vicious, attack strategy of 1988. The 1996 campaigns of Dole and Clinton were rather mild in this sense, and the public judged them more positively than the preceding (1992) campaign.[39] The earlier demands by some for restrictions on the content of media ads have somewhat subsided in recent years, probably fortunately, if we keep in mind our First Amendment values.

THE PRESIDENTIAL CAMPAIGN DEBATES

Debates have become an integral part of the American scene, going back to the Lincoln–Douglas debates before the Civil War. In one of these, Lincoln put Douglas on the spot with his classic point-blank question: "Do you approve or disapprove of the extension of slavery to the new Territories?" Our modern period began in 1960 with the Kennedy–Nixon debates, cited years ago as evidence of the new impact of television. The Roper survey reported then that a high percentage of viewers (44 percent) said they were influenced by the debates, and 6 percent said they voted on the basis of the debates (three to one for Kennedy).[40] In 1976 the Carter–Ford debates were of interest, particularly because of Ford's misstatement that Poland was not under Soviet influence. In 1980 the debates came toward the end of a very closely fought battle between President Carter and Reagan, with Carter ahead early and then losing the battle, conceivably in part because of the debate performance. The public seemed reassured of Reagan's presidential capabilities. One poll reported that "Reagan converted a 4 percent deficit into a 5 percent lead over Carter."[41] The 1984 debates between Mondale and President Reagan came at a time of a sharp decline in the challenger's standing in the polls (a 41 percent approval rating, compared to 56 percent for Reagan). Reagan lost the first debate (October 7), appearing tired and confused, and the press made a major story out of it. Mondale cut Reagan's advantage in half. But in the second debate (October 21) Reagan came back strong and his style and sharp, even clever, responses were thought to have saved him.[42]

In recent presidential campaigns, the debates have fluctuated considerably in importance. In 1988, it is argued, Dukakis' debate performance may not have cost him the election, but it did cost him the opportunity to get back into the race after falling behind for the first time in August. The Bush campaign's attack strategy had had its effect. While the first debate was essentially a draw, the second was a bad story for Dukakis, with his failure to handle with passion the death-penalty question put to him. Polls reported that after Dukakis had cut the Bush lead to two points, Bush took charge and increased his lead to thirteen points after the second debate.[43] So the debate was in a sense a critical moment in the campaign.

In 1992 there was considerable sparring by the two teams (and Perot) over the number and types of debates. The Commission on Presidential Debates favored a different format, and the Democrats supported this. In the end, it was decided to invite Perot, and there was agreement on three different types of debates: the traditional panel of reporters, a sole moderator, and questioning by a group of ordinary, uncommitted citizens (picked at random by the Gallup Organization). This latter "town hall" context appealed to Clinton greatly, since he had been conducting his own such meetings earlier. The questions asked October 15 by the audience of 209 citizens were quite different from those in the other two debates. The debates were seen by close to 100 million people. They did not basically change the lead Clinton had in the polls, but they did legitimize Perot's candidacy. Perot did better than Bush in two of the three debates. He was far behind Clinton in the "town hall" debate (54 percent to 20 percent), slightly ahead in the first, and even with Clinton (at 30 percent) for the third debate. His voter support jumped from 7 percent to the final 19 percent he secured in the election.[44]

In 1996, the debates seemed to be much less relevant to what went on. Dole and President Clinton debated twice, once again in the "town hall" format, but they were not central events in the campaign. The candidates were more civil, the public less worried, the debates perceived as less helpful than in 1992. Thus in 1996, the debates may have had much less influence on the voter's choice.[45]

The debates seem to have a variety of functions, varying by campaign. Obviously, they can attract an immense audience that secures first-hand information about the candidates and their views on issues and their leadership styles. All of this is less mediated by the press than is usually the case, although mediated somewhat by moderators, panels of journalists, and the "town hall." The public's observations about how these elites discourse with each other (intellectually and emotionally and directly) is an important aspect of the campaign. The public can learn about character, perhaps, and about the nature of political conflict, as well as perhaps find a measure of political entertainment. In the "talk show" debates, there is even an element of direct public contact and communication. For the candidate the debates are sometimes an experience to be avoided. For example, Carter's team in 1980 wanted him to avoid the debate with Reagan. Other candidates need debates badly to give them a chance to recoup their losses: Dukakis in 1988 and Dole in 1996. The 1992 debate was a big help for Perot and he was very angry at his exclusion from the 1996 debates. It is quite clear that these debates have become fixed in our campaign culture. Yet they must be seen as only one facet of the campaign. What goes before these September or October debates is in reality of much greater significance. A candidate cannot win on debating skill alone in the last month. He or she can only use the debate to make the final push, the final impression, the final act of convincing the "wafflers" and mobilizing the lethargic. But usually the die is cast much earlier.

THE CAMPAIGN ROLE OF PARTIES AND INTEREST GROUPS

Political campaigns in the United States in the past have often been seen as distinctive; on the one hand, it was claimed, party organizations played a lesser role than previously, while on the other hand, interest groups played a larger role than ever. In our previous discussion of the party organizations in this book, we have presented evidence that the party is still a structure of action, nationally and state and locally, which has reclaimed considerable importance in the campaign process. Our discussion of the local party activists (Chapter 8) documented the increasing activity in campaigns of the county parties and the precinct/ward leaders in the cities. If anything, these local parties were more active in 1992 than in 1980 in registering voters, conducting house-to-house and telephone canvassing, and getting voters to the polls on election day. Thus, a national sample study of county party organizations found 92 percent engaged in these activities in 1992 compared to approximately 70 percent in 1980. And the impact of this activity on the vote has been also documented in over thirty studies (see the latter part of Chapter 8). Parties at the grass roots are very active and effectual today.

The extensive and detailed study of campaigning for Congress by Herrnson reaffirms, to a considerable extent, this basic finding and also reveals the role of political action committees and interest groups in campaigns.[46] We reported some of the findings of Herrnson in our previous discussion of national and state party organizations, particularly spelling out his conception of the national parties now as "intermediary structures" working closely with state and local organizations (Chapter 7). It is well to remind ourselves here of the specific findings of Herrnson on the role of the party organizations in the campaign. His interviews with candidates for the House and Senate asked them to rank seven types of institutional actions as to their importance in eight aspects of campaigning, ranging from candidate recruitment to getting out the vote. Using only some of Herrnson's data, we find that in competitive House contests where an incumbent Congressman or Congresswoman is running, the National Campaign Committees of both parties are ranked high in their involvement in candidate recruitment, in the development of issues for the campaign, and in the proportion of media ads. In conducting the get-out-the-vote drives and providing campaign workers, the local parties are ranked very high in both parties. Indeed, Herrnson concludes, "the above findings suggest that even though local party organizations no longer play a dominant role in many phases of campaign politics, they continue to make important contributions to aspects of campaigning that require direct contact with the voters."[47]

As for the role of interest groups and PACs, Herrnson also provides empirical evidence. These groups have become increasingly important in American campaigns in the postwar period.[48] And the laws regulating money in elections have encouraged the multiplicity of PACs. Candidate strategy, both in fund-raising and in mobilizing public support, is deeply involved with appeals to such groups. Indeed, to some extent parties and candidates work *with* PACs in the campaign. Research strongly suggests that such groups are closely linked to voters' information about, and evaluation of, candidates as well as their voting decisions. Herrnson's study of congressional campaigns provides us with much-needed data on the role these groups play in the campaign process.

In addition to his investigation of the roles of party organizations, Herrnson asked candidates how important unions, other interest groups, and PACs were in their campaigns. Again, his findings are illuminating. Focusing again on incumbent candidates in competitive races, he discovered that for the Democrats, unions as well as other interest groups are ranked as high as local parties in helping with get-out-the-vote drives and in providing volunteer campaign workers. Interest groups were also important in providing assistance in developing media ads. For the Republicans it is interesting that, while unions were rather irrelevant, other interest groups were helpful in providing volunteers as campaigns workers and to a much lesser extent in other ways. The Republicans relied much more on the National Committee and the local party organizations for most of this assistance. As for fundraising, as one might have expected, the Republicans relied heavily on PACs, as did the Democrats, as well as on the National Committees. For the actual conduct of the campaigns, however, unions and interest groups had more important roles in the Democratic than the Republican campaigns.

The very lively presence of interest groups in American campaigns is a significant feature. From research on the relevance of such groups for the vote, we find corroboration of their importance. Arthur Miller's analysis of their role in the 1992 election presents compelling evidence. He identifies the key "Democratic groups," traditional and new (blacks, women, ethnic, environmental, liberal), and the key "Republican groups" (big business, anti-abortion, conservative, Moral Majority, evangelicals). He then analyzes (1) the extent to which they are in *inter*group terms ideologically different, (2) the extent to which they are in *intra*group terms "constrained" attitudinally, and (3) the extent to which these two political group clusters were as such useful in explaining candidate evaluations and the vote. He concludes that "the most striking finding from the multivariate analysis is that the group affect measures . . . were among the strongest factors influencing candidate evaluations and the vote choice."[49]

The implication of this is fairly clear. The information and cues provided to voters by these groups influence them greatly. Further, as candidates are linked to such groups, which is certainly highly probable, these group connections of candidates, as perceived by voters, may harm or help candidates. Being linked as a Republican to the evangelicals may be wise or unwise; being linked as a Democrat to the liberals may also be wise or unwise. In any case, voters perceive these linkages, which constitute part of the candidate image that leads to their vote decision.

AN EVALUATION OF THE CAMPAIGN PROCESS

Campaigns are major phenomena linked to mass-elite communication relationships in democracies. They pose real challenges: *for candidates* trying to get their messages across, *for parties* seeking to mention or expand their power base, *for interest groups* seeking influence on their own or through the party system, *for the media* seeking to perform this reportorial-editorial role in order to gain influence and even respect, and *for the public* trying to make meaningful political decisions. American campaigns are thus multifunctional. They are also rather overwhelming — in duration, cost, complexity of appeals, and the media barrage. The study of these campaigns is fascinating. They teach us a great deal about the American political party system and its culture.

We have attempted to describe and analyze the campaign process from a variety of perspectives in this chapter. Uppermost in our minds is the question: Do campaigns have effects? While it is clear that campaigns vary considerably, it is clear also that there are significant consequences. Only a minority of voters are unaffected. Strategies employed by parties and candidates do make a difference. How the media report the news of the campaign makes a difference. The media ads, negative or positive, can make a difference. The issue "images" candidates communicate to the public in the campaign certainly can make a difference. The presidential debates sometimes can make a big difference. What party organizations do to help candidates, particularly in providing financial support, in assisting with the development of media messages and, above all, in mobilizing the vote, can make a difference. What interest groups do by supporting candidates in a variety of ways can, in our system, make a difference.

Voters thus are subjected to a variety of stimuli and appeals, which may convert only a few of them, considering their basic partisan and ideological predispositions. But these stimuli are very relevant for resolving indecision, reinforcing commitments and/or activating people to vote. Of course, some voters may be confused or "cross-pressured" as a result of the campaign. The early Lazarsfeld study in the forties said social and political cross-pressures occurred for many citizens (one-third of their sample), which led to personal conflicts over the vote decision, which in turn may lead to delay in voting decisions or even decreased interest and withdrawal from participating in the election.[50] Cross-pressures can result from overlapping group affiliations (with different partisan orientations for the groups) or from a conflict of vote intention with partisan predisposition. Janowitz and Marvick's analysis of the 1952 election revealed that a high proportion of citizens were in a conflicted primary group situation so far as politics are concerned (68 percent), and that these people were less inclined to vote (32 percent were nonvoters compared to 7 percent and 20 percent, respectively, for Republicans and Democrats who were not cross-pressured).[51]

Two major concerns, from among many final observations that could be made, emerge from the study of American campaigns. One is the very negative feelings the public has developed about campaigns as now conducted. Actually, this is not just a recent development. Despite the claim that we are now in the age of "electronic campaigns" and the allegedly "postmodern" age of politics, the functioning of campaigns today is very similar to twenty–thirty years ago. Patterson's study of the role of the mass media in the 1976 election resulted in a set of conclusions that might well be echoed today. He saw the media as much more fascinated by emphasizing the competitive struggle between candidates in 1976 than in the forties.[52] Issue positions, he argued, were much more emphasized earlier. Whereas in 1940 50 percent of media attention was given to issues and 35 percent to the fight between candidates, the reverse was true in 1976. It was the "power struggle between the candidates" that took over in the media. He claimed that as a result the voters in 1976 were less concerned about a candidate's policy positions — 34 percent of the respondents talked about policy, while in 1940 it was 67 percent. The "horse race" dominated the media focus. And the media are clearly perceived by the public today as treating campaigns in the same way. The Roper surveys found

83 percent of their 1996 sample agreeing that the "media lead candidates to avoid issues and perform for the cameras." And Woodward's recent content analysis of evening television news from 1972 to 1992 concludes that "the focus on conflict" and "the horse race" is very evident in the primaries, but is also present in reports of the conventions and the coverage during the fall of election years.[53]

A second major concern is the low turnout in our elections, even after all the efforts to mobilize the vote. In 1996 we declined to a vote of only 48.5 percent of the eligible voting population (the latest, 1998, Census Bureau report now says it was 54 percent). Are the negative appraisals of American campaigns possibly linked to low voter turnout? Public cynicism does seem to be high, as is public indifference to election outcomes. The parties actually contacted a higher percentage of citizens in 1996 (29 percent) than in 1992 (20 percent). Perhaps this only kept the turnout of voters from being even lower. One must certainly seriously entertain the possibility that the types of campaigns we run are a contributing factor to public withdrawal from electoral participation. We will discuss this probe further in a subsequent chapter.

In long-range terms, we must be concerned about the impact of the campaign, and particularly the mass media, on the public's view of the quality and effectiveness of the American system. Few studies have directed their attention specifically to this matter. A sophisticated analysis by Miller, Goldenberg, and Erbring of the media's role in the 1974 election and its impact on the public is revealing. The survey of the public in that election was combined with a content analysis of the front pages of the daily newspapers read by these respondents — ninety-four newspapers throughout the country during a sample ten days in October and November preceding the election. The results permitted analysis of the linkage between the type of newspaper content (extent of media criticism of institutions), the actuality and intensity of voter exposure to such media content, and the public's disaffection with the system. This was, then, a rigorous test of the relevance of media coverage.

This research revealed that a few newspapers (6 percent), in their front-page stories, praised our system, a larger proportion (31 percent) were critical, and the remainder (63 percent) were neither of the above. What is significant for our discussion here is the finding that political parties were particular targets of criticism (Table 11.4).

TABLE 11.4 ■ Critical Newspaper Coverage (as a percentage)

Topic of Story	*Percentage of Critical Stories*
Political parties	70
Congress	42
Administration	40
State and local government	34
Supreme Court	25

SOURCE: Arthur H. Miller, Edie N. Goldenberg, and Lutz Erbring, "Type-set Politics: Impact of Newspapers on Public Confidence," *American Political Science Review* 73 (March 1979): 67–84.

This was 1974, of course, and the Democrats were in conflict with President Ford in the Watergate aftermath. Nevertheless, the *relative negativism* concerning parties stands out. What is, finally, extremely significant in this study is the finding that political disaffection of the public was significantly related to newspaper criticism in 1974, even after controlling for education and media exposure. Further, a test was made to determine to what extent the policy dissatisfaction of citizens or their negative attitude toward particular incumbents ("incumbency effect") influenced results. They found that media content had an independent role in influencing public attitudes. The authors conclude, "Readers of papers containing a higher degree of negative criticism directed at politicians and political institutions were more distrustful of government . . . [and] higher levels of exposure . . . [yielded] higher levels of cynicism and feelings of inefficacy. . . ."[54]

The study of campaigns, their management and effects, clearly raises a good many questions, as many as it begins to answer. One last question is, Does all of this help the voter make an intelligent decision, to produce a deliberative, rather than a manipulated, result? V. O. Key posed this question long ago. He said, "the picture of the voter that emerges" from much of the "folklore of practical" politics as well as from "the new electoral studies is not a pretty one." The voter is seen as "an erratic and irrational fellow [sic] susceptible to manipulation by skilled humbugs. . . ." But he argues then that

> voters are not fools. . . . in the large the electorate behaves about as rationally and responsibly as we should expect, given the clarity of the alternatives presented to it and the character of the information available to it. In American presidential campaigns of recent decades the portrait of the American electorate that develops from the data is not one of an electorate strait jacketed by social determinism or moved by subconscious urges triggered by devilishly skillful propagandists. It is rather one of an electorate moved by concern about central and relevant questions of public policy, of governmental performance, and of executive personality.[55]

The issue posed by Key is a basic one. Does the campaign process as we know it contribute to responsible electoral behavior or not? This question we must continue to debate and study, particularly today. Although our campaigns perform many salutary functions, the question is whether in the end they encourage effective citizen deliberation as to the voting decision. The American problem may be campaign overkill for many citizens — too much noise when their voting task is very complex. The problem is certainly not that campaigns are irrelevant.

NOTES

1. Herbert E. Alexander, "Financing the 1996 Election," in *America at the Polls 1996* (Storrs, CT: Roper Center, 1997), 142.

2. The polls tracking the candidates' strengths in the last three presidential campaigns are found in Pomper, ed., *The Election of 1988,* 105; *The Election of 1992,* 145; and *The Election of 1996,* 186, (Chatham, NJ: Chatham House, 1989, 1993, and 1997).

3. This conception of the media is developed in great detail by John R. Zaller, "A Theory of Media Politics: The Struggle Over News in Presidential Campaigns," paper presented as the Converse-Miller Lecture, University of Michigan, April 14, 1997.

4. Based on the Roper Center review of Federal Election Commission Data, *America at the Polls 1996,* 153.

5. William R. Riker, "Campaign Rhetoric," Speech to the American Academy of Arts and Sciences, April 8, 1992, reported in *The Bulletin of the American Academy of Arts and Sciences* 46 (February 1993): 37.

6. Paul S. Herrnson, *Party Campaigning in the 1980s* (Cambridge: Harvard University Press, 1988), chap. 1.

7. Riker, "Campaign."

8. Angus Campbell et al., *The American Voter* (New York: Wiley, 1960).

9. Warren E. Miller and J. Merrill Shanks, *The New American Voter* (Cambridge: Harvard University press, 1996).

10. Lewis Froman, in M. Kent Jennings and Harmon Zeigler, eds., *The Electoral Process* (Englewood Cliffs, NJ: Prentice Hall, 1966).

11. E. E. Schattschneider, *Party Government* (New York: Rinehart, 1942), 33.

12. Ibid., 45.

13. Paul Lazarsfeld, B. Berelson, and H. Gaudet, *The People's Choice* (New York: Columbia University Press, 1948).

14. Angus Campbell et al., *The American Voter* (New York: Wiley, 1964), 51.

15. Angus Campbell, Gerald Gurin, and Warren E. Miller, *The Voter Decides* (Evanston: Row, Peterson, 1954), 11-27.

16. Warren E. Miller and J. Merrill Shanks, *The New American Voter* (Cambridge: Harvard University Press, 1996), 391.

17. Roper Center, *Polls 1996,* 174.

18. Ibid., 177.

19. Ibid., 175.

20. Ibid., 176.

21. John R. Zaller, "A Theory of Media Politics: The Struggle Over News in Presidential Campaigns," paper presented as Converse-Miller Lecture, University of Michigan, April 14, 1997.

22. Pomper, *Election of 1988,* 1996, graph on page 89.

23. See, for example, Shanto Iyengar and Donald Kinder, *News That Matters* (Chicago: University of Chicago Press, 1987).

24. Zaller, "Theory," 1, 22.

25. See, for example, Arthur Miller and Bruce E. Gronbeck, *Presidential Campaigns and American Self Images* (Boulder: Westview Press, 1994), especially chap. 5.

26. Roper Center *Polls 1996,* 181.

27. *The Public Perspective,* Roper Center, (February/March 1997): 4-5.

28. The discussion of the 1996 campaign by Marion R. Just, in Pomper et al., *The Election of 1996,* was extremely useful and heavily relied on for this discussion. See chap. 2, "Candidate Strategies and the Media Campaign."

29. Pomper, *Election of 1996,* 85. The chapters in this book on the strategy of the campaign and the role of the media were relied on heavily for this narrative.

30. Ibid., 123.

31. Again, the very useful piece by F. Christopher Arterton, in Pomper, *The Election of 1996,* provided good background for this summary.

32. Ibid., 124.

33. See Warren E. Miller and J. Merrill Shanks, *The New American Voter,* 1996, for a detailed analysis of policy issues as well as predispositions influencing the 1992 vote, particularly chaps. 11 and 12.

34. Ibid., 349-56.

35. We draw heavily on the discussion of the 1988 campaign by Marjorie R. Hershey in Pomper, *The Election of 1988,* 73-102, 105.

36. See Barbara G. Farah and Ethel Klein, in Pomper, *The Election of 1988,* chap. 4.

37. Quoted by Dianne Rucinski, in Arthur H. Miller and Bruce E. Gronbeck, eds., *Presidential Campaigns and American Self Images* (Boulder: Westview Press, 1994), 146.

38. Ibid., 150.

39. *New York Times* /CBS Poll (October 16, 1996), cited in Pomper, *Election of 1996,* 104.

40. Richard S. Salant, "The Television Debates: A Revolution that Deserves a Future," *Public Opinion Quarterly* (Fall 1962): 341. See also Stanley Kelly, Jr., "Campaign Debates: Some Fads and Issues," *Public Opinion Quarterly* (Fall 1962): 351-66.

41. Pomper, *Election of 1980,* 80-81.

42. Pomper, *Election of 1984,* 100-3.

43. Gerald Pomper, *Election of 1988*, 89-96, 107.
44. Pomper, *Election of 1992*, 93-96, 119-21.
45. Pomper, *Election of 1996*, 90-91, 126-27.
46. Paul S. Hernnson, *Party Campaigning in the 1980s* (Cambridge: Harvard University Press, 1989).
47. Ibid., 104.
48. Jack L. Walker, Jr., *Mobilizing Interest Groups in America* (Ann Arbor: University of Michigan Press, 1994), 19-40.
49. Arthur Miller, "Social Groups as Symbols in America's Sense of Democratic Consensus," in Arthur H. Miller and Bruce E. Gronbeck, eds., *Presidential Campaigns and American Self Images*, 205.
50. Lazarsfeld, *The People's Choice*, 56-64, 68.
51. Morris Janowitz and Dwaine Marvick, *Competitive Pressure and Democratic Consent* (Chicago: Quadrangle Books, 1964), 77. Peter Sperlick's study in 1971, *Conflict and Harmony in Human Affairs: A Study of Cross Pressures and Political Behavior* (Chicago: Rand McNally, 1971) suggests that under certain conditions cross-pressures can activate a person to vote.
52. Thomas E. Patterson, *The Mass Media Election: How Americans Choose Their President* (New York: Praeger, 1980), 30.
53. J. David Woodard, in Arthur Miller and Bruce E. Gronbeck, eds., *Presidential Campaigns and American Self Images*, chap. 6.
54. Arthur H. Miller, Edie N. Goldenberg, and Lutz Erbring, "Type-Set Politics: Impact of Newspapers on Public Confidence," *American Political Science Review* 73 (1989): 80-81.
55. V. O. Key, Jr., *The Responsible Electorate: Rationality in Presidential Voting, 1936-1960* (Cambridge: Harvard University Press, 1966), 4-8.

CHAPTER 12

Money and Political Campaigns: The 1996 Debacle

The year 1996 saw the U.S. public exposed to the most expensive political campaign in our history. Four billion dollars were spent on all the campaigns (national, state, and local). This high cost and the questionable ways in which the money was sometimes raised and spent, as the post-election analyses and investigations reveal, suggest a major crisis in party and candidate finance practices. How irregular, how illegal, and how scandalous the 1996 campaign finance operations were has yet to be finally determined. But we are clearly in the throes of a controversy over what happened, why and how it happened, and how the system should again be reformed. As one scholar put it, "1996 saw the final collapse of campaign finance regulation . . . and, as if on schedule, another scandal began to unfold, causing another embarrassed president to call for yet another round of reform."[1] The former president of Common Cause confirmed this assessment, calling the 1996 campaign "the dirtiest ever" and saying that "campaign finance laws have been replaced by the law of the jungle."[2] Corporations contributed millions directly to parties and candidates in 1996, although we thought they were forbidden by law to do so. The AFL-CIO spent millions on the campaigns from the union treasury, although we thought they were forbidden by law to do so. The presidential candidates, Clinton and Dole, were each given $61.8 million in public funds under the law, but spent more than double that amount despite the understanding they would keep within these funding limits. Foreign contributors gave millions to both parties, a practice prohibited by law. The "soft money" spent in presidential elections alone (funds raised outside the federal limits) was over $100 million.[3] In 1997 the new Task Force on Campaign Finance Reform, including nine scholars expert on the question and chaired by Herbert Alexander, issued a report making many recommendations for change. They said we are now "experiencing a much more dynamic diffuse funding system" in which new political actors "spend money in campaigns that candidates neither raise nor control." These "new realities," they say, raise serious questions about accountability, competitiveness, sources of funds, and enforcement.[4] These and other excesses document "the new reality in the world of campaign finance" that emerged in 1996.

Actually, we have been through this type of crisis before. Indeed, we seem periodically to go through the following cycle: concern about the amounts of money spent on campaigns and how it is raised and mass media exposures and critiques of funding practices, followed by congressional investigations and revelations of abuse; demands for new legislative regulations, followed by new reforms. Then the furor subsides for a while until the cycle begins all over. It happened in the 1890s, the 1920s, the 1940s, and the 1970s. In each period, campaign funding abuses led to reforms, which we subsequently discovered were inadequate or violated, or both. Today we face some of the same questions as in the past, plus some new questions:

1. *The Costs:* Are they too high today? Should individuals, groups, candidates, parties be allowed to spend as much as they wish in the interest of free speech (as the Supreme Court has told us)? Does the differential access of candidates (and would-be candidates) to these funds determine who can run for public office and who can survive? Can costs be cut, for example, by requiring free time from the media?
2. *The Sources:* Who should be permitted to contribute to candidates and parties, and who should be barred? How much should be allowed? How should such money be given, or spent — through parties, interest groups, direct expenditures by individuals, or in other ways? Can we really enforce such limits?
3. *Disclosure:* Is it possible to design and enforce a publicity law that provides adequate and timely information to the voters? Can it assure accountability? Does disclosure solve our real problems or only inform us of what our problems are?
4. *Public Funding:* Is the policy of paying presidential campaign costs out of the public treasury, on the assumption that candidates will observe the limits imposed, working? Is it proper to use public funds for such a purpose? Should we extend this policy to congressional campaigns?
5. *Public Trust:* Is our system of virtually uncontrollable spending in campaigns, with all its excesses, damaging public trust in the system?

These are the major questions we face again today.

Twenty-five years ago we were faced with a political finance mess. That crisis of the seventies, under President Nixon, revealed blatant violations. The Committee to Re-elect the President (CREEP) in the 1971–1972 period secured large sums of money, much of it cash that was never disclosed, which was used to pay for a variety of improper purposes, including hush money payoffs to the Watergate burglars. These slush funds were traced to corporate bank accounts, leading to the convictions of some business executives as well as Nixon aides. These revelations set the stage for the 1974 Federal Election Campaign Act.

A BRIEF HISTORY OF THE REGULATION OF CAMPAIGN FINANCE

The experience of the 1970s was just the latest example of how difficult it has been to deal with the use of money in American politics. To provide an historical perspective, we provide a brief summary for the past century. For most of this century and even earlier, Congress has been struggling from time to time with this problem. Probably our earliest concern was the exploitation by parties of federal employees for money-raising purposes, and this led to the provision in the Civil Service Reform Act of 1883 that federal government employees could not be solicited on governmental premises for funds to support parties or candidates. The *first approach* to regulation was, thus, to dry up one of the sources of campaign funds, federal appointees. But this act was only a beginning and did not really impose any stringent limitations. In the elections of the period from 1888 on, the Republicans particularly, under the financial leadership of Mark Hanna, their national chairman, were able to secure huge sums from wealthy businessmen, as well as from corporations.[5] Standard Oil gave sums ranging from $100,000 to $250,000 in the elections of 1896-1904. This situation raised a hue and cry that resulted in a 1907 act prohibiting the direct contribution of corporations to federal campaigns, an act which essentially has been retained since that date. Again, the aim was to close a source, linked to a desire to limit the direct influence of a particular interest group with ample funds through which they desired to control the presidency. In 1912 Woodrow Wilson, very aware of this danger, directed his finance chairman that "no contributions whatever be even indirectly accepted from any corporation" and that none of the money used in the campaign should come from "three rich men in the Democratic party whose political affiliations are . . . unworthy."[6]

Aside from these legislative attempts to dry up the sources, a *second approach* was that of publicity and disclosure — forcing candidates to reveal the sources of their funds. In 1910 Congress passed a law requiring periodic financial reports by candidates in general elections; in 1911 this law was extended to include primaries and conventions. But there was considerable debate at the time over the propriety and constitutionality of such legislation and its interference with First Amendment rights. The Supreme Court did indeed rule in 1921 that Congress had no authority to regulate primary elections.[7] It was not until twenty years later that the Court reversed this position.[8] In the meantime because of the great public concern, again, about the state of party and campaign finance, Congress passed a new basic law, the Federal Corrupt Practices Act of 1925. This act spelled out the publicity and reporting requirements rather comprehensively for congressional candidates, but not for presidential candidates. All political committees seeking to influence the election of federal candidates and functioning in two or more states were required to report to the Clerk of the House several times a year all contributions and expenditures, including the name and address of every person contributing $100 or more in a calendar year and the name and address of every person to whom an expenditure of $10 was made. Further, all individuals spending $50 or more to influence such federal elections had to file, as did all candidates for both houses of Congress, House candidates with the Clerk of the House and Senate candidates with the Secretary of the Senate. Such reports were due ten to fifteen days before the election and within thirty days after the election. This law remained on the books until superseded by the 1971 and subsequent legislation.

The 1925 Corrupt Practices Act included provisions to deal with the use of political money in a *third way,* by placing *ceilings on the amounts* that might be spent. It limited House candidates to $5,000 and Senate candidates to $25,000 (the exact amount linked to the votes cast in the last election). Certain major expenditures were, however, exempted: personal traveling and living costs; stationery, printing, and postage; distribution of letters, circulars, and posters; and telephone and telegraph service. Although considered largely ineffective, because of decentralization in campaign operations and the loopholes in the law, as well as nonenforcement of its provisions, this approach remained. And in 1940 the Hatch Act placed a ceiling on the amounts national committees could spend: a maximum of $3 million was specified as a permissible aggregate amount. This approach, too, was debated at length, those opposed contending that it led to more auxiliary or separate committees (each of which could spend $3 million!).

The Hatch Act also approached campaign finance in a *fourth way*, by imposing a limit of $5,000 on the amount that any person could contribute in any calendar year to the nomination or election of a candidate to federal office, to an organization working for a candidate, or to the national political party. This prohibition was intended (by some Congressmen) to limit the sources of party funds, or at least to limit the probability that big donors could "buy" candidates. However, contributors who wanted to give more than $5,000 were facilitated by the creation of more committees (to each of which $5,000 could be given). Yet the act did tend to change the big donor's role, while not, however, controlling the total amount the donor could give.[9]

The final major act adopted prior to the 1970s was the Taft-Hartley Act of 1947, which prohibited labor unions (as the 1944 Smith-Connally Act had earlier) and corporations from expending or contributing organizational general funds on behalf of the nomination or election of federal candidates. In a sense the purpose was to ***break the direct financial linkage of interest groups*** to political parties. The way was left open, however, for the creation of voluntary political action committees, to which labor union members could contribute money if they so desired and which could spend money on their own initiative in support of candidates for federal office. This opportunity, plus the propriety of spending certain funds from union treasuries for the publication of political information and for "political education," was upheld by the Supreme Court. This, in a sense, was the precursor of the development in the 1970s of political action committees.

After the Taft-Hartley Act nothing was done, although there was considerable concern about various aspects of party finance. In 1956 Senator Francis Case of South Dakota revealed that he had been offered a $2,500 campaign gift by natural gas lobbyists for his vote on the Harris-Fulbright Natural Gas Bill (President Eisenhower later vetoed the bill). President Kennedy set up the Commission on Campaign Costs, which made an excellent set of recommendations on all the major aspects of party finance reform and also suggested the use of public funds on a matching basis to finance candidates for federal office. But nothing happened during the 1960s, despite more revelations about the improper use of money. Legislation was proposed in Congress; indeed, one act (the Long Act of Senator Russell Long) was adopted and then repealed. This set the stage for the Federal Election Campaign Act of 1971 (FECA), which replaced the 1925 Corrupt Practices

Act. New provisions were included also in the Revenue Act of 1971. FECA 1971 was in turn changed by FECA 1974, which was supplemented by amendments in 1976 and 1979. In addition, Supreme Court interpretations of these acts have modified the law substantially. A summary of the basic elements of the federal legislation may be helpful.

THE PROVISIONS OF FEDERAL REGULATIONS OF PARTY FINANCE AS OF 1996

Since the adoption of the 1974 act, while the basic components of the law have remained intact and were operative as the 1996 campaign approached, major changes have been made as a result of court decisions. That act provided for full public funding of presidential candidates, extremely detailed publicity requirements (administered by a new agency, the Federal Election Commission) and fairly precise limits on contributions and expenditures by candidates, parties, and political groups (PACs). However, the 1974 act was soon amended by Congress, reinterpreted by the Supreme Court, and modified by policy statements of the FEC. In a significant 1976 case, *Buckley v. Valeo,* the Court argued, on the basis of First Amendment free speech rights, that unlimited expenditures by parties and groups were permissible if their advertisements did not "in express terms advocate the election or defeat of a clearly identified candidate." Thus, so-called issue advocacy ads were allowed with no expenditure limits. In 1978 the FEC itself loosened the spending limitations by allowing parties to spend large amounts of "non-federal" money for activities benefiting both federal and state candidates (regulated only by state laws, which were usually permissive). In 1979, Congress amended the 1974 act by permitting unlimited contributions to state and local parties for volunteer, voter registration, and get-out-the-vote drives even if these efforts benefited presidential and congressional candidates. And finally, in the middle of the 1996 campaign (June 1996), the Supreme Court again intervened by reversing the FEC. In a split decision, it permitted state party organizations as well as other political groups to spend as much as they wished as "independent expenditures," independent, that is, of direct candidate support, so long as the spending was not coordinated with the candidate or his/her campaign team. Thus we opened up very wide the opportunities for "soft money" solicitations and "independent expenditures" and in essence overturned specific party and candidate spending limits.

Keeping the above summary of the 1974 party finance law (with revisions) in mind, let us now describe somewhat more carefully the particular provisions of the major sections of that regulation in effect for the 1996 election.

Publicity Requirements

Careful and detailed bookkeeping by political committees and candidates for federal office, or their financial representatives, is now required. This continues and extends the requirements first initiated in 1910 and supplemented in 1925.

Candidates must have one official committee whose treasurer is responsible under the law for reporting. The candidate must keep records of all contributions of $50 or more (by name and date) and must provide detailed reports on all contributions of $200 or more, including the occupation and principal place of business of the contributor; all cash on hand, all money raised by special events (dinners, rallies), all loans and obligations. In short, almost all financial transactions must be carefully recorded. Quarterly reports must be filed every year, as well as ten days before an election and thirty days after an election. Also all contributions of $1,000 or more received in the last two weeks before the election must be reported, up to two days before the election. Every political *committee* spending or receiving money in excess of $1,000 must file with the Federal Election Commission. It must report the names and addresses of all its members, as well as a variety of information about its relationships to other groups, its accounts, the candidates it supports, and of course its contributions to candidates or committees supporting candidates, and its expenditures on its own initiative. *Individuals* are treated like committees if money is spent in order to influence the election of a federal candidate. Thus, those who must report are the treasurer for the candidates committees, all other political committees, including party committees, and individuals — all who contribute to or spend money in connection with federal elections.

Limits on Contributions and Spending

Individuals may directly contribute a maximum of $1,000 to a candidate in a primary campaign and $1,000 in a general election, for a total of $2,000 maximum to any particular candidate. The limit an individual may contribute directly is $25,000 to all candidates and committees in any one calendar year. Individuals may contribute up to $5,000 to a political action committee and $20,000 to a national party committee (see Table 12.1). Individuals may make "independent" expenditures supporting a candidate, but these must not be coordinated with the candidate's operations.

Candidates for the Senate and House may spend as much of their own funds or family fortunes as they wish to (the 1971 act limiting Senate candidates to $35,000 and House candidates to $25,000 was invalidated by the Supreme Court). Presidential and vice-presidential candidates, however, who accept public funding may only spend $50,000 from their own private (personal or family) funds.

TABLE 12.1 ▪ Federal Party Finance Law Contribution Limits (as of 1996)

Contribution Recipient	*By a Person*	*By a PAC*
1. Candidate	$1,000 per election	$5,000 per election
2. Party (national committee)	$20,000 per year	$15,000 per year
3. PAC	$5,000 per year	$5,000 per year
Overall Limits	$25,000 per year	No Limit

SOURCE: Brooks Jackson, in Larry Sabato, ed., *Toward the Millennium: The Elections of 1996* (Boston: Allyn and Bacon, 1997), 227.

Groups have a basic limit of $5,000 in direct contributions to a specific candidate in a particular election but no aggregate limit on the amounts that may be contributed to all candidates and committees. The Federal Election Commission ruled that unions, corporations, and other groups may set up PACs from voluntary contributions and that such PACs could spend unlimited funds in support of candidates, as long as they did this "independently" (without collusion with the candidate's representatives). Further, all PACs of a particular company or union would be treated as a single committee, and the $5,000 limit for direct contributions to a candidate would be applied in this sense.

Party committees have special limits on the amounts they may contribute to particular candidates or their committees, as well as expenditure limits which vary by state depending on the size of the voting-age population. The 1979 amendments provided liberal exemptions to party committees in determining their contributions. For example, payments for campaign materials, legal services, and voter registration drives are exempted.

Miscellaneous regulations require that loans be treated as contributions; no cash contributions of over $100 are permitted.

Public Funding

A "tax checkoff" was allowed (by the Revenue Act of 1971) for all those paying a federal income tax, amounting to a $1 deduction ($2 for those filing joint returns) from their tax as a contribution to a general fund for all eligible presidential candidates. In 1993 this was increased to $3 (and $6). This fund is used to support presidential candidates running in the preconvention and postconvention periods, as well as to support the costs of national party conventions. The tax checkoff takes place every year, but money is paid out only in presidential election years. Over 25 percent of those paying taxes used the checkoff up to 1984, but the percentage has declined since then to only 12 percent in 1997.

Funding for candidates was provided in the preconvention period on a matching basis. A candidate had to establish credibility by first raising $100,000 consisting of private contributions of $250 or less which would add up to at least $5,000 in each of twenty states. Contributions of this size would then be matched by the federal government. The total amount that a candidate in the preconvention period was permitted to spend was $37.1 million in 1996 (up from $13.1 million in 1976). These amounts are adjusted to rises in the cost-of-living index. Eleven candidates qualified for such federal funds in 1996. But Forbes did not accept public funds and could spend as much as he wished of his own personal fortune (over $37 million) and what he collected from others (reportedly over $3 million).

Candidates are subjected to two other requirements, however: (1) they must receive 10 percent or more of the vote in two consecutive presidential primaries in order to remain eligible for federal funds; and (2) if they fail to remain eligible or withdraw from candidacy, they must return the balance of funds received from the government after previously incurred debts are repaid.

Both national conventions could receive help to defray costs, $12.4 million each in 1996 (up from $2.2 million in 1976). A broad interpretation by the Federal

Election Commission permitted certain financial assistance by the host state and city, as well as by corporations, but corporations were forbidden to make direct contributions for this purpose (except that they could donate to host committees).

The law provides now for direct allocations, through a Presidential Election Campaign Fund, to presidential candidates who are nominated by their parties, to be spent in the postconvention period. The major party candidates (those who receive 25 percent of the votes cast in the preceding presidential election) would secure full funding. This amounted to $61.8 million each in 1996. This would be adjusted each election year to the cost-of-living index. No public funding would be provided unless the presidential candidates agree to use only such public funds in the campaign, thus eliminating direct private contributions to presidential campaigns in the general elections period. Minor parties who received 5 percent or more of the vote in preceding presidential elections would be entitled to receive funds after the election as well as qualify for funds in the next election. Perot thus received $29 million in 1996 because he received almost 19 percent of the 1992 vote. The proportion of the maximum amount such a presidential candidate could receive is based on a formula that relates the vote of the minor party to the average popular vote in the presidential elections of the Republican and Democratic parties. After the election, minor party candidates who did not previously qualify for funds could apply and receive the public funds they were entitled to under this formula (if they had received 5 percent or more of the votes). The public funds received by major and minor parties also constitute spending limits for the campaign, except that there is also an allowance for the national committees to spend a limited amount from *party funds* for the presidential campaigns ($12 million for the Republican and Democratic National Committees each in 1996).

Administration of National Party Finance Regulations

The new agency administering the election law is the Federal Election Commission, established in 1974. After a controversial history, it now consists of a bipartisan body of six members appointed by the president and confirmed by the Senate. The members must terminate outside business interests one year after joining the commission. A major effort has thus been made to make it an independent agency. The 1976 amendments gave the commission additional powers to prosecute violations of the campaign finance law and generally increased its monitoring and supervisory roles. The commission aims at voluntary compliance with the disclosure requirements of the law, although it has the responsibility to take action to secure compliance on the basis of sworn, written complaints. Processing and analyzing the disclosure reports present a major task. While the commission initially ran into opposition from Congress, and its earlier regulations (concerning congressional "slush funds" and filing requirements) were rejected by Congress, it has acquired considerable legitimacy now. The FEC has the responsibility to handle civil cases. It works with the Department of Justice in handling criminal violations.

The criminal penalties under the law are considerable. They provide that knowingly violating the campaign law (involving sums over $1,000) is punishable by a one-year jail sentence and a fine of $25,000 or three times the amount of the contribution or expenditure violation. Civil penalties are fines of $5,000 ($10,000 if knowingly committed), or twice the amount involved in the violation of the contribution or expenditure limits. The 1974 act provided also that a candidate who failed to file reports could be excluded from candidacy for the term of that office plus one year; this is no longer true under the present law. In addition, there is still on the books the prohibition on the use of promises of public appointments, contracts, or other benefits for the purpose of securing a person's support for a candidacy — fines of $1,000 and/or one year prison terms are the minimal penalties provided.

INCREASING COSTS: A MAJOR CONCERN

The first major concern, for some people but not for all, is the sharp increase in the amounts spent on all campaigns. The historical trajectory is interesting to keep in mind. Abraham Lincoln spent $100,000 in 1860, Theodore Roosevelt $2 million in 1904, Herbert Hoover $6 million in 1928. Their Democratic opponents did not quite match these amounts: Stephen Douglas $50,000, Alton Parker (1904) $700,000, and Al Smith (1928) $5 million.[10] By 1972 it was beginning to be quite a different story — Nixon spent $62 million, McGovern $50 million. And in 1976, $160 million was spent on presidential campaigns alone.[11]

The basic overall trends in campaign spending are presented in Table 12.2. Alexander calculates that we spent $4 billion on all campaigns in 1996. The big jump, however, was actually in the eighties — a 56 percent increase in constant (adjusted) dollars between 1980 and 1988. Compared to that increase, the 12 percent jump from 1988 to 1996 is small. Similarly, if one looks at the total costs for presidential candidates, one notes that again it was the increase from 1980 to 1988 that was huge (30 percent) in contrast to the 4 percent increase since 1988.

One must remember that the 1974 law set specific limits on presidential spending, which limits were increased from time to time because of cost of living

TABLE 12.2 ▪ The Increase in Campaign Costs, 1960–1996 (in millions of dollars)

Total	*1960*	*1972*	*1980*	*1988*	*1992*	*1996*
All Elections						
Actual Dollars	175	425	1,200	2,700	3,220	4,000
Constant Dollars (1960)	175	301	431	675	679	761
Presidential Elections						
Actual Dollars	30	138	275	500	550	700
Constant Dollars (1960)	30	98	99	127	118	132

SOURCES: Herbert E. Alexander, "Financing the 1996 Elections," in *America at the Polls 1996* (Storrs, CT: Roper Center, 1997), 142 and 145; Paul Beck, *Party Politics in America* 8th ed. (New York: Longman, 1997), 273.

TABLE 12.3 ▪ Changes in Major Party Presidential Spending Legal Limits (in millions of dollars)

Year	*Preconvention*	*Convention*	*Postconvention*
1976	13.1	2.2	25.0
1980	17.7	4.4	34.0
1984	24.2	8.1	47.3
1988	27.7	9.2	54.4
1992	33.1	11.0	65.5
1996	37.1	12.4	73.8

SOURCE: Roper Center, *America at the Polls 1996*, 144.

adjustments or congressional actions. These imposed limits assumed there would be no sudden skyrocketing of expenditures, at least insofar as formal prescribed limits are concerned (Table 12.3). There were, however, changes in allowed spending from 1976 on — indeed, fairly sizable increases.

While the candidates and parties adhered rather strictly to the prenomination and convention limits (convention expenditures allowed were increased to $3 million for costs not met under the earlier limits), the amounts spent after the convention by no means were controlled by the $73.8 million allowances for each party. Alexander's calculations are that the two major parties in 1996 spent over $350 million after the convention, and the minor party candidates another $30 million.[12] If we add the $60 million for the two conventions and the $250 million for the prenomination period, we reach the $700 million figure as the total presidential campaign costs.

The specific expenditures by the presidential candidates who actually announced (there were seventeen) reveals how far we go in public support for these contenders (see Table 12.4). Eleven received public funds, ranging from $350,000 to over $13 million. Steve Forbes accepted no public funds and spent almost $42 million, primarily from his own funds. Perot, the FEC ruled, could qualify for $29 million for the general election campaign, based on his 1992 vote. This he decided to accept (but it was not included in his preconvention receipts or disbursements). It is important to note that nine candidates other than Clinton and Dole were able to mobilize over $150 million to compete in our presidential sweepstakes in order to contest against the two front-runners, four of them challenging, at least financially, fairly successfully. As to whether these 1996 figures are high compared to the past, one should note that the 1988 preconvention battle in both parties produced an expenditure level almost as high: $213 million.

A major question posed of late is whether the presidential campaign fund will in the future be able to support candidates at these levels. The fund, created in 1971, is replenished by the income tax checkoff system that permits taxpayers to "give" three dollars each year from the federal treasury to this fund (it was originally one dollar). In the early days a fairly large percentage of taxpayers used the checkoff, 28.7 percent in 1980. But by 1996 the percentage dropped to 12 per-

TABLE 12.4 ▪ Presidential Prenomination Costs in 1996 (in millions of dollars)*

Candidate	*Total Disbursement*	*Individual Contributions*	*Federal Matching Funds*	*Political Committee Contributions*	*Other*
Clinton	38.1	28.3	13.4	.04	.7
Dole	42.2	29.6	13.6	1.2	.3
Forbes	41.7	4.2	0	—	.03
Buchanan	24.5	14.7	9.8	.02	.01
Alexander	16.4	12.6	4.6	.3	.1
Gramm	28.0	15.9	7.4	.4	5.2
Lugar	7.6	4.8	2.6	.14	.19
Specter	3.4	2.3	1.0	.16	.04
Taylor	6.5	.04	0	0	.004
Wilson	7.2	5.3	1.7	.2	.1
Perot	8.0	.8	0	0	0
Total — All 17 Candidates	234.1	126.5	55.95	2.5	6.7

* The overall spending limit for the pre-convention period if a candidate accepted public funding was $37.1 million.

Note: Candidates not listed are LaRouche, Dornan, Keyes, Browne, Hagelin, and Lamm.

SOURCE: Federal Election Commission, August 31, 1996. See Anthony Corrado, in Gerald Pomper, *The Election of 1996* (Chatham, NJ: Chatham House, 1997), 138–9.

cent. And the FEC has warned that for the next presidential election there may not be enough in the presidential campaign fund to pay all costs, particularly if, as predicted, the election in year 2000 will have many contenders entering from both parties, as well as possibly, again, third-party candidates. The payments from this fund have increased from $71 million in 1976 to $234 million in the last election. This is a matter Congress may have to consider unless public support for the tax checkoff increases.

Are these costs too high? There are two sides to this argument. There are those who deplore these escalating costs and argue that they are much higher than necessary, even illegal. *Newsweek* columnist Robert Samuelson, however, argues that we are "criminalizing" the use of money in politics, that "politics requires money" if people are really to compete and limiting what can be spent would be "curbing free expression." Further, he points out that the $4 billion spent in 1996 was only one-twentieth of 1 percent of the GDP ($7.6 trillion). Above, all he claims, "there hasn't been much evidence of serious influence buying." The "reformers," he says, "have manufactured most of the immorality."[13] The Alexander Task Force on Campaign Finance Reform also is inclined not to be worried too much about the aggregate amounts of spending. They "take the view that money as such, is not evil, and that it buys necessary campaign organization and campaign communication which can help inform the electorate. . . . competitive elections require that candidates communicate with voters, that (this) costs money. . . . We must make sure that the overall level of campaign funding is high enough to support that domestic dialogue."[14] The task force actually recommends raising certain contribution and spending limits, abolishing others, and above all making sure that challengers have a chance for adequate funding.

THE COST OF CONGRESSIONAL CAMPAIGNS

Running for the U.S. Congress has become more expensive also. These costs have increased from a total spent for House and Senate races of $33 million in 1960, to almost $100 million in 1976, to $653 million in 1996. In constant dollars, this represents an increase of 160 percent in the past thirty years. The data for the most recent years (Table 12.5) reveal the great leaps in expenditures in 1992, in 1994 when the Republicans won both houses of Congress, and in 1996 when yet a new high was achieved. It is in these last elections, 1994-1996, that the Republicans have been more dominant in mobilizing support for their congressional candidates.

The average expenditure for House candidates in 1996 was $516,219, compared to $295,602 in 1986; for the Senate the average has increased from $2.7 million to over $3.5 million. The Republican average for the Senate has been higher almost every year, but for House candidates this became so only in 1996.[15] We have come a long way from the days of the seventies, when the total expenditures for House and Senate were much lower. Even in 1980 for Senate and House seats the Democrats reported a total of only $125 million, the Republicans $117 million.

The status of the House candidate — whether an incumbent, a challenger to an incumbent, or a candidate in an open race — reveals great variance in campaign expenditures (Tables 12.6 and 12.7). It is obvious, first, that challengers have the greatest difficulty in mobilizing money and thus have only one-third to 40 percent of the funds available to incumbents and those running in districts with open seats. However, one notes that challengers have been doing much better in securing

TABLE 12.5 ▪ Campaign Expenditures for House and Senate Over the Years (in millions of dollars)

	Midterm Years			Presidential Years		
Expenditures	*1986*	*1990*	*1994*	*1988*	*1992*	*1996*
Total	400.0	403.7	615.4	408.3	528.3	652.6
Democrats	193.8	214.9	301.7	220.0	285.5	303.5
Republicans	206.1	187.8	309.5	187.2	238.3	348.1

Note: These are expenditures for both primary and general elections, for general election candidates.

SOURCE: Roper Center, *America at the Polls 1996*, 149. The data were taken from Federal Election Commission reports, by Rockefeller Institute of Government at the State University of New York and from the Center for Responsive Politics. See also the American Enterprise Institute's *Vital Statistics on Congress, 1995-1996*, edited by Norman J. Ornstein, Thomas E. Mann, and Michael J. Malbin; cited in Roper Center, *America at the Polls*, 1997.

TABLE 12.6 ▪ Average Expenditures for House Candidates by Status of Candidate (overall average)

Status	*1996*	*1994*	*1992*	*1990*	Percent Increase *1990–1994*	Percent Increase *1992–1996*
Incumbent	678,556	561,441	594,699	422,124	33	14
Challenger	286,582	240,183	167,411	134,465	79	71
Open seat	647,336	585,991	588,722	443,129	32	10

SOURCE: Adapted from Roper Center, *America at the Polls 1996,* 150.

funds in recent years than have other candidates — an increase of over 70 percent since 1990, far better than incumbents or open seat candidates (Table 12.6). Of particular significance is that this improvement of financial support for challengers occurred in both parties, but at different elections. The Republicans demonstrated a phenomenal increase in 1994 (104 percent compared to 35 percent for the Democrats), linked no doubt to the Republican success in electing seventy freshman Congress members that year. Yet as significant is the Democratic recovery in 1996, an increase of 124 percent over 1992 (80 percent over 1994), while the Republican average actually decreased by 13 percent 1992–1996 and 1994–1996. This probably helped the Democrats in their defeat of twelve Republican 1994 first-termers (Table 12.7). As Corrado points out, however, in his analysis of the financing in 1996, most of the Republican 1994 first-termers saved themselves in 1996 by very successful early solicitation of funds, besting their Democratic challengers by over $400,000 in the average in amounts contributed. He points out that these first-term Republicans, on arriving in Washington in early 1995, were warned and urged to begin fund-raising at once and were aided by Gingrich and other Republican leaders in securing early PAC commitments. As a consequence, the Republican 1994 first-term incumbents had by the end of 1995 already raised five times more than their Democratic challengers, over ten times more from PACs.[16]

There thus seems to be some reality to the observation that the capacity of challengers to raise additional funds, and the capacity of incumbents, especially recent incumbents, to compete effectively in fundraising with their challengers, is

TABLE 12.7 ▪ Percentage Change in Average Expenditures for House Candidates, by Party and Status

	Percent Change				Dollar Basis (in thousands of dollars)			
	1990–1994		*1992–1996*		*1990*		*1992*	
Status	*D*	*R*	*D*	*R*	*D*	*R*	*D*	*R*
Incumbent	46	14	-5	35	427	414	622	553
Challenger	35	108	124	-13	131	134	144	276
Open seat	2	14	15	5	548	538	561	617

SOURCE: Adapted from Roper Center, *America at the Polls 1996,* 150.

associated with individual candidate successes and failures, as well as with potential and actual party turnover. There remains some controversy over this question of who is benefited the most.[17] The reality still is that challengers usually raise much less than incumbents.

THE ROLE OF CITIZEN CONTRIBUTIONS

In all the years of campaign spending, individual citizens have been a major source of the money that candidates and parties receive and spend. The percentage of American adults who report contributions to a party or candidate has fluctuated over the years, but it has rarely been below 8 percent. In the 1960s it was over 11 percent; it was over 10 percent in 1972 but in recent years has declined: 1980 (8.0 percent), 1988 (8.7 percent), 1992 (6.6 percent), but 1996 (7.9 percent).[18] While these percentages may seem small, if one calculates them as a percentage of the adult population, the total number giving may be surprising. The number of contributors has increased: calculated at three million in 1952, twelve million in 1972, and possibly close to sixteen million today.[19] And these individual contributions are very important, both for presidential candidates in qualifying for public funding (needing contributions of $250 or smaller in twenty states), as well as for congressional candidates, for whom individual giving is the backbone of their fund-raising operations. In 1992, House candidates raised $310 million, 49 percent of which came from individual contributions; Senate candidates that year raised $181 million, 66 percent of it coming from individual contributions. In 1996, of the $220 million raised by Senate candidates, 63 percent came from individual contributions; of the $440 million raised by House candidates, 55 percent came from individual gifts.[20] Remember that federal contribution limits are $1,000 per person per candidate. Few people can give that amount, or even $500. As a matter of fact, an unpublished study found that since 1972 the average number of $1,000 gifts in each presidential campaign was only 4,308.[21] The implication is that amassing the amounts needed in these federal elections requires a tremendous effort and a tremendous response from a very large number of people making small contributions.

THE BIG MONEY

While direct individual contributions are of great importance, for some time they certainly have not been sufficient. Even the regular contribution limits under the 1974 act permitted those with big money and motivated to spend it on the campaign to do so. Remember that individuals could give $5,000 to a PAC and $20,000 to a party committee per year. PACs, in turn, could give $5,000 per election to a candidate and $15,000 to a party committee. Even in the past, corporation executives could legally give much money personally in these ways to candidates and party committees, by encouraging family members and friends to donate to many candidates and party committees. Theoretically, of course, the individual overall limit was $25,000 a year. Yet this was not enough. And the parties found they could

"Big Money" in Campaigns: Much Reform Action but Little Change

A recent Speaker of the California State House of Representatives, Democrat Jesse Unruh, has provided American politics with a major dictum when he coined the phrase "money is the mother's milk of politics." Such a phrase highlights the role of special interest groups and their "big money" in American politics. In fact, this "special interest money system," argues Brooks Jackson, "aims to buy access to and influence over lawmakers (and) . . . is taking the place of votes and diluting the birthright of ordinary citizens: government for and by the people." Hence, there has been a continuing impulse to reform campaign finance in the American political process. Yet despite reform, the role of money has remained constant.

Frank Sorauf in his book, *Inside Campaign Finance,* agrees that in the initial years of party politics, "the parties dominated finance because they dominated the campaigns themselves . . . much of the party-centered campaigning needed no cash, it rested heavily on services volunteered or bartered for some party-controlled favor . . . when cash in large sums was needed, the parties went to men of wealth, the storied fat cats of party and campaign lore" (2-3). Later, when parties declined somewhat, campaigning became more specialized and expensive. In this changed political context, big money and special interest groups became critical, and so did the need for more reform.

There are many instances of expensive campaigns in 1992. Third-party candidate H. Ross Perot spent nearly $73 million of his own money to run for president. In 1968 when George Wallace ran on his third-party ticket without having his own money, he relied on Texas oil billionaire Bunker Hunt to secretly bankroll his efforts. In the summer of 1968, Hunt "sent his lawyer to Montgomery to meet with . . . [the Wallace people] at the Jefferson Davis Hotel. Hunt's representative brought with him a large briefcase filled with hundred-dollar bills; in later years, the total was put at $250,000 to $300,000 dollars."

To stop the infusion of big money into campaigns has been virtually impossible. We have tried various reforms to little effort. Sometimes candidates refused big money, and some have been elected despite this. Two Wisconsin senators illustrate this well. Senator William Proxmire worked out an agreement with his campaign rival(s) not to spend more than $25,000 dollars. Senator Russell Feingold won reelection in Wisconsin in 1998 even though he refused big money and PAC contributions.

There have been many investigations into the role of big money and many proposals for reform. The Alexander Task Force 1997 Report is discussed in this chapter in detail. Another effort is that of the Pew Foundation, which created a task force in 1998 chaired by Professor Larry Bartels of Princeton. The Twentieth Century Fund investigated the problem in 1984. We should not forget that already in 1907 we passed a national law that forbade corporations to contribute directly to political parties, seeking thus to control the fat cats and big money. But today after the many task forces and more legislation, and investigations following the 1996 campaign, we are essentially still faced with this same basic problem that has confronted us for almost a century.

SOURCES: Brooks Jackson, *Honest Graft: Big Money and the American Political Process,* updated ed. (Washington, DC: Farragut Publishing Company, 1990).

Frank Sorauf, *Inside Campaign Finance* (New Haven: Yale University Press, 1992).

Dan T. Carter, *The Politics of Rage: George Wallace, The Origins of the New Conservatism, and the Transformation of American Politics* (Baton Rouge: Louisiana State University Press, 1995).

Report of the Task Force on Campaign Reform, *Campaign Reform: Insights and Evidence* (Princeton: Woodrow Wilson School of Public and International Affairs, September, 1998).

use the liberalized interpretation of the law by the courts, and to some extent by the FEC, to raise much more money.

In the seventies, Nixon's personal attorney went to prison for selling an ambassadorship for $100,000, and twenty-one corporate executives at that time were convicted for illegally contributing corporate funds of $100,000 each. These were violations of the law. Today, big money is mobilized openly, but with more finesse. On January 24, 1996, the Republican National Committee raised $16 million in one night in a grandiose fund-raiser in Washington, DC. One investment banker boasted he sold $2 million worth of tables. He said, "These people buy into the agenda because it's going to create the kind of atmosphere in the country that's going to allow them to prosper."[22] In May the Democratic National Committee raised $12 million with its fund-raiser, a "National Presidential Gala." The audience included representatives of corporations, listed as contributors, many lobbyists, labor unions, and wealthy persons. "It was a situation ripe for scandal," says Brooks Jackson.

Both parties offered special access to big donors. The Republican chair promised that those raising $250,000 would get lunch with Bob Dole and Newt Gingrich; $45,000 would only lead to breakfast with Gingrich. The Democrats promised each donor of $100,000 four "annual events" with the president and vice president, plus annual retreats with party leaders. Both parties were taking in much more money than in 1992, and the big increases in giving were in "soft money," often from corporations and labor unions, which were essentially unregulated. The "nonfederal" soft money in 1996 amounted to $89 million for the Democrats and $94 million for the Republicans, more than three times the amounts of such money raised in 1992. The Democrats transferred 56 percent of this money to the state and local parties, the Republicans 46 percent. Corporations giving to the Republicans included Philip Morris (over $2 million), Nabisco (almost $1 million), Atlantic Richfield, Brown and Williamson Tobacco, U.S. Tobacco, Chevron, General Motors, and Amway. Corporations giving to the Democrats were Joseph Seagram ($1.1 million), Walt Disney ($866,000), Communication Workers of America ($770,000), MCI, AFSCME, and AT&T. These large gifts to the national parties were used to help the party and its candidates in a variety of ways. Watergate seemed to pale in comparison to this explosion of fund-raising. But with Watergate there was a coverup; the 1996 extravaganza was more in the open, though not completely.

Unique to 1996 was the foreign money scandal. It began with the discovery by the *Los Angeles Times* of a $250,000 contribution to the Democrats that was traced to a South Korean bank through a Los Angeles company. The DNC had to return the money. Then it was found that an Indonesian couple grateful to Clinton for receiving them, gave $450,000 from funds originating outside the country, an illegal act. One person, John Huang, a DNC fundraiser connected to an Indonesian conglomerate, was responsible for this and other illegal contributions. A $140,000 fund-raiser held at a California Buddhist temple was also implicated. In all, the Democrats had to return more than $3 million in such foreign donations. Dole also had his troubles, but not with foreign contributors. One of his financial vice chairs was fined $6 million for securing $120,000 in illegal corporate donations to the Dole primary campaign.[23]

Lack of access to big money, some people feel, is one of the problems would-be candidates face. To run for president, at least $30 million is necessary ($37 million was allowed in 1996) for the long period before and during the primaries. Studies show that only a small number of Americans will give $1,000. One must search for small contributors, but it is essential to tap a seam of bigger donors. Brooks Jackson thinks that Lamar Alexander lost the 1996 Republican presidential race because his money ran out. He couldn't even raise $20 million, whereas Forbes spent over $37 million of his own money and Dole raised over $30 million. Alexander ran out of money when his campaign was gaining on Dole.

The other issue concerning big money is whether it really purchases access and even influence. There is no doubt that big money gained access — to the White House (even to the Lincoln bedroom) and to many parties and dinners with Democratic leaders. The big givers to the Republicans also certainly were given access — $45,000 got you breakfast with Gingrich; $250,000, and you were joined by Dole. Raising money with a guarantee of particular types of access to leaders, while legal, is certainly "unseemly," as one Senator put it. Whether this leads to meaningful influence is another matter. Most observers deny it and debunk it. It is difficult to link contributions to leaders' actions.

There is no doubt that money may help to keep certain "interests" more salient than others in the minds of lawmakers, to predispose them to support such interests, and to communicate more frequently with them. And occasionally it may be more productive. As one big giver put it, "these people buy into the agendas," which one can interpret in many different ways! As the Supreme Court put it, "money is speech," and "speech" can be productive. An example that emerged in the post-1996 investigation concerns the case of Roger Tamraz, who was promoting a plan to build an oil pipeline in Central Asia. To get support for this, he gave $300,000 to the DNC. He was invited to six social functions at the White House, but no private meeting with Clinton, and no government support for his venture. When asked why he did it, he said "access," which he said to him meant that his rivals, including major oil companies opposing him, knew he had "it.""Access" is its own reward. When asked if he got his money's worth, he responded, "I think next time I'll give $600,000."

THE ROLE OF PARTIES IN CAMPAIGN FINANCE

The role of parties has been a big issue for years. In the past, political scientists complained that the way we regulated the solicitation and use of money in campaigns resulted in the weakening of parties. PACs were facilitated, but parties were not. National party committees were limited under the Hatch Act (1940). And under the public funding provisions of the FECA of 1974 the funds were given to candidates, not to the parties. The argument was that we should change the law in order to give parties a larger role, as mobilizers and controllers of campaign funds.

With the new court decisions in the Buckley and Colorado cases, plus congressional amendments to the 1974 act, that has changed. The parties (national, state, and local) in recent elections are playing a more significant role. The reason is that parties are no longer limited by the restrictions imposed by the 1974 act (i.e., the specific limits on direct contributions to party committees and the requirement that expenditures by party committees be from such "hard money"). Court decisions now allow parties to raise "soft money" (money donated by corporations, unions, other groups, and individuals in unlimited amounts for a variety of purposes). The court decisions permit these "soft money" funds to be spent by the parties in a variety of ways: for transferal to the states and local organizations for "party building" and voter-registration drives; for joint federal and nonfederal campaigns on behalf of the party; for fund-raising and administrative costs (especially to raise more "hard" and "soft" money); and for direct support for local and state candidates.

Two new developments that use soft money should be noted. One has come to be called "independent expenditures," permitted by the Colorado court decision. Parties under this decision can set up new divisions in their organizations (called by the Republicans the "I-team") which are legally distinct from the regular organization, and whose function is to spend unlimited funds on behalf of candidates. This is allowed as long as it is not spent in "coordination" with a candidate on his or her campaign organization. The national senatorial and congressional campaign organizations could, and did, set up such operations. This new development, in effect, meant that FECA limits were no longer relevant.

The other 1996 innovation was "issue advocacy ads." These are political ads used by an individual or political group or party that express support for a program, or policy, or issue, which may be related to a candidate but which does not "expressly" advocate "the election or defeat of a particular, identifiable candidate" (the FECA description). Both parties participated heavily in such "issue" campaigning in 1996. It is claimed that the initial stimulus for such ads was the strategy of the AFL-CIO, which used such ads in its attempt to defeat Republican candidates in 1996. Early in the campaign it announced its intention to spend $35 million to influence congressional elections, most of it on grassroots programs. The money would come from union treasury funds, defended because this effort would be called a program to "educate" the public about issues important to working people. Both parties decided to use this issue ad strategy in the presidential election campaign after the primaries and in congressional campaigns.[24] In the final analysis, the money spent on the ads, over $40 million estimated for the Democrats, and probably a similarly high amount for the Republicans, was not prohibited by, or counted against, FECA limits.[25] The parties thus enhanced their financial participation in the campaign considerably.

Financial reports on party spending in 1996 reveal the tremendous increase in the role of parties (Table 12.8). In both "hard money" and "soft money" expenditures there has been a considerable rise. But the increase in "soft money" expenditures, for the purposes described earlier, are truly staggering — over 200 percent increases for both parties. The national party organizations have, as a result, worked with states and local parties to strengthen the party role at all levels.

TABLE 12.8 ■ The Increased Financial Involvement of the Party Organizations (in millions of dollars)

Type of Transaction	*1996*	*Percent Increase over 1992*
A. "Hard money" expenditures by the three national committees		
Republicans	400.0	58
Democrats	198.6	35
B. "Soft money" spent by the three national committees		
Republicans	149.0	224
Democrats	117.5	257
C. "Soft money" transferred by national parties to state and local parties		*1992 Amount Transferred*
By Republicans (RNC)	47.8	5.3
By Democrats (DNC)	53.9	9.5

SOURCE: Anthony Corrado, in Pomper, *The Election of 1996* (Chatham, NJ: Chatham House, 1997), 151–53.

This development, associated with the innovation in types of expenditures which have emerged recently, is now a matter of some controversy. The Task Force on Campaign Finance Reform debated this issue and recommended "removing all limits on the amount that parties may contribute to or spend in support of any of their candidates." This resulted in a dissenting opinion by three scholars on the task force, who feared this would result in party organizations operating as "financial holding companies," as well as result in other problems, including the problem of getting adequate disclosure. The task force did recommend abolishing "soft money" and also reforming the use of monies for "issue advocacy" as well as "independent expenditures." But in the last analysis the task force spoke out in favor of the role of parties, while the dissenters wished to reduce the parties' role.[26]

The overall role of parties in campaign finance has assumed a greater importance, as Table 12.9 reveals. The amounts have increased from $206 million in 1980 to $881 million in 1996, a fourfold jump. The 1992 period saw a particularly large increase. The Republicans continue to do much better than the Democrats in the aggregate, but the proportions have changed: in 1980 the $37 million the Democrats raised was 18 percent of the party total; in 1996 it was close to 40 percent. Overall, congressional spending (Part B of Table 12.9) is surprisingly low in relation to the total party effort. Sorauf noted this in his study with Wilson in 1990, commenting that the money congressional candidates received from the parties ranged from 6 percent to 20 percent. Actually, in 1996 it was closer to 5 percent. As Sorauf noted then, one must remember that we are reporting only the dollar receipts and contributions of parties to congressional candidates. The services the parties provide to their congressional candidates in office facilities, supplies, access

TABLE 12.9 ▪ Increase in Party Participation in the Financing of Campaigns (in millions of dollars)

Type of Transaction	*1980*	*1984*	*1988*	*1992*	*1996*
A. Total contributions and "coordinated" receipts of the major parties					
Democrats	37.2	98.5	127.9	192.5	332.3
Republicans	169.5	297.9	263.3	316.0	548.7
Total	206.7	396.4	391.2	508.5	881.0
B. Spending by parties in congressional elections (in millions of dollars)					
Democrats	3.0	7.8	11.2	19.7	17.9
Republicans	12.2	17.9	17.8	26.4	22.6
Total	15.2	25.7	29.0	46.3	40.5

SOURCES: Frank J. Sorauf and Scott A. Wilson, 1990, "Campaigns and Money: A Changing Role for Political Parties?" in L. Sandy Maisel, ed., *The Parties Respond* (Boulder, CO: Westview Press, 188, 1990), 190; Roper Center, *America at the Polls 1996*, 154. Based on data from the Federal Election Commission.

to polls, research, and assistance in planning campaigns, which are not necessarily included in the figures used here, may rationalize this gap somewhat. We know now that the national parties are working more closely than ever with congressional candidates and state and local parties, providing assistance which may not show up in these monetary calculations.[27]

THE PACs AND CAMPAIGNS

Political Action Committees have existed for over forty years as major actors in our campaigns. Created by unions, corporations, and many other types of groups with both particular and general political objectives, funded by voluntary contributions from individuals (or other groups), they seek to channel money to their favored targets among the parties and candidates. They thus supplement and channel in a major way direct individual contributions. And although there was at one time deep concern about the role of PACs, in fact the 1974 FECA was relatively tolerant and generous toward them, allowing $5,000 contributions by PACs per election directly to candidates, whereas individuals were limited to $1,000. Since the court decision opening up campaign funding by allowing independent expenditures and "issue ads," the role of PACs has broadened considerably. The number of PACs has grown from about 600 in the early seventies to 4,430 registered in the 1996 election. Most of their contributions go to congressional campaigns, very little to presidential. The increase in funds given to House and Senate candidates has jumped dramatically over the years, as the following data reveal.

PAC Contributions to House and Senate Candidates over Time

Year	*Total Congressional Receipts (in millions)*	*Total Receipts from PACs*	*Percent from PACs*
1974	73.9	11.6	15.7
1976	1043.8	20.5	19.6
1994	740.6	178.5	24.1
1996	659.6	192.0	29.1

One can see that both the dollars candidates mobilize and the PAC dollars given to them have increased in tandem, and quite impressively.[28] However, the percentages, while also increasing, have not even reached 30 percent yet. PACs thus continue to play a smaller role than parties and individual contributions, despite all our concern about them.

The history of PAC involvement in campaigns goes back to the 1930s when labor unions contributed to the Democratic campaigns, as the corporations had earlier, before forbidden to make direct contributions by the 1907 act. The unions were prohibited from such contributions for the war period by the Smith-Connally Act of 1943 and then by the Taft-Hartley Act of 1947, which forbade "any corporation whatever or any labor organization to make a contribution or expenditure" related to any aspect of the national electoral process. Labor unions then began to set up political action and political education committees using voluntary contributions from members, but carefully avoiding using union treasury funds. Alexander Heard has documented the early involvement of labor PACs: in 1952 they disbursed $1.8 million and in 1956 $1.7 million. He compared these amounts to those which the officers and directors of the one hundred largest corporations personally gave to national candidates in those years ($1.0 million in 1952 and $1.9 million in 1956).[29] Over 90 percent of the corporate money went to Republicans, while the Democrats received most of the labor money. These labor funds from PACs increased constantly in the sixties to $5 million by 1968. These PACs were reorganized and accepted by the 1974 FECA. The big increase in the number of PACs occurred between 1974 and 1976 — a 63 percent increase, and then another 95 percent increase by 1978. This was due primarily to the establishment of PACs by corporations, trade associations, cooperatives, and other non-labor groups. Between 1974 and 1978, corporate PACs increased 55 percent, unions only 5 percent, and other groups 45 percent.[30] It was clear in those earlier days that, as the chair of the U.S. Chamber of Commerce put it, "The prevailing attitude is that PAC money should be used to facilitate access to incumbents."[31] The basic goal was then, as it is today, to try to establish a financial association, if not a relationship of influence, with those already in power, not necessarily to expend money on those aspiring to power.

PAC contributions to congressional candidates vary considerably by their status (incumbent, challenger, open seat). We noticed earlier that challengers generally fare worse than the other two types in their capacity to secure funds (Table 12.6). How do the contributions from PACs fit into this picture? Normally PACs have given much more money to incumbents. This was true in 1978 (PACs gave 57

TABLE 12.10 ▪ Proportionate Contributions of PAC Money to Congressional Candidates by Type of Candidacy, 1995–1996 (in percent of total PAC donations)

Congressional Chamber	*Incumbent*	*Challenger*	*Open Seat*	*Democrats*	*Republicans*	*Totals for House and Senate*
House	58.6	10.6	9.1	38.9	40.0	78.7
Senate	9.4	3.2	8.7	7.8	13.3	21.3
Both houses	68.0	13.8	17.8	46.7	53.3	100.0

Note: The total PAC dollar contributions to House and Senate candidates for the two major parties was $191.5 million. PACs also gave to "other parties" a total of only $500 to Senate candidates and $728,663 to House candidates.

SOURCE: Adapted from Anthony Corrado, in Gerald Pomper, *The Election of 1996* (Chatham, NJ: Chatham House, 1997), basic tables, 157, 161.

percent of their $34 million to incumbents) and even more so in 1996 (68 percent of $191.5 million).[32] The actual proportion of PAC money that went to different types of congressional candidates in the last election campaign (1995-96) are presented in Table 12.10. From this one can see that the House incumbents are the real beneficiaries (58.6 percent). And even House challengers do slightly better than Senate incumbents. We can emphasize these differences by presenting the data for 1995-96 a different way.

1996 PAC Contributions for Both Major Parties

	Total Receipts (in millions)	*Total PAC Money (in millions)*	*Total PAC Money for Incumbents (in millions)*	*Percent of Total PAC Money Given to Incumbents*
House	$436.2	$150.8	$112.9	74.8
Senate	218.0	40.7	18.1	44.4

One should note also that the two parties share PAC money fairly evenly overall, but the Republicans recently have been doing better than the Democrats, a clear reversal of the earlier trend when Democrats outdistanced Republicans considerably. (See Table 12.11.)

This shift in PAC contribution patterns occurred in 1995. Apparently the Republican House leadership let interest groups and lobbyists know that the Republican expectations for PACs would change, now that the Republicans had control of both chambers. A memo was reportedly sent to the PACs, business PACs in particular, on behalf of the new Republican leadership, which warned that "legislative interests of those who help the congressional enemies of business and don't support our friends, could be affected negatively. . . . if ever there was a time for the business community to help its friends, it is now."[33] A shift did indeed take place. A big dinner was held by House Republicans who invited their lobbyist and PAC friends to come and pay their respects to the new leadership. It was called "The New Majority Dinner." It was rated "the most successful event in the 129-year history of the National Republican Campaign Committee." The results were soon apparent in 1995, during which Republican incumbents raised $26 million in PAC money compared to $11 million in 1993 (while the Democrats slumped from $25.5 in 1993 to $18.5 in 1995).[34] A final summary comparison of what happened in the reversal of PAC giving between 1994 and 1996 is presented in Table 12.11. The decline in PAC giving to Democrats is clear, for both House and Senate, while PAC giving to Republicans increased 75 percent for their House candidates. Incumbents still got most of the money in 1996, but the incumbents were even more Republican than Democrat. Corporate and professional PACs gave over 80 percent of their money to incumbents, while the unions gave 63 percent to incumbents.[35]

PAC money was also important in the congressional contests between the seventy-one Republican first-termers (elected in 1994) and their Democratic challengers. The Democrats were able to defeat a few of these Republicans (twelve) and PAC money may have helped the Republicans in winning some of the others. The early fund-raising by the Republicans in 1995 led to a huge margin in PAC funds for Republicans — an average of $139,000 for the seventy-one Republican first-termers compared to an average of only $12,500 for the Democrats — by the end of 1995.[36]

What is the impact of PAC money? The labor leaders argued that their large increases in campaign spending were effectual. Some of their targeted Congress members were defeated (four in particular), but more importantly, they felt that their issue ad campaigns had pushed Republicans to accept a minimum-wage increase and to back away from their desire to cut Medicare and Medicaid. On the other hand, Republicans protested the extent of labor's involvement in the campaign and said they would press for campaign reform so that union dues would no longer be used without permission in campaigns.

On the larger question of whether PAC money means not only access but actual influence on legislative policy decisions, the evidence is not convincing. The very careful study by Grenzke of 172 House members and 120 PACs concluded that PAC giving by itself was not closely linked to legislative behavior. PAC money is often given in the hope of encouraging a representative to support a particular

TABLE 12.11 ▪ The Party Shares of PAC Money as They Changed

	Democrats		Republicans	
PAC Money	*House*	*Senate*	*House*	*Senate*
1993–94	88.1	23.7	43.7	23.0
1995–96	74.4	15.0	76.3	25.7
Percent change	–16.0	–36.7	+74.6	+11.8

SOURCE: Anthony Corrado, in Gerald Pomper, *The Election of 1996* (Chatham, NJ: Chatham House, 1997), 157,161; Jackman, in Sabato, *Toward the Millennium: The Elections of 1996* (Boston: Allyn and Bacon, 1997), 250–52; Federal Election Commission Reports.

policy and also to maintain a sympathetic position revealed earlier. The causal relationship is thus not clear: Does money induce legislative behavior, or does behavior attract money? Grenzke quotes from interviews with PAC leaders that indicate most do not delude themselves. They admit "our contributions are too small to change a vote," or "they [Congress members] behave as they would anyway, and the money comes after," et cetera. She concludes from her rigorous statistical analysis, in which she controlled for other factors (such as the electoral vulnerability of the member, the political leanings of the district, the power of the member over the PAC's agenda, and the member's ideology) with this observation: "contributions (by PACs) generally do not maintain or change House members' voting patterns." What PACs do is get access, which may be important for them. And she sees "access" as a "hearing": "Money may facilitate an opportunity to present one's case, and in the absence of conflicting testimony, the member may change his or her position as a result of the meeting." But she then reveals that on the average the House members in her study received contributions from 138 PACs.[37]

PACs have become key actors in our campaign finance. They do not dominate, however. They are defended as legislative channels for funds, and for groups (and individuals within groups) to participate in the party and electoral process. Their role has been somewhat expanded of late. They do seek access and influence, but there is no evidence they have any significant control over policy. While campaign reform is again being discussed, there is no strong evidence that we are disposed to restrain PAC involvement.

CONCLUSIONS: MAJOR CONCERNS AND WHAT REFORMS ARE NEEDED NOW?

Today, after one hundred years of repeated efforts at regulating the use of money in campaigns, we seem to face the same concerns as before. The costs of our campaigns are very high, there are loopholes in the laws that allow contributions and expenditures never intended, the campaign playing field is not level for those in power and those out of power and challenging those in power, certain interests (corporations and unions particularly) are more directly involved than we ever wanted, and the affluent buy candidate status easily while able leaders cannot compete for lack of money. Money, lots of money, is indeed "the mother's milk of politics."

Despite a fairly strict set of regulations adopted in 1974 in the aftermath of Watergate, politicians aided by the Supreme Court have devised ingenious techniques and approaches to circumvent the law, to raise and spend as much money as they like, almost without limits, or despite the limits. As a consequence, in the aftermath of the 1996 election we began to realize that we now have a new, complex, technologically advanced, wide-open system of campaign finance in which everyone seems liberated to collect and spend at will, sometimes without adequate disclosure or accountability to the public. The old regulations seem totally inadequate to cope with this ingenious, even devious, system of campaign finance.

There are two ways of reacting to this state of affairs, two schools of thought. One is that our campaign finance system is in a shambles, scandalous, an embarrassment, and a threat to the integrity of the electoral process. Not only Common Cause, but also many members of Congress, have made such evaluations during the recent investigations. On the other hand, there are those who are more sanguine. They believe in the First Amendment rights of free speech and applaud the Court's dictum, paraphrased as "money is speech." They argue for very few limits on the use of money in campaigns, do not worry about the escalating costs, and oppose most legislation, except a carefully crafted public disclosure law requiring periodic reports on who collects, who contributes, and who spends money. In the words of *Newsweek* columnist Robert Samuelson:

> Money, they say, is corrupting politics. It isn't. Campaign spending isn't out of control or outlandish . . . contributions have not hijacked legislation. . . . campaign finance laws are so arbitrary and complex they invite 'criminality' or its appearance. Rep. John Doolittle of California sensibly suggests abandoning all contribution limits and enacting tougher disclosure laws. The best defense against the undue influence of money is to let candidates raise it from as many sources as possible — and to let the public see who's giving. That would be genuine reform.[38]

For another columnist, George Will, the only regulation we need is "no cash, no foreign money, full disclosure."

We have in this chapter documented certain key features and developments in campaign finance in recent years. Some of these are salutary and do not necessarily invite criticism as much as others. Many people will support the Supreme Court's position that people should be allowed to spend as much money as they want to in connection with campaigns. The increased role of national party committees in collecting money and sharing it with state and local parties can be seen as a positive development for believers in the party system. Since we expect interest groups to be involved with the electoral process, in contrast to the British system, the fact that PACs play a role in our system does not jar the sensibilities of most Americans. While the media costs are high, absorbing close to 50 percent of all candidate and party expenditures, the public is certainly given a considerable opportunity through the media to be informed about the candidates and issues and also about who is contributing and spending.

Yet many of us have nagging concerns about what is going on in the world of American campaign finance. We sense that the system is "too rich." The costs are high, too high for many who aspire to a career in political life; the costs may also be considered unnecessary. Why should not the media, especially television, provide free time for candidates and parties, as is the case in Britain, since the resource television exploits is a public resource?

Linked to that worry is the feeling that the wealthy control the electoral process too much. "Big money" can buy a candidacy, as Forbes did in 1996 with $40 million from his personal fortune. Or big money can deny a candidacy to another person. Big money can lavish millions on the Republican and Democratic National Committees and thus is awarded "access" if not "influence." Big money can spend unlimited amounts on "issue advocacy ads" that are worded cleverly (to come within the law) to support a candidate of their choice. Without "big money," our campaigns would never reach the $4 billion level they do today. It is important to remember that those in power get the bulk of the money, thus preserving the incumbents in office. Those in power, national campaign committees, distribute the money to others, like-minded, aspiring candidates who they can accept into the power structure. Those in power in the parties work with the PACs to make sure the candidates they see as most winnable and deserving receive such interest-group funds. And, finally, those in power make, and will make, the campaign regulation rules to suit their own purposes. The symbiosis between "big money" and "big politics" can be alarming if one reflects on it.

The unfairness element in campaign finance can be troublesome. It is apparent in the financial disadvantage faced by challengers of incumbents in Congress. Although our analysis showed it is possible through almost superhuman efforts for challengers sometimes to mobilize enough money to win (as was the case of most of the first-term Republican House members running for reelection in 1995–96), normally incumbents get most of the money (as they did for the rest of the House members in the last election). The finance playing field is just not level. Presidential aspirants are also disadvantaged. It is reported that in 1995 Jack Kemp withdrew from the Republican prenomination battle when his advisors told him the number of hours a day he would have to spend telephoning donors asking for the money he would need. Lamar Alexander did do the fund-raising work, but not enough. He had to bow out after raising $20 million. In another sense, there is a clear advantage for certain interest groups with large sums of money to be heard in the electoral process and gain the ear of legislators, while other groups are excluded. As Schattschneider observed long ago, our pluralist system functions with an "upper class" tendency, the organized pressure group system is very small, the range of interests represented is narrow, and "the people with the greatest needs" are not the most active and are not taken into consideration by those organized groups playing a major role in politics.[39] Thus, lower class interests, unorganized or poorly organized, may not be reflected in this financial system.

A third type of unfairness is found in the treatment of third parties. The public funding system for presidential candidates is rigged in favor of the two major parties. Third parties have to get at least 5 percent of the vote and then would receive an allocation of public funds proportional to their vote, *but only after the election.* Attracting funds for a third-party candidate is difficult, and this delay in matching federal funds increases the campaign burden. One must remember the other obstacles faced by third-party presidential candidates, such as the need to qualify to get on the ballot in all fifty states by securing the requisite number of signatures in each state (estimated cost at $1.6 million in 1996). The elimination of third-party candidates from presidential debates in 1996 is another example of inequity. But certainly the formidable prospect of raising millions to compete with the two "rich" major parties discourages third-party candidates for president as well as Congress (there is only one independent Congressperson in the House, Bernard Sanders of Vermont).

A further nagging concern is the "hidden" money raised and spent, particularly "soft money" and "independent expenditures." In 1996 there were certain outright illegal contributions, only discovered as the result of journalistic investigation, particularly into the foreign contributions. Both parties had to return millions when these illegal contributions were revealed. Most people would admit that these innovations in fund-raising in recent elections need much better disclosure through better reporting requirements, as well as very visible publicity and action to curb these excesses. Our system is not as transparent and "proper" as we would like it to be.

Finally, one notes certain "irrational" aspects of the current system. So-called issue ads are permitted if they do not ask people to "vote for" or "elect" a candidate, although they praise a candidate for president and his behavior on certain types of legislation. For example, they are permitted to say: "President Clinton says, get it done. Protect our values"; or they can talk about "the Bob Dole story" or oppose "President Clinton's spending on illegal immigrants," so long as the magic words "vote for" or "elect" are not used. This is sheer irrational license. Just as it is gross license to forbid by law (1907, 1947, 1974) corporations and unions from contributing to candidates and parties directly from their company or union treasuries, but then permit them to use such funds from their treasuries as "soft money" expenditures. And to approve the unlimited use of funds for candidate A or party B, so long as those spending the funds do not "coordinate" this spending with the candidate or party, is sheer naïveté. Obviously we are tolerating a lot of what used to be called "honest" illegality, or outright dissimulation, in our system under the present regulations.

What is to be done? There is a clamor for reform, from President Clinton, many members of Congress, the press, scholars, and the public. Can reform be achieved? Many are dubious that those in power, who thrive under the present system, will be willing to make the radical changes that are necessary. How should we go about the reforming of our present system? Over the years a set of principles have emerged which should guide our reform efforts. The two basic assumptions underlying these principles are that campaign contributions and expenditures should be focused on assisting the citizen in making an informed voting decision, while also providing the citizen with a meaningful opportunity to participate in campaigns. These are the primary and central purposes, not safeguarding the political security of incumbents, or the dominant influence of the wealthy, or the political access of PACs, et cetera.

With these assumptions in mind, we propose the following basic principles, or requisites, to guide our next efforts at campaign finance reform:

1. Strict responsibility and accountability for every person, group, PAC, candidate, and party organization for all funds contributed, received, and spent in connection with all primaries and general elections. *The Responsibility Principle.*
2. Complete public disclosure of all amounts, sources, and purposes for which money is spent in campaigns, so that voters may have timely and

adequate access to the financial information relevant for their voting decisions. Yet it must be recognized that disclosure does not rectify problems. *The Full Disclosure Principle.*

3. Greater equalization of opportunity for all candidates to acquire the funds necessary to compete effectively. *The Equalized Competition Principle.*
4. Continued support for First Amendment rights of citizens and groups to contribute and spend, within reason, to further their political goals. *The Free Speech Principle.*
5. Commitment to the strengthening of the parties' role in the campaign process, so that party organizations at all levels can mobilize the necessary funds to support candidates as well as build party institutions, so that they can function as central structures of our system. *The Strong Parties Principle.*

To put these principles into effect and remedy the current abuses in the use of money in campaigns is no easy task. The Supreme Court has helped open up the campaign finance process and this, then, puts constraints on reform. It is hoped that the Court will agree to the imposition of reasonable limits. The parties controlling Congress and the presidency have recently been loathe to pass significant reform legislation, since they have achieved their power and status through the present, relatively unregulated, system. Yet past Congresses, as the result of either pressure "from below" or their own initiatives, have adopted new party finance reform. The moneyed interest and its interest groups will certainly lobby against any major change, yet in 1907 and 1947 Congress passed laws forbidding corporations and unions to contribute directly to campaigns from their corporate assets. We hope that, with enough will, and pressure, and some idealism, our elites can and will act.

What specific measures for reform should be considered? Obviously there are many suggestions. We present here some of the major proposals advanced by the Task Force on Campaign Reform, chaired by Herbert E. Alexander, which we feel provides us with new and creative thinking on the subject.[40] They also should provoke the kind of debate among academicians and politicians necessary to achieve a better campaign finance law. The task force actually submitted thirty-nine recommendations. We select ten major proposals from that list. The task force takes the position that "campaign finance has entered a new era and we must now confront the new realities." Fundamental changes have occurred, and "the cumulative effect of these changes has been an exploding fragmentation of the campaign and the involvement of new groups in it." Consequently the "regulatory regime of 1974 . . . seems increasingly outdated." Here are ten of the task force's carefully considered recommendations for new reforms.

1. *Liberalization of Limits on Contributions.* Based on the need in a democracy for the "presence of multiple candidates mounting credible campaigns," and a desire "to maximize the competitiveness of elections," they recommend that the limits on individuals' contributions to candi-

dates for federal office should be raised from $1,000 to $3,000 per candidate per election; that the $20,000 limit on contributions to parties be abolished; that the total amount allowed for individual contributions per annum should be increased from $25,000 to $100,000, and all of that could be given to a party if the individual so desires. It is hoped that increasing these limits will curtail independent expenditures now allowed by the courts and redirect this money to candidates and parties. (Report, pages 8, 12-13, 20-21)

2. *Elimination of Expenditure Limits.* Spending limits on congressional candidates should be eliminated, because limits "undermine competition," restrict free speech (as the Supreme Court has said), fragment fund-raising, encourage "subversion," and are difficult to enforce. Four members of the committee dissented from this proposal. (Report, pages 18-19, 28-30)
3. *Abolition of Soft Money.* "Soft money should be abolished," because this money evades limits, is often misused, and, above all, is not easily accountable. "This means that the national level party committees should be barred from tunneling soft money to state and local party committees and that federal candidates and their operatives should be barred from soliciting such funds." (Report, pages 5-6)
4. *Increased Financial Role of Parties.* In a desire to strengthen the role of parties and their links to candidates, it is proposed that the limits on the amounts parties and their committees can contribute to congressional candidates be eliminated. Further (as stated in Number 1), individual contributions to parties now limited to $20,000 annually should be boosted to a new, recommended maximum of $100,000. This, it is suggested, may provide some relief for the elimination of "soft money" as well as force local and state parties to be more independent. Three of the task force members dissented from this recommendation. (Report, pages 20-21, 31-32)
5. *Expanded Role of Volunteer PACs.* The role of special interests and PACs is viewed benevolently. It is proposed that there be "no aggregate limits on the amounts candidates may accept from all PACs." Further, the current limit of $5,000 which an individual may contribute to a PAC should be increased to $6,000. However, early in the report the task force takes the position that "unions, corporations, and trade associations should be barred from paying for issue advocacy with treasury funds obtained from union dues and corporate profits." Only voluntary contributions should be used for this purpose. (Report, pages 7, 22)
6. *Seed Money for Congressional Candidates.* In the interest of assisting candidates challenging incumbents, a unique proposal for "seed money" is presented for launching "a credible campaign." The contribution limit for individuals should be $10,000 (rather than the proposed $3,000) for the first $100,000 raised by House candidates and for the first $500,000 raised by Senate and presidential candidates. (Report, page 12)

7. ***Liberalize and Expand Public Funding.*** The public funding system for candidates for the presidency needs some technical improvements. On some problems we cannot do much through national regulation. Fund-raising under the "front loading" nature of presidential primaries is difficult to deal with at the national level. We can do something about other problems: tax checkoffs may provide inadequate funds for the next presidential year and should be raised to five dollars; the threshold for qualifying for matching funds should be "indexed for inflation"; state-by-state limits on spending in the prenomination phase should be eliminated — they "inspire subterfuge." We should raise the limits on spending in the prenomination period to equal that in the postnomination period ($61.8 million in 1996); the limits on spending by presidential nominees after the conventions, evaded in the past, can be reinforced if we eliminate "soft money." The task force endorses "a system of partial financing" of candidates for the U.S. Senate and House in general election campaigns. Candidates who accept public funding should agree to contribute no more than $50,000 of their own funds. (Report, pages 14–17)
8. ***The Millionaire Problem.*** Their concern that "millionaire candidates" undermine "electoral competition" and "increase public cynicism about the role of money in campaigns" led the task force to (1) suggest that the Supreme Court reverse its "reading of millionaires' First Amendment rights," (2) suggest that the Court view candidates' utilization of their own fortunes as "self-contributions" and thus subject to contribution limits, and (3) hope that their other recommendations, such as the elimination of limits on the amounts parties can spend, will eventually impose some control over "big money" self-financed candidacies. (Report, pages 13–14)
9. ***Regulate Disclosure, Publicity, and Enforcement.*** Disclosure regulations are one type of reform developed in the past that has been successful. What is needed now both at state and federal levels is a tightening of disclosure procedures "both organizationally and statutorily." "Disclosure is most effective when the information is readily available and summarized in ways that make the data meaningful and readily accessible." Electronic filing of disclosure reports should be mandatory for all committees receiving or spending above a reasonable threshold. All disclosed information relating to finance in elections should be made available "in the form of press releases, direct access to agency computers, optical imagery systems, and over the internet." The Federal Election Commission has not been able to properly enforce disclosure requirements. Its bipartisan structure of three Democrats and three Republicans, with four votes required to issue a ruling, "stymies enforcement" and "virtually guarantees administrative and policy gridlock." The FEC, if reconstituted with an odd number of commissioners selected less for their partisanship, would be able to function more effectively. (Report, pages 9–10)

10. ***The Constitutional Free Speech Issue.*** A major controversial issue with which the task force grappled was whether the Supreme Court should be urged to allow limits on political speech (and thus on campaign expenditures), if Congress finds such limits necessary "to preserve the integrity of the electoral system." Reopening this constitutional question, which was settled in the *Buckley v. Valeo* case in 1975, resulted in a strong dissent by one task force member (Michael J. Malbin). It is a basic position, to which the other task force members revert at different points in their report, as a critical issue to be resolved. (Report, pages 3, 14, 25–26)

These reform proposals are indeed innovative in many respects. They evolve as specific recommendations to achieve basic principles the task force is committed to, similar in many ways to the five guiding principles we articulated earlier. Heavy emphasis is on accountability, competitiveness, and the regulation of campaign finance in such a way as to maximize informed electoral decision making. These are idealistic, although controversial, goals and objectives. For some they may appear too idealistic, unrealistic, and problematic. They do provide us with a new approach to the reform problem we face in America and an agenda on which we would do well to reflect carefully.

NOTES

1. Brooks Jackson "Financing the 1996 Campaign: The Law of the Jungle," in Larry J. Sabato, ed. *Toward the Millennium: The Election of 1996* (Boston: Allyn and Bacon, 1997), 256.
2. Fred Wertheimer, "The Dirtiest Election Ever," *Washington Post,* November 3, 1996, Cl (cited by Jackson in Sabato, *Millennium,* 226).
3. Herbert E. Alexander, 1997, "Financing the 1996 Election," in *America at the Polls 1996,* (Storrs, CT: Roper Center, 1997), 143.
4. Task Force on Campaign Finance Reform, "New Realities, New Thinking," Citizens Research Foundation, University of Southern California, 1997.
5. This review of the early legislation is based in part on the useful summary found in Herbert E. Alexander, *Financing Politics* (Washington, DC: Congressional Quarterly, 1984), 61–69.
6. Originally quoted in Eugene H. Roseboom, *A History of Presidential Elections* (New York: Macmillan, 1957), 316. Cited by Alexander, "Financing," 67–68. The three men were Belmont, Morgan, and Ryan.
7. *Newberry v. the U.S.,* 256 U.S. 232 (1921).
8. *U.S. v. Classic.* 313 U.S. 299 (1941).
9. See Alexander Heard, *The Costs of Democracy* (Chapel Hill: University of North Carolina Press, 1960), 347–50, for an analysis of the effects of the Hatch Acts.
10. Louise Overacker, *Money in Elections* (New York: Macmillan, 1932), 71n, 73.
11. Herbert E. Alexander, *Financing the 1976 Election* (Washington, DC: Congressional Quarterly Report, 1979), 166–67.
12. Alexander, "Financing," 143.
13. *Newsweek,* October 6, 1997, 53.
14. Task Force on Campaign Finance Reform, ii, 4, 11.
15. Roper Center, *Polls 1996.* 150.
16. Anthony Corrado, "Financing the 1996 Elections," in Pomper, *The Election of 1996,* 1997, 161–62.
17. See, for example, the debate by Gary C. Jacobson, Donald O. Green, and Jonathan S. Krasno, in the *American Journal of Political Science* 34 (1990): 334–72. Jacobson argues that "a challenger's level

of spending can make as much as a 12 percentage point difference in votes" (357). Green and Kresno do not so much dispute this as assert that "access to financial resources continues to rank as one of the chief advantages of incumbency" (371).

18. National Election Studies. See Steven J. Rosenstone and John M. Hansen, *Mobilization, Participation, and Democracy in America* (New York: Macmillan, 1993), 61.

19. Early figures from Alexander, *Financing Politics* (Washington, DC: Congressional Quarterly, 1980), 81.

20. Roper Center, *America at the Polls 1996*, 153; Corrado in Pomper, *The Election of 1996*, 155-60.

21. Brooks Jackson, in Sabato, *Millennium*, 230.

22. This description of some of the 1996 financial activities draws heavily on Brooks Jackson, in Sabato, *Millennium*, 242-48; also see Anthony Corrado, in Pomper, *The Election of 1996*, 151-55.

23. Brooks Jackson, in Sabato, *Millennium*, 246-48; also Corrado, in Pomper, *The Election of 1996*, 153-55.

24. Ibid., 145-50, 162-64.

25. Brooks Jackson, in Sabato, *Millennium*, 236-40.

26. Task Force Report, see dissent, 31-32.

27. See the excellent article by Frank Sorauf and Scott Wilson as a source for Table 12.9, 191.

28. Sources are Robert A. Diamond, ed., *Dollar Politics: The Issue of Campaign Spending* (Washington, DC: Congressional Quarterly Press, 1973); Common Cause, *1974 Congressional Campaign Finances*, 1976; Federal Election Commission Reports, 1995 and 1996.

29. Alexander Heard, *The Costs of Democracy* (Chapel Hill, NC: University of North Carolina Press, 1960).

30. *Congressional Quarterly Weekly Report.* March 17, 1973, 577-87; April 1978, 26, 30; Federal Election Commission *Record* 7, No. 3 (1981), 11.

31. Quoted in Dan Nimmo, *The Political Persuaders: The Techniques of Modem Campaigns* (Englewood Cliffs, NJ: Prentice Hall, 1970), 153.

32. For 1978, FEC press release of June 29, 1979; for 1996, Anthony Corrado, in Pomper, *The Election of 1996*, based on FEC report as of November 25, 1996, 157-61.

33. Quoted by Brooks Jackson, in Sabato, *Millennium*, 250.

34. Ibid., 251.

35. Ibid., 253.

36. Corrado, *The Election of 1996*, 162.

37. Janet M. Grenzke, "PACs and the Congressional Supermarket: The Currency is Complex," in *American Journal of Political Science* 33 (February 1989): 1-24.

38. Robert J. Samuelson, "Making Pols into Crooks," *Newsweek*, October 6, 1997, 53.

39. E. E. Schattschneider, *The Semi-Sovereign People* (New York: Holt, Rinehart and Winston, 1960), 31-34, 105.

40. The Task Force submitted its report in 1997 after holding meetings in July to October 1996. Aside from Alexander, the other members of the Task Force were: Janet M. Box-Steffensmeier, Anthony J. Corrado, Ruth S. Jones, Jonathan S. Krasno, Michael J. Malbin, Gary Moncrief, Frank J. Sorauf, and John R. Wright. The report is published by the Citizens Research Foundation of the University of Southern California, March, 1997.

CHAPTER 13

Parties and the Election Process

Elections are in a very real sense a major test for political parties in democracies. After all, their leadership selection activity, their organizational effort, their intensive campaign work, and the election all tell parties and their candidates where they stand vis-à-vis the public. Since parties must win elections (or demonstrate that they have the potential for winning) if they are to be taken seriously, elections force a discipline on parties that cannot be matched by any other aspect of democratic life. Although there are many ways to look at elections, certainly a central perspective is their relevance for testing the strength, viability, and power potential of the parties.

For some time in the United States there have been signs of a decline in the public's involvement in elections. In the 1996 presidential elections we witnessed a low point in voter turnout (48.8 percent), the lowest level of voting in presidential elections since 1924. Further, there are some indications that citizens are less active in campaigns than before, although they are participating more in politics in other ways (such as directly contacting officials and joining political interest groups). This poses what Brody twenty years ago called "the puzzle of political participation in America."[1] At a time of increasing education and efforts to make the task of voting easier, why this decline in voting? How serious are these developments, what causes this decline, and what are the implications for our system? If elections are supposed to be central to effective democratic government and act as vehicles for social, economic, and political change, does increasing citizen apathy for electoral engagement hinder the achievement of these objectives? These are questions we need to keep in mind as we analyze change in electoral behavior.

Scholars have over the years seen elections as fulfilling different functions in democratic systems. Elections are *expressions of the interest of citizens* in the political system and its leadership selection processes. By voting, citizens may be responding to a desire to be involved in politics, to have some say in what decisions are made, and to a "sense of efficacy," a conviction that ordinary persons can be effective in the system. Nonvoters may not necessarily be negative about voting; they may stay home out of a confident feeling that the system is functioning satisfactorily. And not all voters are optimistic that they can influence political decisions. Yet the act of voting does represent for many an interest in politics, if only at a minimal level.

Elections are *communication opportunities for citizens.* In one sense, people can merely communicate their preference for one set of leaders over another set. In a more radical sense, elections may represent opportunities for people to protest, even to "revolt" against those in power and the policies they stand for and have adopted. President Lyndon Johnson said, "The vote is the most powerful instrument ever devised by man for breaking down injustice."[2] As a vehicle for political action, the voting act is vital. It may be an act of reprisal against leadership, communicating at the same time a desire for a leadership of a different type or one which behaves differently. George Bernard Shaw claimed that a general election enables the people, if they choose, to do what the French did in their revolution — they "overthrew one set of rulers and substituted another."[3]

Elections may be institutionalized *representative mechanisms.* In offering the process of voting for particular candidates with particular socioeconomic backgrounds, committed to particular objectives linked to perceived interest-group demands, the election may be an instrumentality by which people, in groups or individually, see themselves as placing *their representatives* in public office. There are, of course, other ways that these groups may seek to implement group objectives, through lobbying and direct contact with those in public office, but the election itself may be a major means to channel the demands of groups and to make them legitimate, because with an election one achieves a breakthrough into the formal decisional process itself.

Elections provide the basis for *power acquisition and power rationalization.* Joseph Schumpeter wrote that an election is the "institutional arrangement for arriving at political decisions in which individuals [those who would be leaders] acquire the power to decide by means of a competitive struggle for the people's vote."[4] This refers, of course, to a democracy in which presumably there is freedom for rival leadership cadres to compete for power. Elections are one means by which people can obtain power, but in a democracy it is a crucial and distinctive means which distinguishes democracy from other forms of power acquisition. Thus, the election process is vital for the acquisition and exercise of power in a democratic system.

There are those who would argue that elections in democratic societies are a sham or ritual or merely manipulative exercises by the elite in power. Some years ago, the Soviet newspaper *Pravda* ridiculed Western elections as "falsifications" and "legalized swindling," a caricature which most of us do not appreciate.[5] Most scholars see elections as meaningful institutions performing important functions. Above all, they have a major relevance for political parties in our system. Through elections, the public is expressing an *interest* (or lack of interest) in parties, *communicating* something about or to parties, designating which party should *represent* its interests, and providing support for, or *giving power* to, one set of party leaders (and taking power away from another set of party leaders).

THE ELECTION SYSTEM IMPOSES CONSTRAINTS ON PARTIES

The special nature of the American election system should be kept in mind as the factors determining voting and nonvoting in the United States are examined. First, it is important to remember that our voter registration procedures are *voluntaristic*, requiring every person to take the initiative to go to the local clerk of elections to register at age eighteen or thereafter. In Europe the government assumes the responsibility for registering voters. The job of the voter and the party is more burdensome here.

A second feature is the "long ballot" in the United States. Others have called it the "bedsheet ballot," by which is meant that we elect many different types of officials at many different levels of government, and we do this frequently (that is, for short terms in office). The job of the citizen, weighing the merits of many different candidates for many different offices, is thereby difficult. It is no wonder that after elections, when the votes are analyzed, there appears the phenomenon of "ballot fatigue," the tendency for fewer people to vote for those at the bottom of the ballot (for register of deeds, for example) than for the candidates at the top of the ballot (for president or U.S. Senator or governor). The heavy task this imposes on the parties is obvious. They must mount and finance and coordinate, if possible, a great variety of campaigns simultaneously, and they must communicate effectively with the public the merits and policy positions of all their candidates so that they maximize their chances of electoral success. It is a far cry from the European system, in which the voter goes to the polls to elect a member of parliament or (at a different date) a member of a local municipal or county council — but no one else. The ballot in such a system is small, the election issue is focused, and the task of the citizen and the party is much more simplified.

A third critical feature is the "winner-take-all" single-member district (SMD) concept of elections in the United States, in contrast to the proportional representation system which exists in most European countries, except Britain. The candidate who secures the largest number of popular votes in a state or district wins the election. The task of parties, therefore, is magnified, since they must win a plurality or they get nothing.

There are other obstacles to voting in the United States that make the parties' task of getting out the vote onerous. Election day is not the national legal holiday it is in many European countries, where workers have the day free and can vote at any time. Further, the polls are open longer in many countries. It has been suggested that in the United States voters might be permitted to vote over an entire twenty-four hour period, thus increasing the time span in which they can participate. Absentee voting is difficult in many states and could be simplified greatly.

Finally, the length and expense of American campaigns certainly place a burden on the parties and also on the citizen who is exposed for months to party propaganda that may alternately confuse, bore, and anger him. British parliamentary elections take less than one month (technically, twenty-one days). Our presidential campaigns last at least nine months (from the February primary in New Hampshire to the first Tuesday after the first Monday in November). Our off-year election campaigns to Congress last at least four months (July to November) and often longer. Add to this the great cost of the campaigns and the need to raise huge sums of money during the campaign when energy should be devoted to mobilizing votes, and one can see how the parties' task is complicated and multifaceted in the United States.

Some argue that ours is a "low-turnout culture," one in which voting participation is not emphasized nor valued, the way it is in Europe. This may well be, and the subsequent analysis of data will seek to examine this argument. But it must be admitted that the characteristics of our election system make involvement in the elections a demanding activity, both for individual citizens and for the parties.

CHANGES IN OUR ELECTION SYSTEM

Our election system has changed nationally in a variety of ways over the years. In addition, there have been changes adopted by individual states. First, by the Civil War, the right to vote was extended to all white males by elimination of property and taxpaying qualifications, and to blacks by the Fifteenth Amendment in 1879. In 1920 women were enfranchised and in 1972 the Twenty-sixth Amendment reduced the voting age to eighteen. The legal and extralegal requirements to effectively disfranchise blacks in the South after the Civil War were gradually removed, by court decision and by the Voting Rights Act of 1965. These included the poll tax (forbidden by the Twenty-fourth Amendment), literacy tests (outlawed by the 1965 act as amended in 1970), the white primary (declared unconstitutional by the Supreme Court in 1944), and periodic re-registration (repealed by all but three states by 1972).[6] Research reveals that in states with a poll tax, literacy tests, and periodic registration, blacks were "35 percent less likely to vote than their counterparts in states without these [qualifications]."[7] Following the Voting Rights Act of 1965, the federal government sent officials into southern counties and parishes to see that blacks were registered, and that plus the actions of civil rights organizations (SNCC, NAACP, CORE, etc.) had a tremendous impact on black voter participation.

Voter registration laws were adopted by states throughout the country from the late nineteenth century on. States differ in the difficulty of registering to vote. The number of days before the election required for registration varies; some states permit it to be done on election day. The place of registration varies: in the county building, or shopping centers, or at home via house-to-house registration, or even by postcard or when getting a driver's license. The 1970 amendments to the Voting Rights Act attempted to ease these regulations somewhat — for example, by declaring that registration should not be required more than thirty days before the election. Residency requirements have been eased also (the Voting Rights Act declared a maximum of thirty days for presidential elections, extended by the Supreme Court now to all elections).

Aside from such changes in the suffrage and the conditions of voting, our system has also undergone other changes, some of them increasing the burden of the voter, as indicated earlier. The adoption of the Constitutional amendment in 1913 for the popular election of U.S. Senators is one such change: The changes inspired by the Progressive movement earlier in this century — referendum, initiative, recall, short terms of office, the direct primary — have complicated the task of the party and voter considerably. Obviously our campaigns (discussed in detail in Chapters 12 and 13) have changed greatly. Citizens now are exposed to very frequent election contests that are very confrontational, expensive, and saturated with media coverage. The election is a recurring and intense focus of our democratic culture.

While many changes in the way we conduct elections have occurred, certain aspects of our election system have not changed. We still retain the electoral college system for selecting the president, and the single-member-district system with the plurality vote in electing the members of Congress and state legislators.

In the study of this highly complex election system and the electoral behavior that occurs and is expressed through it, two major features stand out: the political parties have adapted to the system and learned how to work within and through it; and the public, unfortunately in recent years, seems to be less willing to participate in this election system, even though we have made it easier to do so. We shall discuss the problem of increasing nonvoting as we proceed with our analysis.

VOTING PARTICIPATION AND NONVOTING

The decline in voter turnout in American elections since 1960, and in a longer-range perspective since the 1890s, *may be* a significant indicator of the public's acceptance of parties, the campaign process, the election system itself, and the performance of governmental leadership. There are, however, alternative explanations of low aggregate participation. That there has been a decline, however, cannot be easily refuted. Figure 13.1 reveals the ups and downs, as well as the periods of greatest and least turnout, in presidential elections. American citizens who were eligible were attracted to the polls in large numbers before 1900 — close to 80 percent or more participated.

After the realigning election of 1896, it appears that a decline set in that was only temporarily checked in 1916 when President Woodrow Wilson was reelected. The decline continued through the World War I years and the arrival of suffrage to women in 1920, lasting until 1924 (when the all-time low of 49.3 percent was reached). During the 1920s and early 1930s, turnout picked up, reaching the high point of 62.1 percent in 1940 at the time of World War II. The small proportion of absentee soldiers voting contributed to the low of 55 percent in 1944. The Dewey-Truman contest motivated even fewer voters proportionately in 1948 and the turnout dropped to 53.5 percent. After 1960 another period of decline began, culminating in the 1980 turnout of only 54 percent of the eligible voters. By 1996, turnout had fallen even further, to less than 49 percent. Thus, since the 1890s there have been three major periods of decline: from 1896 to 1924, a 30 percent decline; from 1940 to 1948, a 9 percent decline; and from 1960 to 1996, a 15 percent

FIGURE 13.1 ▪ Fluctuations in Turnout in Presidential Elections, 1876–1996

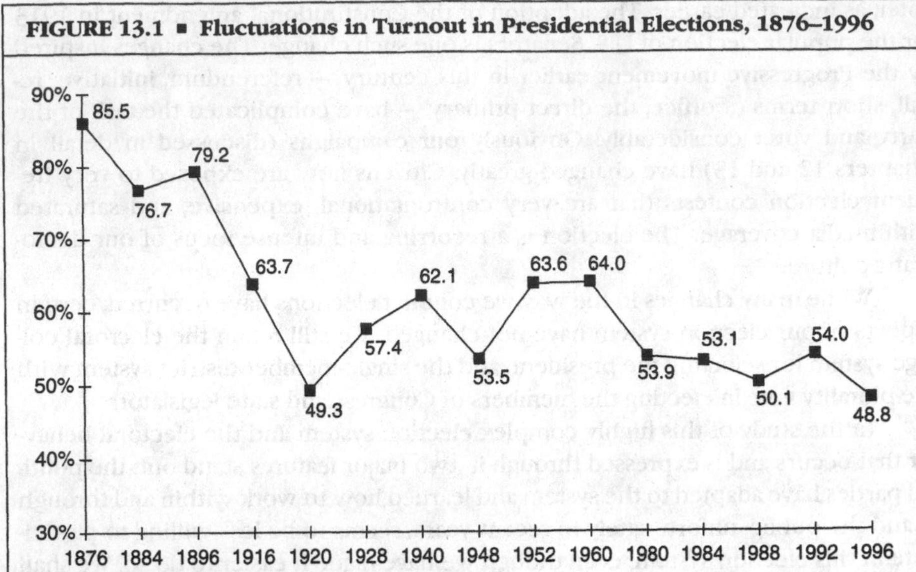

SOURCES: U.S. Bureau of the Census, Statistical Abstract of the U.S., 99th ed. (Washington, DC, 1978), 520. For earlier data see also Robert E. Lane, *Political Life* (Glencoe, IL: Free Press, 1959), 20–21; and W. D. Burnham, "The Changing Shape of the American Political Universe," *American Political Science Review* 59 (March 1965): 11, fig. 1. Steven J. Rosenstone and John M. Hansen, *Mobilization, Participation, and Democracy in America* (New York, Macmillan, 1993), 57. Roper Center, *America at the Polls 1996,* 34. There is controversy over the exact turnout in recent years and how it should be calculated. See Paul Beck, *Political Parties in America* (Chatham, NJ: Chatham House, 1997), footnote 1, 418.

decline. The high level of voter interest of the latter part of the nineteenth century has never been restored.

In comparison with other countries, U.S. voting turnout percentages are low. For recent elections they compare as follows.[8]

Country	*Percent*
Italy	87
Sweden	87
Britain	78
Netherlands	78
Japan	73
India	59
Germany (West)	77
France	69
Canada	69
Israel	77
Switzerland	46

Even at our recent high point in 1960, the U.S. is below most of these countries, by five to twenty-three percentage points, and recently the difference became greater.

In elections below the presidential level, turnout is often much lower. Those participating in elections for the House of Representatives vote at the 35 to 40 percent level. The turnout in gubernatorial, county, and city elections can be still lower, with fewer than 30 percent of the voters participating.

The drop in turnout is significant, it has reached a low of about 15 percentage points below the 1960 level of participation. The decline can be demonstrated dramatically if one looks at the change in the actual vote cast in recent years:

Total Votes Cast in Presidential Elections (in thousands)

Party	*1984*	*1988*	*1992*	*1996*	*Percent Decline or Increase*
Republicans	54,450	47,946	39,102	37,869	−30.5
Democrats	37,574	41,016	44,908	45,629	+21.4
Two-party total	92,024	88,962	84,010	83,498	−9.2

Even with the Perot vote in 1996 (just under 8 million), there still is an aggregate decline in the vote, attributable particularly to the decline in votes cast for the Republicans. Thus, despite an increase in our population and the size of the eligible electorate, the total vote cast is lower. There obviously is a large core of regular nonvoters, and it is increasing. What is particularly surprising is that this occurred while the southern vote (the ten states of the "Solid South") increased their vote in recent years from 21.5 million in 1984 to almost 24 million in 1996). The South used to be the region of low turnout, but no longer. Variations by states are more pronounced now, more so than by regions. Thus, for example, among the big states California had a 700,000 vote decrease in 1996 (over 1992) and New York a decrease of over 400,000; on the other hand, Florida increased 500,000 and Texas over 400,000. Particular conditions of local politics are responsible for these variations in turnout.

The decline in voter participation is important for what it tells us about the lagging political interest of citizens, the inadequacy of candidates and party efforts and performance, and, above all, for its link to critical personnel and policy outcomes. As Lijphart put it, "The overall weight of evidence strongly supports the view that who votes and how people vote matters a great deal. Indeed, any other conclusion would be extremely damaging for the very concept of representative democracy."[9]

GROUP DIFFERENCES IN VOTER TURNOUT

Who votes and who does not vote tells us something about what categories of citizens are attempting to have a voice and what citizens are disinterested, who cares about exercising democratic rights and who does not care, and, thus, who may be represented by political leaders and who may not be represented. As V. O. Key put it long ago, "the blunt truth is that politicians and officials are under no compulsion to pay much heed to classes and groups of citizens that do not vote."[10]

While voting participation has declined overall in the past three decades, to the alarm of many, one must realize that this record of participation varies by groups and categories of citizens, and that it varies over time for these groups. Thus, national election studies of self-reported voting in presidential elections in the critical period of early decline, 1960–1976, found these differences for particular groups.[11]

Group	*Percentage Change in Voting Turnout*
Religion	*1960–1976*
Catholics	-19
Protestants	-7
Union status	
Union families	+2
Non-union families	-13
Age group	
21–24	-5
33–52	-10
69–72	-7

Although turnout decline did pervade most socioeconomic groups of citizens, the phenomenon, and its extent, was not universal and must be understood in terms of the political status and orientation of these groups at particular historical points. When unions were on the rise in the sixties and early seventies, their members showed no decline in voting. Catholics had a high turnout in 1960 (74 percent) in Kennedy's election, but numbers declined after that. Voting turnout decline, contrary to expectations, was higher for the middle-aged than for either young or older voters. The old argument that decline in turnout is attributable primarily to the apathy of younger citizens is not borne out by the data.

It is particularly interesting to go back to the earliest national election studies, 1948 and 1952, and compare the rates of self-reported turnout fifty years ago with today (Table 13.1). Here we see that in the low-participation years preceding the 1960 election, the percentages for some groups, such as educational groups, were very similar in 1948 and 1996. The turnout differential between those with elementary schooling and college education was 24 percent in 1948 and 31 percent in 1996. Age cohort differences were also similar. As for the religions groups, the Catholics had high turnout in the fifties and nineties, while the Protestants had a somewhat better record in 1996 than in the fifties. One should note also the decline in all of these groups in the seventies, as illustrated by the data in Table 13.1 for the 1972 election year.

In a long series of studies, it has been demonstrated that usually the better educated and more affluent citizens have a higher voting participation record than those with a lower socioeconomic status (SES). That, according to Verba and Nie, is "the standard model" of the linkage between SES and the vote.[12] The differential turnout between high and low SES in the early ears was 25 to 30 percentage points. While it is argued that in years of high national turnout this bias (vote differential)

TABLE 13.1 ■ Over-Time Comparisons of Voting Turnouts in Presidential Elections for Selected Groups (percentage of self-reported vote)

Group	*1948*	*1952*	*1972*	*1992*	*1996*
Education					
Elementary school	56	62	44	50	54
High school	67	80	56	73	67
College	80	90	74	89	85
Age					
Under 35	56	68	58	68	60
Over 35	67	77	75	82	78
Religion					
Protestant	57	72	53	76	79
Catholic	80	85	65	82	80

SOURCE: Angus Campbell and Homer C. Cooper, *Group Differences in Attitudes and Votes* (Ann Arbor: Survey Research Center, University of Michigan Institute for Social Research, 1956), 19–37; Warren E. Miller and J. Merrill Shanks, *The New American Voter* (Cambridge: Harvard University Press, 1996), 39–69, Appendix B. National Election Studies.

in favor of those with high status is not as great as in years of low natural turnout, yet the difference is always considerable. Socioeconomic status and age are two variables closely linked to the vote in the United States. In Dalton's analysis for 1992, the correlations with the vote are .36 for educational level and 2.1 for age: those with higher educational achievement and those who are older are significantly more likely to vote. (The correlation for gender in 1992 was .00).[13]

There are some exceptions, however, to the turnout decline phenomenon: blacks, southerners, and women all increased their turnout since the fifties (Table 13.2). The sixties brought a sharp increase in the turnout of African Americans, particularly with the passage of the Civil Rights Act of 1964 and the Voting Rights Act of 1965, both of which led to federal government pressure concerning discrimination against blacks. In the decade from 1960 to 1970, there was a tremendous increase in voter registration in southern states (Table 13.3). Blacks were supported

TABLE 13.2 ■ Increase in Voter Turnout for Blacks, Southerners, and Women (in percent)

Category of Voter	*1952*	*1960*	*1972*	*1980*	*1984*	*1992*	*1996*
African Americans	33	53	65	67	66	70	68
Residents of Southern and border states	35	50	44	50	45	67	—
Women	60	—	62	—	—	76	75

SOURCES: National Election Studies. For early data see Angus Campbell and Homer C. Cooper, *Group Differences in Attitudes and Votes* (Survey Research Center, Institute of Social Research, University of Michigan, 1956). For later data see Miller and Shanks, *The New American Voter* (Cambridge: Harvard University Press, 1996), 39–69, Appendix B.

TABLE 13.3 ▪ Percentage of Voting-Age Population Registered to Vote in Southern States

	Non-White		White	
State	*1960*	*1970*	*1960*	*1970*
Mississippi	5.2	71.0	63.9	82.1
South Carolina	13.7	56.1	57.1	62.3
Alabama	13.7	66.0	63.6	85.0
Georgia	29.3	57.2	56.8	71.7
Virginia	23.1	57.0	46.1	64.5
Louisiana	31.1	57.4	76.9	77.0
Arkansas	38.0	82.3	60.9	74.1
North Carolina	39.1	51.3	92.1	68.1
Florida	39.4	55.3	69.3	65.5
Texas	35.5	72.0	42.5	62.0
Tennessee	59.1	71.6	73.0	71.6
Averages	29.7	63.4	63.8	72.2

SOURCES: *Historical Abstract of the U.S., 1971* (Washington, DC: Government Printing Office, 1971); original report, *Voter Registration in the South,* prepared by the Southern Regional Council, Voter Education Project; see Frank Feigert and M. Margaret Conway, *Parties and Politics in America* (Boston: Allyn and Bacon, 1976), 84–85.

by the civil rights movement in both the South and the North. Contrary, then, to the normal trend in voting participation in the country, the utilization by blacks of their voting rights increased greatly. Further, there was a modest increase in southern turnout which, as we indicated earlier, continued to increase in recent years. Women equaled men in turnout by the eighties, with no real decrease since that time.

EXPLANATIONS OF VOTER TURNOUT DECLINE

A variety of theories, some with considerable supporting evidence, has been advanced to explain the paradox of decline in voting in the United States and our low turnout compared to other countries. It is both a comparative problem and an over-time problem. What is so puzzling is that general theories of social and economic change, while suggestive, are only partially successful, because other countries also have experienced such change and yet have maintained voting participation at a high level, while the United States has not. "Economic development," "the mass society," "cognitive sophistication," the "aging of the party system," and similar types of theories may explain other system phenomena in modern societies, but may not be as useful in explaining comparatively low voting levels in the United States. We are forced to look at the political culture of the United States, at our "low-voting-turnout culture," and ask why and how such a "culture" has come about.

THE ELECTORAL SYSTEM AND ITS RESTRICTIONS

We begin with a look at certain features of our electoral institutions as they are linked to low turnout. As discussed earlier, the American system makes heavier demands on citizens than in other countries, or, to put it differently, makes voting more difficult. The first problem is our registration system, which under our decentralized form of government is decided at the state level and varies by state. Although liberalized recently by some states, essentially it differs in that we do not have in the United States an automatic system of registration, as in most other countries. The U.S. citizen has to make the special effort to register, usually thirty days before the election, while in other Western democracies one is automatically eligible and registered upon reaching a certain age and the government has the responsibility of preparing the voting register. In these countries, this reduces the "cost" (in effort and time) to the individual for voting. There has been much research on the effects of our registration system on voting, not all of it convincingly demonstrated. Burnham argued that the adoption of our registration system resulted in a 17 percent decline in turnout from 1896 to 1916.[14] Other scholars have claimed that the easing of registration in certain states (registration on election day, post card registration) has led to small turnout increases in those states. Generally, comparative analysis estimates that the vote increases could range from 8 to 15 percent if we moved to the European automatic registration system.[15] This may be so, although we have not tried it in the United States. In explaining the decline in voting turnout in the United States between the 1960s and the 1980s, Rosenstone concludes that more liberal registration laws have only improved turnout by 1.8 percentage points, because there were too many other, more powerful, factors that were responsible for nonvoting.[16]

However, there is clear evidence that registration laws, and their administration, in the south were major obstacles to black voter turnout. Blacks had low participation in the fifties, only 33 percent but that number increased to over 60 percent in the sixties and seventies (Table 13.2). Rosenstone and Hansen have shown that 44 percent of this increase was due to the easing of registration laws. For blacks (and probably other minorities such as Hispanics), registration was a major impediment.[17]

A second feature of our system that is identified as restrictive is the single-member-district (SMD) arrangement. It is argued that a proportional representation system (PR) encourages more participation and would boost turnout by up to 12 percent. Yet not all PR elections in other countries have high turnout; it depends on the country and the type of election. A third feature with an alleged negative effect is the frequency of our elections. These increase the "cost" of voting to the citizens, the need to take the time to go to the polls twice or three times, or more, a year. Fourth, the timing of our elections is mentioned as a deterrent — weekend voting in Europe is associated with a possible 6 percent increase in turnout.[18] Finally, it is argued by Lijphart that compulsory voting as in some countries, today or in the past (Belgium, Netherlands, Austria, and Australia, for example) should be considered as a way to enhance the importance of elections, stimulate citizen interest and desire to become more informed, and implement the norm of the citizen's duty to vote. While this interesting recommendation should indeed be debated, in the future as it has in the past, it is clearly not a practical objective in the United States today.

There are two obvious observations: (1) simplifying and easing the task of voting should certainly be encouraged in the United States (in such ways as the "motor voter" law passed in 1993 that requires states to provide citizens the opportunity to register to vote when they apply for a driver's license); but (2) on the basis of most recent research, while there are negative effects of our registration system, generally one can say that these restrictions are not today the primary explanation for turnout decline in the U.S.

Generational Explanations for Low Turnout

Empirical research has clearly shown that age is linked to voting; few would say that age is completely unrelated, and other factors explain it more satisfactorily. The disagreement is over how to conceptualize that relationship of age to voting. The following theories have been advanced:

1. Early socialization to politics definitively determines one's predisposition to vote or not vote, and this decision persists throughout one's life. There is minimal support for this theory.
2. The "life-cycle" is determinative; that is, voting is a function of age and changes with age for all people. One's age determines a person's integration into society. Therefore, all people are less likely to vote when young, more likely in middle age, even more likely as senior adults, and less likely in later years. This theory appeals to many.
3. The historical generational period in which one enters young adulthood (age 17–25) and first becomes eligible to vote exposes one to a particular set of political events, which influences one's attitudes toward politics and determines one's orientation to voting. Historical periods have distinctive characteristics for each country (for example, the New Deal and post–New Deal period in the United States). As one adult group is replaced by another adult group over time, and since their socialization experiences can be different, the level of turnout can change, positively or negatively.
4. The "life experience" theory argues that what matters is not the particular period in which a person comes to political adulthood, but the particular experiences of each individual citizen. As people acquire more political contacts, knowledge, skills, and resources, they can become more interested in politics develop more positive attitudes and evaluations about parties, politics, and elections, and this can lead them to be more participative, including using their vote more frequently.[19] (However, one must concede that such experience may also induce a negative reaction).

We know that there are great differences in voting rates by age. In the 1952–1988 national elections, the averages of self-reported voting were youngest, 42 percent; young adults (25–44), 60 percent; middle-aged, 70 percent; older adults, 75 percent; most elderly, 65 percent. The basic question is, Which of these theories best explains why these disparities continued to appear?

The historical generational conception is illustrated by the work of Miller and Shanks. They divide the electorate into three historical-era groups, based on the year they were first eligible to vote: before 1928, 1931–1964, 1968 to the present. They thus have three groups: pre–New Deal, New Deal, and post–New Deal. They present a wealth of evidence to demonstrate that the decline in voting was essentially a post–New Deal phenomenon (Table 13.4). The drop in turnout begins in 1968 and continues for the post–New Deal generation throughout the seventies and eighties, increasing slightly in 1992. There is a 23 percentage point difference between the New Deal and Post–New Deal generations in 1968, and while that narrows to an average of 16 percentage points later, it continued consistently lower. The contextual argument Miller and Shanks present as an explanations is, as one might expect, a long narrative about the political and social turmoil that young adults in the post–New Deal period were exposed to. These included the racial crises of the late fifties and early sixties, leading to the federal government's intervention in the South and the passage of the 1964 and 1965 Civil Rights and Voting Rights Acts. Also, the urban violence in the North, the political assassinations of the sixties, the Vietnam War and opposition to it, the tumultuous Democratic convention in Chicago in 1968, the emergence of the new "counterculture," Watergate, and eventually Irangate were all part of the social upheaval of the times. This entire period from Eisenhower to Reagan is characterized as not an "era of good feeling," but an era that would, and did, induce a rejection of politics by young adults. While the New Deal period is seen as one of great progress and positive politics, the successive years were seen as a time of "post–New Deal shocks" leading to withdrawal by many from belief and participation in the system. Other data presented in this

TABLE 13.4 ▪ Contrasts in Voter Turnout by Generation: Prior to, During, and After the New Deal (percentage of self-reported voting among non-blacks)

Political Generation	*1960*	*1968*	*1976*	*1980*	*1988*	*1992*
Pre–New Deal	83	74	70	71	66	*
New Deal	81	80	80	80	80	82
Post–New Deal	*	57	61	62	65	73
Total cases	1,676	1,254	2,148	1,237	1,552	1,938

* Too few cases for analysis.

SOURCE: Warren E. Miller and J. Merrill Shanks, *The New American Voter* (Cambridge: Harvard University Press, 1996), 51.

study demonstrate that it was those with a lower education (grade school) who had particularly low turnout (ranging from a high of 41 percent in 1972 to a low of 27 percent in 1980 and 1992). The authors conclude: "The national decline was largely result of the non-Black electorate having replaced its older cohorts (of the New Deal period) . . . with a host of new additions" who voted at a much lower rate.[20] It was not "aging" but generational replacement that is seen as the key explanation of turnout decline.

The other major hypothesis about the role of age in turnout decline is the "life experience" theory, as utilized by Rosenstone and Hansen. The term "life experience" as they use it embraces a variety of conditions: socioeconomic status (resources), sense of political efficacy, positive support or affect toward parties and candidates, and social involvement (social contacts, community integration, et cetera). All of these are part of the accumulating "experience" that comes with growing older and with exposure to the system of electoral-party politics, including contacts by parties; it is not defined by an historical era of socialization. This experience matures in some persons but not in others. And when these types of experiences occur, and lead to *positive* orientations toward parities and elections, they determine voting participation, as well as other forms of political participation. So goes the argument. Rosenstone and Hansen demonstrated how each of these can have an effect. They find these "resources" to have a positive relationship to voting, while other variables are negative and depress participation (legal restrictions, other demographic variables, certain types of minority status, for example). But one "cause" or "source" of voting decline that stands out is *party and social movement mobilization* of citizens to vote. Their probit analysis concludes that 54 percent of the decline in voter turnout in presidential elections from the sixties to the eighties is attributable to a decline in party and social movement effort to contact citizens about voting, 17 percent is due to "a younger electorate," 9 percent to "weakened social interest," and the remaining 20 percent to a decline in sense of efficacy and "weaker attachment to and evaluations of" the parties and candidates.[21] Thus they feel they have identified the particular conditions responsible for voting turnout, or for failure to vote, by citizens as they become older. Aging by itself is not the relevant cause; the political experiences one is, or is not, exposed to as one ages provide the critical explanation.

Attitudes and Beliefs as Factors Explaining Voter Turnout

Implicit in much of the foregoing discussion of the causes of nonvoting, particularly the generational differences, is the assumption that certain types of attitudes or orientations to politics play a very important role. In the early classic study of participation, Verba and Nie use a model that emphasizes certain attitudes such as the belief that politics is important and participation is a civic duty, and strong identification with parties. Their model includes also the "social circumstances" of citizens that are conducive to, or inhibit, the taking on of such attitudes, which result in either voting or nonvoting. The model they use in simple form is:[22]

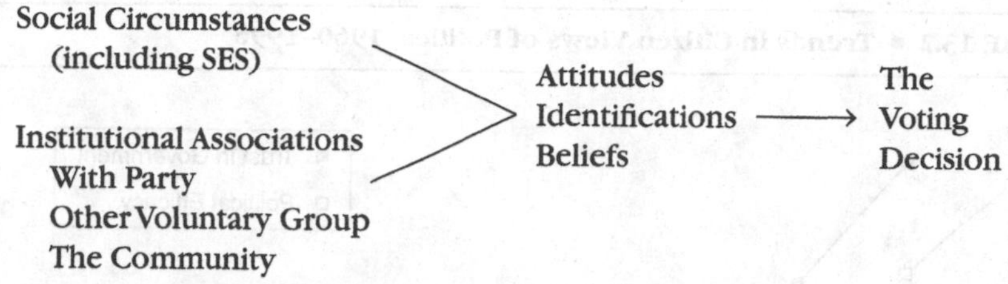

Thus, attitudes are central to any theory of voting, yet there is often not complete clarity about what specific attitudes are relevant, or about the way that attitudes are linked with other influences to explain turnout. For example, Rosenstone and Hansen disagree with those who have argued that a sense of civic duty to participate is significantly associated with voting. That the sense of civic duty has declined in recent years should suggest a relationship, but in fact the "impact on the probability of voting in presidential elections" is small (about 6 percent). Similarly, trust in government is discussed as a possible determinant. It has sharply declined, from over 70 percent in 1960 to between 20 percent and 30 percent today. And the public has become increasingly cynical about the government's responsiveness "to what people think."[23]

But both of these attitudes toward government do not seem to be related to voting participation. Rather Rosenstone and Hansen's research argues that three other types of citizen orientations are, in their opinion, significant: (1) a sense of political efficacy ("people like me don't have a say in what the government does"); (2) "affect toward the parties" and concern over which party wins the election; (3) strength of party identification. There has been a decline in all three factors in the last thirty years, and all three are correlated with a decline in the vote.

There has been much research on the importance of the lack of trust in government, as well as on the lack of a sense of efficacy, as two orientations possibly related to the vote. Both distrust and inefficacy have shown a considerable increase over the years, although in the eighties and nineties trust and efficacy have both increased (see Figure 13.2). The most important for explaining voting turnout seems to be the sense of efficacy. This was illustrated by an analysis of voting and nonvoting in the 1972 elections (Table 13.5). While there was a 9 percent decline in turnout for those least trusting of government, the real attitude factor seems to be the lack of a sense of efficacy, even among those who were not cynical. It was, therefore, not lack of trust which led to nonvoting but a feeling that voting participation was not worthwhile, that the individual citizen was not heard and could not be effective through political involvement such as voting. Rosenstone and Hansen in their analysis of turnout from 1956 to 1988 also confirm the importance of efficacy and reject the argument that trusting citizens are more likely to vote.[24]

Party identification (conceived as psychological attachment to parties) is a good example of a very important voter orientation that seems clearly related to voter turnout (Table 13.6). As we reported earlier (Chapter 5), there has been very little decline in self-reported voting by strong Democrats and Republicans.

FIGURE 13.2 ▪ Trends in Citizen Views of Politics, 1960–1996

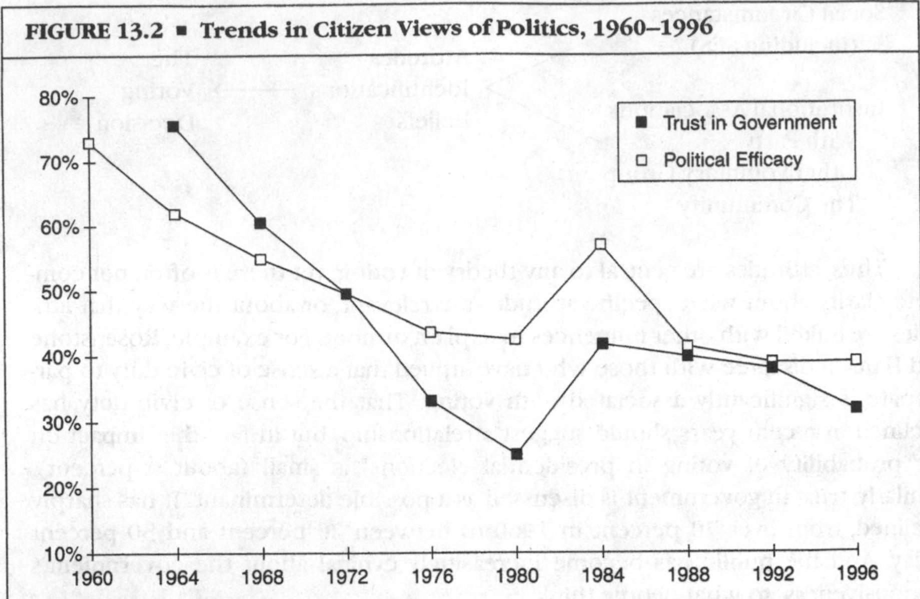

SOURCES: University of Michigan SRC/CPS/NES; see also Norman H. Nie, Sidney Verba, and John R. Petrocik, *The Changing American Voter* (Cambridge: Harvard University Press, 1976); Arthur H. Miller, "Political Issues and Trust in Government: 1964-1970," *American Political Science Review* 68 (September 1974): 951-72.

But weak identifiers and independents (including independents "leaning" toward Republicans) have been more likely to stay at home on election day. From 1960 to 1988 independents dropped from 77 percent to 50 percent, a decline of 27 percentage points. For "independent Republicans" the decline was almost as great, followed by weak Democrats: strong attachment and commitment to a political party clearly seems to have inhibited nonvoting. A separate analysis of African Americans from the fifties on reveals a consistently high turnout rate for strong Democrats,

TABLE 13.5 ▪ Distrust of Government and Political Inefficacy: Which Linked Most to Nonvoting (presidential election, 1972)

Sense of Political Efficacy	Level of Trust in Government (percent nonvoters) *High*	*Low*	*All Respondents*
High	11	16	13
Low	41	38	39
All Respondents	26	35	

SOURCE: Arthur Miller and Warren Miller, "A Majority Party in Disarray: Policy Polarization in the 1972 Elections," *American Political Science Review* 70 (September 1976): 753-78.

TABLE 13.6 ■ Party Identification Strength and the Vote (turnout rates in percent)

Party Identification	*1960*	*1964*	*1972*	*1980*	*1988*	*1992*	*1996*	*Percent Change 60-96*
Strong Democrats	85	82	80	84	80	85	77	-8
Weak Democrats	79	73	71	65	63	73	62	-17
Independent-Democrats	72	71	72	69	64	73	58	-14
Independents	77	62	53	54	50	60	44	-33
Independent-Republicans	87	84	76	75	64	72	66	-21
Weak Republicans	88	85	79	77	76	77	70	-18
Strong Republicans	89	92	87	88	89	88	89	0

SOURCE: National Election Studies.

only slightly lower for weak Democrats. Today strong Democrats vote at the 80 percent level or higher.

The bottom line for Rosenstone and Hansen is that a weakened feeling of efficacy explains 9 percent of the decline in voting after the 1960s, while weaker attachment to parties explains another 11 percent of this decline. In a special analysis of the increase in black voter turnout between the 1950s and 1968–1972, efficacy was not an important factor in the overall analysis (explaining only 1 percent of the increase), but stronger attachment to parties was important (explaining 13 percent). Yet, for blacks three other key factors explaining their increased turnout were of primary significance: easing of registration laws (explaining 44 percent), an increase in "personal resources" (24 percent), and mobilization by the civil rights movement and through party contacts (18 percent).[25]

The Role of Parties in Voter Turnout

Throughout our discussion on the determinants of nonvoting, we have analyzed and evaluated a variety of explanations: socioeconomic status, institutional restrictions, generational differences, and attitudes and orientations toward politics. There remains controversy over the utility of these theories, but the considerable research done in recent years has helped us gain a better understanding of the problem and its possible causes. The role of parties as major actors in the election process recurs in much of this research. Much emphasis is placed on the relevance of parties as organizations, as cadres of activists, as groups with which people identify, and as vote mobilizers. As already indicated, the extent of "mobilization" explains much of the increase or decrease in voting turnout, a mobilization in which parties are very important actors, along with candidates, social movements, interest groups, and the mass media. Yet a major question remains, "What do parties (or these other actors) do specifically that influences whether people vote, what is communicated to citizens during this "mobilization" that either activates them or turns them off of politics?

Nonvoting: An American Paradox!

Since Richard Brody in 1978 emphasized the "puzzle" of nonvoting, our turnout in presidential elections has declined further, to 50 percent or lower. And scholars have discussed a variety of possible explanations, attitudinal (weak party identification, low sense of efficacy, and so on) and institutional (the failure of parties, candidates, and groups to mobilize voters). One important explanation is that the special features of the American electoral system are responsible, particularly the preelection personalized registration system, which we have but which does not exist in other democracies. The argument is that from the latter part of the nineteenth century to 1920 turnout fell sharply because of strict registration laws. After 1920 low turnout was still partly due to the restrictions and barriers imposed on voters because of this registration system. Certainly registration laws in the South restricted access to the polls for African Americans. The analysis and evidence presented in this chapter documents the role of registration in discouraging voting participation.

Some states have been experimenting with revisions of the registration requirement. North Dakota does not require it at all. Other states make registration easier, even allowing registration by post card. The 1993 Federal "Motor Voter" Law permits registration at the time that the driver's license is renewed. However, there seems to be evidence that still today a small percentage of the decline in voting is due to the registration requirements.

One state, Oregon, has gone even further than other states recently to the voting process itself: Voting-By-Mail. In December 1995 and January 1996 Oregon held a special primary election and special general election to elect a new senator to replace Senator Bob Packwood, who had resigned because of scandal. These were the first efforts to use a vote-by-mail system on a statewide basis.

The question was raised whether such changes in the election system did improve turnout. The research by Michael Traugott after this Oregon election suggests that it may have some effect on turnout. He conducted a survey in which he interviewed a sample of 1,483 adults by telephone (with a 60 percent response rate) and discovered that indeed there was some increase which could be attributed to the vote-by-mail system. "In the case of Oregon," he concluded, "these increments were small, typically ranging from 2 to 10 percentage points." The point of his and others research on the election system is important — simplifying and easing the voting act can have some salutary effect.

Other innovations may follow — voting by personal computers, touch-tone phones, or television hook ups? However, despite such evidence most scholars would still say that in the final analysis positive attitudes toward the system and effective voter mobilization are much more important if we really are to get Americans to exercise their voting rights.

SOURCE: Michael Traugott, "An Evaluation of Voting-By-Mail in Oregon," special report prepared for the Workshop on Voting-By-Mail, Washington, DC, 1997.

One basic distinction to keep in mind in answering this question is the past performance of parties, the parties' historical record for good or bad, and, on the other hand, the perceptions of the current actions of parties in the present campaign, during which the immediate mobilization is occurring. Every election year, each potential voter is the target of a barrage of campaign stimuli attempting to influence current images of the parties and candidates. But, as Fiorina reminds us, we all have "retrospective" images of party and candidates' performances that also determine behavior.[26] The decision to vote or not may have a retrospective meaning that may be triggered by party actions in the campaigns, as well as by a response to immediate campaign stimuli about current party and candidate records and appeals.

Another key distinction to keep in mind is that the mobilization by parties (or other actors) may have a positive impact on the citizen or may lead to indifference, rejection, even outright hostility. Contacts with potential voters may be counterproductive, attitudinally and behaviorally. We know of party contacts that have led to increased action by the opposition party. These contacts may also produce a backlash against the mobilization of certain groups (for example, the reported response of some southern whites to the mobilization and participation of blacks). In our earlier Detroit study we found that up to one-fifth of those highly exposed to the two parties were, despite such contact, cynical about government and negative about parties. Although those not exposed to party contacts were much more negative about parties and government, one wonders to what extent party contact led to the negative orientations for the 20 percent.

Mass media coverage during a campaign can also induce negative public attitudes. A study of the 1974 election combined a survey of the national public with a content analysis of the ninety-four newspapers in the cities in which these citizens lived. The authors of that study discovered that 31 percent of the coverage was critical of the political system (70 percent of the coverage of parties was negative), and only 6 percent praised the system. Above all, they found a close link between readership of the newspapers and the dissatisfaction of citizens with politicians and political institutions, even after controlling for education level and media exposure. The authors concluded, "Readers of papers containing a higher degree of negative criticism directed at politicians and political institutions were distrustful of government [and had] higher levels of cynicism and feelings of inefficacy."[27] There are other scholars who have documented the poor image of parties that has appeared in the media. Thus the media, parties, candidates, and other actors can have an impact that does not necessarily "mobilize" support for government or a strong desire to be politically involved. People may not get out to vote as a result of these different types of mobilization. Further, these different types of mobilization may to some extent work at cross-purposes. Conceivably, the media may negate the work of the parties.

When we examine the evidence about the role of parties in increasing or decreasing turnout we should keep in mind the type of analytical model suggested by the research on this subject. That model is presented in Figure 13.3. Party efforts to get out the vote are only one type of mobilization, and this mobilization is

FIGURE 13.3 ■ Factors Linked to Voting

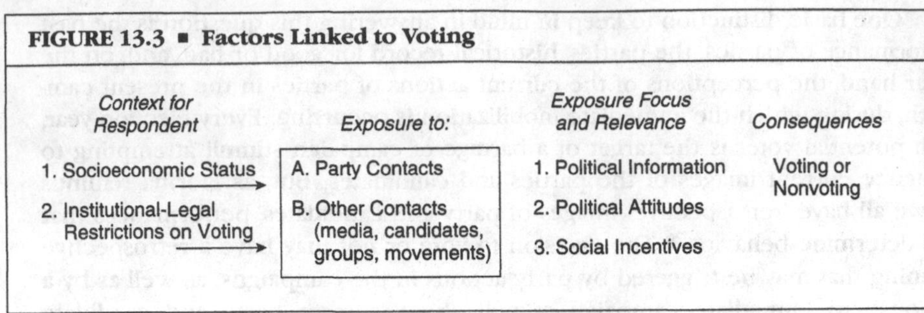

constrained by two basic contextual conditions. These contacts in turn affect citizens' knowledge, political attitudes, and social incentives relative to political involvement, which then leads to the vote decision.

The evidence for certain aspects and relationships in this model is fairly clear-cut and agreed upon. We know a great deal about the contextual situation for the individual — the differentials in social economic status and resources and the variations in the United States (and comparatively) on the legal restrictions on voting. The exposure data are also fairly well understood. We know that mass media exposure to the campaign is high; over 70 percent tuned in to television. We have good data over time on the extent to which the public has been contacted by the parties and candidates in each election, ranging from 12 percent in 1952 to 29 percent in 1972, a decline to 20 percent in 1992 and then an increase to 29 percent in 1996. In addition, we have good data on the different ways in which the public participates aside from voting and how this has varied over the years.[28] What we do not have very specific data on is the precise extent to which people have been contacted in the campaign by candidates, nonparty groups, and social movements. We know that 20 percent of the public report that they attended political meetings or rallies and 16 percent worked for a party or candidate, but the extent of direct mobilization by candidates remains elusive. Similarly, the extent of mobilization in each election by social movements is also elusive, although there are some data on the civil rights "activities" and "events" to which citizens were exposed in earlier years.[29] To what extent "mobilization" as used in the Rosenstone and Hansen analysis incorporates these civil rights activities is unclear.

That part of the model dealing with the link between party contacts (as well as other contacts) and attitudes has interested us for a long time. We are still, however, in a state of disagreement over the precise relationships. Long ago, in our study of the 1956 election, we found that citizens highly exposed to the parties' organizations were (a) more knowledgeable about politics and (b) more supportive of parties as relevant structures for political action.[30] However, this is of course not conclusive evidence that party contact produced greater knowledge and support, although it is highly suggestive. Some recent research questions the relevance of party contacts for certain attitudes, such as trust in government, or belief in governmental responsiveness, or a sense of caring about election outcomes (perhaps as a measure of civic duty).[31] The argument is that while there is some effect, it is minimal. On the other hand, strong party identification, a strong sense of political efficacy, and positive affect toward parties are orientations affected by party mobilization, and these play a major role in any model explaining voter turnout.

One other type of effect of party effort, in addition to attitudes, is the relevance of contact for providing information to potential voters about the election (registration, voting booth locations, who the candidates are, offers of assistance in voting, et cetera). Such effects can be considerable. It is a finding that was already identified and demonstrated earlier. In 1962, Converse noted the importance of "information flow" to voting participation.[32] Rosenstone and Hansen reemphasize the importance of this in getting out the vote.

A final effect of mobilization, it is alleged, is social involvement relevant to the election. The argument is that party contacts "engage social networks," encouraging get-togethers and neighborhood parties, and these reinforce one another's voting intentions. "People participate because people they know and respect ask them."[33] The evidence for much of this is not available, but it does make sense. We do know that those who are contacted by parties are more likely to persuade others to vote and to themselves work for campaigns (although this may be tautological). Strong confirmation that party contacts themselves induce such "socializing" in family groups or work groups needs yet to be clearly demonstrated.

The basic point of recent research should be kept front and center: mobilization, particularly mobilization by parties, is a critical, even major, determinant of voting or nonvoting. In the Rosenstone and Hansen analysis, lower mobilization accounts for 54 percent of the *decrease* in voter turnout from the 1960s to 1988. Increased mobilization accounted for 18 percent of the *increase* in black turnout between the 1950s and 1972 (easing voting restrictions accounted for 44 percent). Lower mobilization accounted for 34 percent of the *decrease* in black turnout from 1972 to 1988.[34] The primary factors that explain the decrease in voter turnout are presented in Table 13.7.

The evidence in support of this theory or model is, as we have said, suggestive, even compelling, and sometimes less definitive and perhaps speculative. But the role of mobilization generally, and party contact specifically, seems central to

TABLE 13.7 ▪ Factors Explaining Decline in Voter Turnout, 1960s to 1980s (turnout declined 11.3 percent)

Factor	*Percentage of Decline Explained*
A younger electorate	17
Weaker attachment to parties	11
Weaker sense of political efficacy	9
Weaker social involvement	9
Decline in mobilization	54

SOURCE: Steven J. Rosenstone and John M. Hansen, *Mobilization, Participation, and Democracy in America* (New York: Macmillan, 1993), 215.

the explanation. While this is so, one must not forget that party contact efforts may not be productive. Party contacts can activate citizens, inform them, and get them involved in the campaign, or these contacts can be completely discounted or ignored and be counterproductive: they may boomerang. After all, a large proportion of citizens who are contacted do not vote (at least 25 percent, while a larger proportion not contacted do vote (over 50 percent). A political campaign is for some a confusing, conflictful experience, and what the parties do and how they communicate with voters may either help resolve this confusion or exacerbate it. If the latter, the citizens may take refuge in nonvoting. The parties in their mobilization efforts may play a double role, both encouraging and discouraging . The irony of the 1996 election was that voter turnout declined to 48.8 percent while party contact efforts increased over 1999. This suggests that the puzzle of nonvoting is still with us.

THE CONSEQUENCES OF NONVOTING

There is something troubling about a high level of nonvoting in a democracy. It suggests that people have given up on the government, see their role in the system as irrelevant, feel that elections produce little change. This sounds like the thinking of the Haitian woman interviewed by an Associated Press journalist in 1996 concerning their election: "Voting doesn't do any good . . . nothing comes of voting." In a developing country people were led to expect "results," but they say, "We voted, we got nothing. We voted again, and we got nothing. Why should we vote now?" Are many Americans saying the same? The recent apparent decline in the vote of Afro-Americans, after they increased their vote dramatically up to the mid-seventies, suggests that some may have lost faith that the system can produce "results." Why vote? While it is important to keep in mind that Americans participate in politics in many ways besides voting — working through community groups, contacting their representatives directly, writing letters, and contributing money to political candidates and causes — still the implications of nonvoting should be considered seriously.

Historically Americans have had considerable faith in the effectiveness of the electoral process, believing that voting is central to the proper functioning of a democratic system. Our level of turnout in the nineteenth century underscores that. It provided a means by which the voter could express himself in relation to authority. "Voting is the foundation stone for political action," said Dr. Martin Luther King, Jr.[35] It is supposed to be a way in which the citizens can act against, or in support of, the system. Also, elections and voting are supposed to restrain and influence elected officials (and possibly also appointed officials) in a democracy. Long ago, Samuel Adams was quoted as saying. "Where annual elections end, tyranny begins."[36] Or, to put it differently, democracy is a system in which elected leaders are always on probation. The assumption underlying these assertions of faith is that voting makes a difference, voting can be effective. By the same token, nonvoting and changes in the pattern of voting can have significant impact. What are the consequences of increased voting or of nonvoting?

The answer to this question is not easy. A great deal of research has been directed at the question of leadership responsiveness to constituency opinion and the extent to which this is due to the actual or anticipated results of elections. Research reveals that members of Congress do feel that they will be judged, if not immediately, at least eventually, by who they vote in the House, even though they also sense that much of the public is apathetic or ignorant about policy matters. At least 80 percent believed that the electoral verdict in their districts "had been strongly influenced by the electorate's response to their records and personal standing."[37] Another study of Congress concluded that many congresspersons fear that a "wrong" vote may be used as a weapon against them by the opposition in an election.[38] Still another study of California legislators discovered that the roll-call votes of representatives (California senators in this case) were more clearly correlated with the opinions of constituents as the election approached (a correlation of 41) than after the election (a correlation of .07).[39] There does indeed appear to be an electoral connection! Although it is true that incumbents are more regularly returned to office than ever before (close to the 90 percent level), it is also true that over 50 percent are reelected at least once with a margin of less than 5 percent. The potential for defeat for a congressperson out of touch with his or her constituency is, therefore, real. And all members of Congress can now remember 1994.

Verba and Nie's study, *Participation in America,* has revealed that in communities where there is a high level of participation, including voting, the "concurrence" between leaders' views on issues is closer to that of citizens than in communities in which there is a lower level of political activity. This is particularly so when there is a strong consensus among citizens in their positions on issues. This leads Verba and Nie to conclude that "where citizens are participants, leaders are responsive."[40] Implicit is the inference that elections and voting make representatives aware of the possibility of defeat, force them to be more aware of the views of the public (particularly as the election approaches), and presumably lead to legislative votes that are more representative of public desires than would be the case if the electoral process was a formality only.

The decline in voting leads some scholars to be truly alarmed about the consequences for our democracy. As we know, turnout at the local level is usually no more than 20 percent of the eligible voters. Even many of the most educated and knowledgeable people never vote in local elections, to say nothing of those less educated and with lower incomes. This led one scholar of American cities to warn that "the level of voting participation [in local elections] is so low as to question seriously the validity of representative elections."[41]

While nonvoting in the United States is a generalizable trend for almost all classes and categories of citizens, it particularly results in the exclusion of the disadvantaged, the least educated, and the poor. As Miller and Shanks demonstrated, as turnout declined during the post–New Deal period, it was the least educated who were more likely to be excluded (or excluded themselves). The decline in voting, comparing the New Deal and post–New Deal periods, varied as follows by education level:[42]

Decline in Voting Turnout by Level of Education (in percent)

Education Level	*New Deal 1952-1956*	*Post-New Deal 1984-1988*	*Percent Change*
Grade School	62	33	-29
High School Graduate	80	54	-26
One Year or More of College	89	84	-5
Difference	27	51	

Verba and Nie revealed that, comparatively, citizens of lower class and lower status were much more responsible for the decline in voting in the United States than in European countries, where on the average the difference in participation between those with high and low socioeconomic status was only 10 percent or less.[43] Thus, a great bias is introduced into our system when voting participation declines. In a sense, the leadership selection process becomes more elitist and upper middle class.

There are other alleged consequences. It is argued that in the United States lower turnout works to the advantage of the conservative forces, the Republicans. This seems to make sense based on our knowledge of which social classes and groups support the two parties. Yet we can always find exceptions. The Republicans won decisively in 1968 and 1972, when turnout was relatively high, and lost in 1996, when it reached a very low point. Nevertheless, comparative studies do fairly consistently document the finding that "left of center parties" do better at the polls when turnout increases.[44]

Basically, perhaps, the major effect of nonvoting over time is cultural. It contributes to citizen behavior patterns, as well as a set of beliefs and expectations about citizens' sense of obligation to participate in politics. The norm of citizen duty to vote no longer is salient in the society. This may be the result of social and economic change, such as the increasing marginalization of certain citizens economically and socially, as well as increasing disillusionment among these people about the performance of the political system. In Europe and several non-Western countries, the norm of voting is high today. It was also high in many of these countries in the past. And it has continued consistently high from year to year. Significantly, and somewhat ironically, this has been so in these countries without the type of party mobilization we conduct in the United States. With the exception of Britain, such personal-contact mobilization has never been a part of their campaign operation. Examples of this continuity in the belief in the norm of voting are many. The United States fell from a turnout of close to 80 percent in the latter part of the nineteenth century to 49 percent in 1920, then increased to 64 percent in 1960, and has now fallen to under 50 percent in 1996. In contrast, France had over 80 percent turnout in the twenties and thirties and remains at 70 percent today. Similarly, before the Nazis came to power, in Germany the voting turnout was 79 percent to 88 percent, and today remains at 77 percent. Similar percentages are true for Austria and the Netherlands also (even after compulsory voting was ended in 1970). Britain has always had a high norm and remains at the 78 percent level today. Unfortunately, in the United States the norm of voting has been unstable and

uncertain over the years, never firmly accepted, despite all our democratic ideas. The erosion of this voting norm in the United States may be so ingrained and far advanced that it will be very difficult to reverse.

NOTES

1. Richard Brody, "The Puzzle of Political Participation in America," in Anthony King, ed., *The New American Political System* (Washington, DC: Enterprise Institute, 1978).
2. *Newsweek,* August 16, 1965, 15.
3. Quoted by Walter Lippmann in *The Phantom Public* (New York: Macmillan, 1930), 59.
4. Joseph Schumpeter, *Capitalism, Socialism and Democracy* (New York: Harper and Row, 1950), 269.
5. *Pravda,* February 27, 1955, section A.
6. For an excellent review of these southern efforts to keep blacks from voting, see Steven J. Rosenstone and John M. Hansen, *Mobilization, Participation, and Democracy in America* (New York: Macmillan, 1993), 196-209.
7. Ibid., 200.
8. Russell Dalton, *Citizen Politics,* (Chatham, NJ: Chatham House, 1996), 45.
9. Arend Lijphart, "Unequal Participation: Democracy's Unresolved Dilemma," *American Political Science Review* 91 (March 1997): 5.
10. V. O. Key, Jr., *Southern Politics in State and Nation* (New York: Vintage Books, 1949), 527. Quoted in Ibid., 4.
11. NES studies were relied on here plus several other studies, although not all of these agree on the exact percentages. One must realize that these are *self-reports* of voting, which students agree are considerably inflated, by 15 percentage points or more, compared to census bureau reports.
12. Sydney Verba and Norman Nie, *Participation in America* (New York: Harper and Row, 1972), 132.
13. Russell Dalton, *Citizen Politics,* 59.
14. Walter Dean Burnham, "The Turnout Problem," in A. James Reichley, ed., *Elections American Style* (Washington, DC: Brookings Institution, 1987), 113. Cited in Rosenstone and Hansen, *Mobilization, Participation, and Democracy in America,* 206.
15. Arend Lijphart, "Unequal Participation," 7.
16. Rosenstone and Hansen, *Mobilization, Participation, and Democracy in America,* 214-15.
17. Ibid., 220.
18. See Lijphart, "Unequal Participation," 7-11, for a review of the studies and evidence on these points.
19. For a discussion of these theories, see Rosenstone and Hansen, *Mobilization, Participation, and Democracy in America,* 137-141.
20. Miller and Shanks, *The New American Voter* (Cambridge: Harvard University Press, 1996), 50.
21. See Rosenstone and Hansen, *Mobilization, Participation, and Democracy in America,* chaps. 5 and 7, for the argument and evidence they use to defend this hypothesis.
22. Verba and Nie, *Participation,* 18-22.
23. Rosenstone and Hansen, *Mobilization, Participation, and Democracy in America,* 146-50.
24. Ibid., 150.
25. Ibid., 214-22.
26. Morris P. Fiorina, *Retrospective Voting in American National Elections* (New Haven: Yale University Press, 1981), 6.
27. Arthur H. Miller, Edie N. Goldenberg, and Lutz Erbring, "Type-set Politics: Impact of Newspapers on Public Confidence," *American Political Science Review* 73 (1979): 67-84, 80-81.
28. Rosenstone and Hansen, *Mobilization, Participation, and Democracy in America,* 57-63; Verba and Nie, *Participation,* 79-81.
29. See Ibid., 188-93. They use the data collected by Doug McAdam, *Political Process and the Development of Black Insurgency, 1930-1970* (Chicago: University of Chicago Press, 1982) (footnote 44, page 190 in Rosenstone).
30. Samuel J. Eldersveld, *Political Parties: A Behavioral Analysis,* (Chicago: Rand McNally, 1964), 453-55, 494.
31. Rosenstone and Hansen, *Mobilization, Participation, and Democracy in America,* 146-56.

32. Philip Converse, "Information Flow and the Stability of Partisan Attitudes," *Public Opinion Quarterly* 26 (1962): 581. Also see Eldersveld, *Political Parties*, 494.

33. Rosenstone and Hansen, *Mobilization, Participation, and Democracy in America*, 176.

34. Ibid., 215–23.

35. King also said the biggest step Negroes can take is in "the direction of the voting booths." See "Civil Right No. 1 — The Right to Vote," *New York Times Magazine*, March 14, 1965, 26. Quoted in Donald R. Matthews and James W. Prothro, *Negroes and the New Southern Politics* (New York: Harcourt, Brace and World, 1966), 11.

36. Quoted in *The Federalist*, No. 53.

37. Warren E. Miller and Donald E. Stokes, "Constituency Influence in Congress," *American Political Science Review* 57 (1963): 48.

38. John Kingdon, *Congressmen's Voting Decisions* (New York: Harper and Row, 1973), 59–60.

39. James H. Kuklinski, "Representativeness and Elections: A Policy Analysis," *American Political Science Review* 72 (1978): 173. Also Duncan McCrae, *Dimensions of Congressional Voting* (Berkeley: University of California Press, 1958).

40. Verba and Nie, *Participation*, 335.

41. M. Gottdiener, *The Decline of Urban Politics* (Beverly Hills: Sage Publications, 1987), 20.

42. Miller and Shanks, *The New American Voter*, 66.

43. Sidney Verba, Norman Nie, and Jae-on Kim, *Participation and Political Equality* (Cambridge: Cambridge University Press, 1978), 151.

44. Lijphart, *Unequal Participation*, 5. His observations are based on a study conducted by Alexander Pacek and Benjamin Radcliff, "Turnout and the Vote for Left of Centre Parties: A Cross National Analysis," *British Journal of Political Science* 25 (1995): 137–43.

CHAPTER 14

The New Technologies: How the Parties Have Adapted

Twenty to thirty years ago scholars were decrying the decline of parties and linking it to the arrival of new technologies, or the new "style" in campaigns, and the powerful new media by which candidates could communicate directly with the voter. In the early eighties, technology presumably relegated parties to the sidelines. Thus in 1984, Rosenstone advanced his theory of major party failure, that

> technological innovations have permitted candidates to be increasingly free of political parties. Independent-minded politicians who were once unwilling to embark on third party campaigns without the help of an already existing locally based party can now take the plunge more readily.[1]

Further, the contention was made that one could explain third-party success by the same reasoning:

> The technological and political change that allowed Jimmy Carter to win the presidency also allowed men like George Wallace and John Anderson to make relatively successful independent challenges without preexisting organizations.[2]

Thus technology, it was alleged, affected both the major and third parties alike. And this has resulted in the irrelevance of parties, it was claimed:

> Technological changes have allowed for increasingly sophisticated political strategies all without the aid of a political party. Little more than a computer, telephone, a postage meter, and a consultant are needed. Direct mail places a candidate in contact with potential supporters. Polls to gauge the pulse of the electorate have replaced the candidate's need for local party organizations.[3]

The argument led to the concept of "candidate-centered politics," which presumably now characterized our system. In 1984, Crotty also argued that

> the role of the political party in campaigns has given way to the technology of television-centered campaigns built on polls and run by media and public relations experts. The evolving politics is a candidate-centered, technocratic exercise in impersonal manipulation. . . . television . . . has supplanted the political party as the main conduit between candidate and voter.[4]

In the earlier postwar period, there was no evidence yet of the dominant role of television. When scholars in 1952 studied the impact of television, they concluded that: "at least in Iowa, the presence of television had a minor net effect, if any, upon the 1952 election."[5] Another observer concluded: "television served as a *substitute for* rather than an *addition to,* the other mass media in areas where it was available."[6] Moreover, when the pioneering voting studies came out as late as the mid-fifties, the power and influence of television was seen as on par with the other mass media, newspapers, and radio.[7] But soon television began to play a more major role in campaigns, as the 1956 and 1960 elections demonstrated. While more people followed the campaign via radio and newspaper in 1952 than via television, by 1956 television was the major medium for 74 percent, increasing to 87 percent by 1960. And it has been dominant ever since.[8] The new media and new technology threatened to take over the parties' functions. Thus, the key question is, how did the parties respond to, and finally adapt and adjust to, this challenge, and survive?

Before we look at how parties adapted and adjusted to the new technology, we must recognize that "technology" is an umbrella term. There is, in campaign politics today, a multiplicity of technologies, and it is essential that we desegregate this umbrella term so as to separate out the major components of this modern technology in American politics.

In the past, political parties used the best campaign technology available. Such procedures as "torchlight parades, speeches, bell-ringing and cannon fire," as well as "party rallies, strong speeches, barbecues, picnics, brass-bands, firm handshakes and mass meetings"— in fact, a combination of the crowd-pleasing aspects of the old-fashioned circus, carnivals, and county fairs to stir and mobilize the partisan followers and political nonbelievers alike. In sum, political parties took their messages and candidates to the people in the best technologies of the day. Hence, it should come as no surprise that, in order to stay in touch with the people of contemporary times, they have employed the latest technologies of this era. However, fast-track, ambitious, and enterprising candidates of today were quicker to exploit the recent technologies than political parties. But parties have indeed adapted and adjusted.

Today the modern technologies can be grouped into four basic categories: (1) professional campaign managers, (2) pollsters, (3) advertising specialists, and (4) public relations (media) consultants. We will discuss each in detail.

THE PROFESSIONAL CAMPAIGN MANAGER

In most circles, campaign manager is simply called the "hired gun." Traditionally, the role now played by campaign managers was undertaken by political parties' precinct captains and the candidates' loyal staffers.[9] Here is how it worked. The parties' grassroots managers — precinct captains, ward bosses, leaders of local political clubs, as well as racial and ethnic group organizations and local party activists — knew their political districts and bailiwicks. They knew the voters and their streets, in both the neighborhoods and across the entire community. For the grassroots managers, these were the people whom they not only had done favors for over the years, but had continually worked with year in and year out. Moreover, they knew how to get these people to the polls on election day. Hence, a linkage would be formed between the candidate's staffers and these early grassroots managers of the political parties. It was up to the staffers to provide as much "walk-around money" (wam) as possible, along with party patronage, to motivate these party managers and get the necessary turnout on election day.[10] However, the central problem with this system was that it was labor intensive, led to corruption, and attracted the attention of reformers. Eventually, progressive reforms targeted corruption and party patronage and reduced their effectiveness.

Stepping into these gaps, vacuums, and poorly coordinated linkages were the professional campaign managers, men and women who develop a strategic plan for the overall campaign and employ in that plan an adroit use of polling, computers, television, direct mail, e-mail, faxes, and money[11] "to elect their candidate(s) to public office."[12] These technologies mean that they can, particularly by television, put the candidate directly into nearly every household and in front of the voter. On this point, Robert Agranoff writes: "Television has become the surrogate party worker, the vehicle for conveying candidate style, image, and issues."[13] Others have said, "These technological modes of campaigning have replaced the precinct or party worker in many places."[14] With these new technologies, the professional campaign managers are no longer making "seat-of-the-pants" judgments.[15]

About the rise of these campaign managers, Stanley Kelley, in his early book on the subject, noted that they initially started in the mid-thirties in California when Clem Whitaker, a newspaperman, and Leone Baxter created the firm of Whitaker and Baxter to handle issues involved in initiative and referendum elections of the state.[16] Eventually the success of this firm spawned competitors, and the concept gradually spread throughout the country. The professional campaign manager operated to bring candidate-centered organization into "the competition for control of the offices of government."[17] And they aimed to do it effectively.

How? Kelley again: "Through [the] media it is possible to communicate instantaneously with people up and down the social scale and widely scattered in space. Publicity can be given to particular events, issues and personalities, and a truly national attention can be focused on a problem."[18] These were the initial findings about the new technological device as of 1956.

Writing even in the 1980s, Larry Sabato said that "political consultants are sometimes as important as voters and candidates in determining the conduct and outcome of elections.[19]

> [Within] two decades . . . political consultants had become a campaign standard across the United States, and not just for major national and statewide contests. State races for lesser offices and U.S. House seats, and elections for local posts and even judicial offices, frequently had the services of one or more consultants.

> The number of consultants has skyrocketed along with the demand for their services. As late as 1960 there were relatively few full-time professionals in the field; twenty years later there are hundreds — thousands if local advertising agency executives specialized in politics are counted. . . . Today the average modern professional manages more campaigns in a year than his predecessors did in a lifetime.[20]

Overall, professional campaign managers have come to master the "state of managerial technology" and have at their disposal more knowledge about the "art and science" of campaigning than do either politicians or political candidates.[21] And this is due to the fact that modern campaigns are now too complex for politicians and candidates to learn how to run in the midst of the campaign. With all of the emphasis on winning, and a lifestyle built around the power, glory, and rewards of politics, it is not surprising to discover that "the majority of candidates for public offices turned their campaigns over to the experts."[22]

But for all their mastery of managerial technology, the new experts come at a price: "Political consultants, answerable only to their client — candidates, and independent of the political parties, have inflicted severe damage upon the party system and masterminded the modern triumph of personality cults over party politics in the United States."[23] This problem notwithstanding, candidates for political office continue to use them.

Enter William J. Clinton. Even as a congressional candidate in Arkansas in 1974, he effectively used campaign consultants and managers to great effect. In his initial campaign in the third congressional district in Arkansas, candidate Clinton had his strategy "outlined" to him by Jody Powell, an aid to Jimmy Carter, the former governor of Georgia who was heading the Democratic National Committee's 1974 congressional team. Of this matter of local campaign managers reaching out to national consultants, Clinton's biographer writes: "Powell came to Fayetteville to advise the Clinton campaign for a few days in the early fall. . . . He left behind a seventeen-point memo."[24] The document was used by the manager to direct the campaign.

As Clinton progressed up the political ladder in Arkansas, so did his efforts to attain the very best campaign managers and political consultants. In fact, on the eve of his initial gubernatorial campaign, Clinton, acting on the advice from his campaign manager and President Carter's pollster, Patrick Caddell, went beyond the realm of state politics. "Early that fall of 1977, Clinton's chief of staff, Steve Smith, placed a call to a young political consultant in New York named Richard [Dick] Morris, who had some novel ideas about how polls could be used to shape rhetorical arguments in campaigns."[25] After that initial meeting, Clinton would have his professional campaign manager work very closely with Morris and use his advice in 1982, 1994, and 1996 to achieve his stunning electoral comebacks. With consultant Morris wedded to his campaign manager, Clinton would become known as the Comeback Kid.[26]

THE POLLSTERS

It all started in 1824. "It was a newspaper, the *Harrisburg Pennsylvanian,* that in 1824 published the first political poll in the United States. The electorate of Wilmington, Delaware, was surveyed on presidential preferences, with Andrew Jackson winning the straw ballot two to one over the eventual electoral college winner, John Quincy Adams.[27] A tradition of polling known as the "straw ballot" polls was born. After this local newspaper began the process, numerous other papers picked up the technique, until by the 1880s nearly all of the major papers were conducting such polls. Following the newspapers, the magazines, weekly and monthly ones, began the process. The most famous of these, the weekly *Literary Digest,* conducted a straw ballot poll in every presidential election from 1916 until 1936. But in 1936, the *Digest*'s twenty years of polling came to an end with the disastrous prediction that incumbent president Roosevelt would be defeated by Governor Alf Landon.[28] Although that year the *Digest* mailed out ten million ballots (down from the twenty million ballots it had mailed out in the 1932 election[29]), it had followed a sampling procedure in which people with telephones, automobiles, upper-crust social-club memberships, and voter registration cards were surveyed. This sampling procedure gave the poll an upper-class Republican bias in the depression-ridden thirties and led to its incorrect prediction.[30] Yet prior to the 1936 election, the *Digest*'s polls had correctly predicted the previous five presidential elections (1916–1932). And it was these correct predictions that captured the political eye, interest, and imagination of the chairman of the Democratic National Committee, James F. Farley. Farley was not only a cabinet appointee of President Roosevelt — the Postmaster General — but his campaign operative. And it was due to Farley's attention to the *Literary Digest* polls that the polling technique came to be used by the major political parties.

As a consumer and follower of the *Digest* polls, Farley told the chairman of the Republican National Committee that he thought the 1932 poll "was a very reliable barometer of political trends."[31] He further stated in 1932 that "the *Literary Digest* poll is an achievement of no little magnitude. It is a poll fairly and correctly conducted."[32]

On April 30, 1935, three years after he made these remarks, Farley conducted a secret poll for Roosevelt based on the sampling principle of the *Digest* polls. Since Roosevelt wanted to keep the sponsor of the straw vote polls secret, Farley and Emil Hurja, the chief statistician and executive director of the Democratic National Committee, mailed out about 150,000 straw ballots accompanied by a cover letter that told the recipient that it was being conducted by the *National Inquirer* magazine.[33] Although no such magazine existed, some 21 percent of the ballots were returned, and the DNC based its strategy for the 1936 election to considerable extent upon these polling results.[34] Thus, via the *Literary Digest* polls, the Democratic party had used a poll to help recapture the 1936 White House. Polling as a technological instrument for political parties was born.

Ten years later, in 1945, the Republican candidate for U.S. Senate, Jacob K. Javits, became the first candidate to pay for a privately conducted poll to help him win the seat. And as he saw it, the poll made him quite successful.[35] The dam had now broken and polling for national political candidates was fashionable. George Gallup had used the poll instrument to help his mother-in-law win the secretary of state position in Iowa in 1932.[36] Thus, it could now be used by both statewide and national candidates. With these candidate usages, the stage was set for the candidate-centered era in American politics that emerged three to four decades later.

After these initial applications of the polling techniques by political parties and candidates, not only did polling move out of the pre-scientific era, beginning in 1936 with the rise of Gallup, Roper, and Crossly polls, but the polls themselves

Franklin Roosevelt and an Early Straw Vote: 1935

In 1935, President Franklin Delano Roosevelt publicly ignored the very popular Governor and Senator (held both positions simultaneously) Huey Long of Louisiana, who had announced presidential aspirations. Privately the president authorized the Democratic National Committee to commission a secret poll (dubbed in that day and time as a straw vote because of its unscientific sampling method) to determine if a Long candidacy would have an impact on the president's reelection chances.

The real reason that Long was being taken seriously was not because of his electoral base in Louisiana and possibly the South. Long, after securing his base in Louisiana, moved his campaign organization to the neighboring state of Arkansas and supported a woman, Hattie Caraway, for U.S. Senator. Against a male opponent and tradition, Hattie and Huey prevailed. After this success, Long in 1934 created the "Share Our Wealth" movement. A year later it claimed 27,000 chapters with 7.7 million followers.

The objective of the DNC straw poll was primarily not to determine where Long was strong, although that was a concern, but which states he might "throw or shift" to the Republican party. The DNC, under the guise that it was conducting a poll for the fictitious magazine, *National Inquirer*, drew a sample 100,000 people from forty-eight states. The names were drawn from the telephone books in each state. Concerned that this sample of telephone owners in the 1930s might be skewed to the rich and wealthy, while many of Roosevelt's supporters came from the "working and lower" classes, the DNC contacted AT&T to get information on which income groups owned phones. Although AT&T assured the DNC that about one-third of its telephone owners were working people and that in the South 3 to 5 percent were African Americans, the DNC took no chances and mailed out an additional 49,742 ballots to "relief recipients." However, when only 21 percent of these later mail ballots were returned, the final analysis and interpretation relied upon the initial group of 100,000.

The results of the poll revealed that Long would get more than 36 percent of the vote in Louisiana, 10 percent of the vote in Mississippi, Arkansas, Utah, South Dakota, Wyoming, Washington, Iowa, Maryland, and Nevada. More importantly, Long's support was large enough to possibly shift five states: Colorado, New York, Illinois, Ohio, and New Jersey. Long was clearly a threat. But before the DNC could put together a strategy to counter Long, he was assassinated late in 1935. Nevertheless, the DNC conducted three more straw polls for FDR.

Here, then, is an early example of the faith in national polling, however unsystematic, for determining campaign strategy.

SOURCE: Emil Hurja, "Supplementary Memorandum on Interpretation of Hearst Poll," (September 7, 1932) in the Emil Hurja papers, Box 63 Franklin D. Roosevelt Presidential Library, Hyde Park, New York.

diversified into several categories. First there is the *benchmark poll,* also called the "baseline" survey, that is conducted early in the pre-campaign period (often a year or more before the election debate); the benchmark is a "planning, not a prediction poll."[37] The second type of poll is "one or more *'follow up surveys'* [which] are conducted to probe more deeply into topics of concern."[38] These polls are designed to build on the findings of the initial benchmark polls. Thirdly, there are the ***trial heat polls,*** which are a series of questions or a question about which candidate will win the election. "Trial heat questions group candidates together in hypothetical match ups and ask citizens for whom they would vote in that hypothetical pairing."[39] Then come panel surveys. These are polls that re-interview at least half of the people in the initial benchmark polls. Such polls are conducted to see how opinion shifts during the election and/or between elections.[40] However, to probe more deeply into the likely voters, vote choice decision, and voting intention, pollsters conduct "focus group" sessions with a few individuals "selected randomly to reflect age, sex, race, economics or life-style characteristics brought under the leadership of a trained discussion leader to talk generally about the campaign and the candidates."[41] As the campaign heads into its final month, weeks, days, and hours, "tracking polls can be created to follow the candidates over the final month, over each week, or over every day to see how the electorate is responding to the candidate, his or her issues, or political ads." Tracking polls are designed "to monitor closely any late shifts in support."[42]

There is one new innovation in polling. All of the above polls are preelection polls, which are based upon the electorate's intentions. At CBS, Warren Mitosky created the exit polls, interviews with voters as they leave polling places. And being "polls of actual voters," they "circumvent the ending problem faced by preelection surveys of determining who actually will vote." "Exit polls are quickly tabulated. Almost instantaneous predictions and descriptions of election outcomes are possible."[43] And with the use of this latest innovation, the networks in 1980 were able to predict the outcome of the election prior to the polls closing on the West Coast and in Hawaii. Today, as a consequence of that reality and subsequent congressional investigations, the networks have agreed to hold their predictions until all of the voting polls have closed.

Thus, the new technology of polling has created a wide variety of preelection and postelections polls, as well as voting-day polls known as exit polls. Since parties at the national and state levels have been able to employ polls to their advantage in presidential elections, such a technology has been a boom to aspiring and enterprising candidates.

Polls and polling helped make potential candidates actual candidates and forecast their possible and potential success. Polls and polling helped these candidates plan their campaigns and remake and shift their strategy as the campaigns progressed. In the pre-primary campaign stage, as well as in the primary stage, where the national parties did not intervene, polls helped candidates become self-starters and self-recruiters. Ultimately this meant that candidates did not need to rely upon parties to launch their campaigns. Thus candidates used polls and polling to relieve parties of their recruitment and mobilization functions.[44]

However, facing this loss of role and function, parties have adapted and adjusted. Today, parties conduct polls and offer the results to their candidates. Moreover, parties conduct surveys on their own image or lack of it. In fact, parties shape and reshape themselves based on what the polls tell them. In point of fact,

Clinton's comeback in the Arkansas gubernatorial election of 1982 and his reelection bid of 1996, after the Democrats loss in Congress in the 1994 election, were based on party polls and their interpretation by political consultant Richard (Dick) Morris.

As one writer put it, "of all the presidents, Clinton has been the biggest and largest consumer of polls." He writes: "The political importance of public approval and the difficulty of maintaining it help explain why Bill Clinton's staff immediately went to work on a plan to keep their boss on top of the polls, while staying in tune with Congress and his own bureaucracy."[45]

Pollsters now direct all aspects of the candidate-centered electoral process. They direct and guide not only the candidate and his message, but the national organization as well. "Carter's pollster Patrick Caddell proposed in a memo that the incoming administration act quickly to 'Carterize' the Democratic National Committee and the national party and determine qualifications for the national chairman."[46] Caddell's memo urged:

> It is clear that if the DNC is going to be "Carterized" and made a political wing of the White House, that requires a chairman who is a loyalist, and essentially a Carter insider. I suspect that any move that brings in someone who has an independent constituency and other political interest is going to result in an inability of the DNC to really carry out the functions we need. I think that the DNC chairman's selections be along the lines of someone who is both a loyalist, an insider, and a person who is willing to take directions from the Governor's political and personal staff.[47]

And when Kenneth Curtis did not take orders well from the White House, he was "quickly replaced by a far more obliging national chair."[48]

However, developing strategy for the national organization is just one of the many roles of pollsters. They also advise candidates how to relate to their constituency group and coalition partners. For example, Clinton pollster Stanley Greenberg urged Clinton in 1992 and 1996 to "de-market the party to African Americans." On this matter, Robert Smith writes: "After the 1984 election the Democratic party commissioned several studies to determine why the party lost the presidency. One was conducted by Stanley Greenberg, Clinton's 1992 pollster. It pointed clearly to the party's identification with Blacks as the source of its problem."[49] And based on the Greenberg findings, Clinton developed a strategy "to appear not beholden to blacks and a hard-line New Democrat on issues of welfare, crime, and Jesse Jackson."[50] In a word, the pollster has "gradually outgrown the role of mere technician, emerging from the cocoon a full-fledged strategist who influences campaign activity in a wide variety of areas."[51]

THE POLITICAL AD SPECIALISTS

The ad specialists are the technocrats who create the political spot. They design and craft a public image for the candidate in thirty- and sixty-second spots. And from these image merchants, the candidate is both packaged and sold to the larger constituency. Yet this is only one-half of the specialists' job. The second role is to package and demarket and unsell the opponent and the opposition. Currently this is known as negative advertising. Hence, the ad specialist engages in both *positive* and *negative* advertising. And the mix can be varied depending on the circumstances.

Political commercials, like regular commercials, mix style with substance and image with reality. They combine the fake and deceptive with the true and real. Both soft- and loud-selling ads are made to have their candidate supported by the electorate. Thus, the ads for new candidates go after name recognition. The candidate is made known to the electorate. A personal image is fashioned to link the candidate to the dominant traits and sentiments in the electorate, or at least in the target audience. The candidate image is so fashioned in the political commercials that the voters will be both motivated and stimulated to participate.

Beyond the desire for name recognition, the commercial needs to address the issues. Here polls, focus groups, and past voting patterns help the candidate say what the electorate wants to hear and/or feels. Carefully exploiting a variety of themes, slogans, and symbols, the political commercials make it appear to the electorate that the candidates recognize the public's "issue" concerns. The candidate seen in the political commercials has the same values, beliefs, and attitudes as do the people themselves. It is beyond clever, it is a way to program the voters. Hence the candidate and the issues are nudged into one package for the voter. These are the positive ads. They are usually combined with the negative ones.

Negative ads permit the candidate to undo the competition. Such ads deliberately create the impression that those values, beliefs, and attitudes that the electorate wants to see coming from its government cannot possibly be achieved with the election of the opponent. The negative political commercials suggest that the opponent cannot possibly bring anything meaningful to the political process. Negative political commercials reduce the opponents to nonpolitical leaders and candidates. They become political "un-citizens." Such ads coupled with the positive ones bolster the candidate's chances of victory and the opponent's chances of defeat. Two well-known examples are President Bush's Willie Horton ads and Senator Jesse Helm's "white hand" ads.

But this is just the story for the newcomer, the first-time political candidate. Political commercials designed for the reelection contest are indeed different. The problem here is not name recognition, but a political track record put under assault and harsh criticism. Long track records, as used in political commercials, can become a burden. In 1982 Clinton faced the same problem in his attempted comeback in the Arkansas gubernatorial election as he had in his defeat in the 1980 contest. President Lyndon Johnson faced such a situation with his potential reelection bid in 1968. Presidents Carter and Bush had track records that were carefully exploited against them, and they lost even though each was an incumbent president. Hence, political commercials for incumbents or candidates for reelection are designed to win voters back and/or attract a new target audience.

To overcome "unpopular decisions" or the "breaking of past campaign promises" as well as questionable records of yesteryear, political commercials have adopted a well-known and successful technique known as the "apology" strategy.[52] In such commercials, the candidate acknowledges past mistakes, errors, shortcomings, and weaknesses and promises to atone for them. There will be no political errors and problems in the future. These ads tell the electorate that the incumbent has gotten their message loud and clear and will follow it slavishly in the future. In some reelection bids, these clever ads have helped incumbents overcome poor track records.

A slight twist on the "apology" political commercials are those ads that call errors and unpopular positions in the past "tough decisions" that had to be made. As such, these unpopular actions should be applauded and respected, not derided. They deserve reelection.

Overall such political commercials, whether they are using some version of frankness or truthfulness as a basis, literally turn error and mishap and shortcoming into virtue, and the political compromises into positive values. Incumbency becomes a focus of heroism.

Although political commercials for first-timers and incumbent candidates differ, as would be expected, the role of race in political commercials since the civil rights era has been nearly enduring in its staying power and presentation. The passage of the 1965 Voting Rights Act and the 1964 Civil Rights Act placed race on the public's mind and into the political process.[53] The Voting Rights Act made it possible for African-American political candidates to have an electoral base in the community, but it also made it possible for them to run in biracial campaigns. And it is the biracial campaigns that have attracted the negative political commercials.

Keith Reeves found that:

> The disposition reality is this: black office-seekers who compete in majority-white settings in the main are unable to attract consistent widespread support because *race perniciously influences both the tenor of their electoral campaigns and their outcome.*[54]

To ensure that race impacts such campaigns, negative racial ads and commercials surface. Examples are the 1983 Chicago mayoralty contest, with its white candidate slogan ("Epton before it's too late"); the 1988 presidential campaign (Bush's Willie Horton Ads); the 1990 Senatorial race of Jesse Helms (where white hands appeared); and the two gubernatorial contests of Los Angeles Mayor Thomas Bradley. All of these high-profile cases, and there are many others, suggest the huge impact that negative racial commercials can have upon the contest and the outcome of the elections. These high-profile cases offer further clues and suggestions to future candidates in biracial contests. The re-entrance of African Americans into politics, then, has stimulated the use of race in both biracial and uniracial contests. Bush versus Dukakis is an example of using negative racial ads as tools to defeat the opposition.

Thus, political commercials are part of the new technology used in political campaigns that the national political parties as well as their state counterparts have used, for better or worse.

THE PUBLIC RELATIONS SPECIALISTS

The fourth area of new technology is that of public relations developed by, as some people call them, the media specialists. These are the specialists who have more to say about the nature, scope, and type of relationship that the candidate will have with the public.[55] They help develop the overall game plan for the campaign, including both the strategic elements as well as the tactical ones. And they help craft the candidate's image presented to the public. It is this image through which the public sees the candidate's personality and competence.

With an image crafted, the specialist implements the campaign plan and schedules public appearances, refining issue positions and shaping press releases. The releases coordinate positions, pictures, and words. Reagan did this successfully in 1984. His campaign introduced a new theme every ten days to two weeks and keyed the president's speeches to it. Situations that might have distracted public attention, or confused it, were carefully controlled. The Clinton campaign in 1996 adhered to this practice as well.[56] The public relations specialist puts it all together by coordinating much of the candidate's contact with the public and the media. Ideally, no stone is left unturned. The public relations people try to orchestrate all aspects of the candidate as he or she is presented to the public. These specialists "make the candidates seem real to the voters," testifying to the candidate's concern for the public they are addressing.[57] And like the other specialists in campaign organization, these specialists use every form of modern technology to improve their art and science. Computer technology is extensively used. Arterton describes how Clinton's specialists used computer technology to coordinate the media campaign with differential voter interests in geographic areas.

> It began by superimposing media markets of a map on the United States. Week by week, each media market was ranked in terms of the number of persuadable voters in the market weighted by the Electoral College voters and the perceived strategic importance of the states reached in the market. The resulting map, in which the media markets were arranged on an eight-point, color-coded scale, quickly revealed where the campaign needed to place its emphasis in travel, field organization and media buys.[58]

With computer technology, the public relations specialist can target certain areas for special effort.

Public relations experts focus media attention on the candidate. Political candidates all over the country have been forcefully and painfully reshaped by consultants — changed diets, new coiffures, refurbished wardrobes, and voice lessons, to say nothing of new ways of walking, different speaking tones, and other emphases linked to the specifics of the campaign strategy. Professionals attempt to make their candidates physically handsome and more attractive, because "ugly folks get [fewer] votes."[59] Due to his loss in 1960 to the more attractive John F. Kennedy, Richard Nixon lost weight, kept a constant suntan for color television, and shaved several times a day to alter his "shadowy look."[60] In 1976, Jimmy Carter "slimmed down twenty pounds, had plastic surgery to correct sagging eyelids, let his hair grow longer, and purchased a number of 'made for TV suits',"[61] while in 1996,

Senator Robert Dole was "improved" by being associated with younger people, in different places, in his political commercials. Hence, public relations specialists use their technology not only for interactive purposes, but also to "improve" the individual candidate as well. These specialists work to sell the candidate to the public, to "improve" the product so that it will be attractive and voter friendly.

Overall, modern campaign technology, which includes (1) management specialists, (2) pollsters, (3) advertisers, and (4) public relations/media specialists who combine their skills and talents with the aid of computers, faxes, e-mail, and the Internet, is impressive. Some argue that this technology permits the political candidate to sidestep political parties and enter directly into the electoral fray, leading to a candidate-centered politics. But others would claim that parties have responded to this new technology by incorporating it into their organizations, activities, and functions.

PARTIES: ADAPTING TO MODERN CAMPAIGN TECHNOLOGY

Clearly, modern campaign technology has created a "personality cult" in American campaigns that personalizes and glamorizes American politics. Enterprising political candidates in the past "saw skills and technologies possessed by consultants that were unavailable from the parties." Back in the eighties, David Broder argued that "the availability of professionals for hire to individual campaigns has been one of the things which enabled candidates to operate independently of parties."[62] It was some time before parties recognized the impact and influence of these new campaign technologies and began adjusting to them. In recent years the parties nevertheless made the adjustments to them. The Republicans adjusted first and the Democrats eventually followed suit. Both parties have developed specific ways in which they employed and deployed these modern campaign techniques in the behalf. Beginning with William Brock, who took over as chair of the Republican National Committee in 1977, and Tony Coelho who in 1981 took over the chairmanship of the Democratic Congressional Campaign Committee (DCCC), much party adaptation to, and incorporation of, modern technology was accomplished.

The RNC under Brock's leadership built a state-of-the-art headquarters where it installed a television studio, computers, a direct-mail program and fund-raising and polling operations. The National Republican Congressional Committee (NRCC), "using techniques borrowed from catalogue merchants and mail order sellers," began sending millions of letters to people whose names appeared on likely-to-vote lists, to lapsed Republican donors, to registered Republicans in prosperous zip codes, and to subscribers to conservative magazines. And people responded well. The NRCC alone had so much money its biggest problem was figuring out how to spend it.[63]

In point of fact:

> In the two-year cycle leading up to the 1980 elections, the NRCC raised $26.8 million, more than double what it had achieved two years earlier. Its income was thirteen times as great as the DCCC's. Then Democratic Committee considered itself success-

> ful if it could afford to donate $5,000 to the campaigns of its most threatened incumbents and influential committee chairmen, who viewed the contributions as entitlement of office.[64]

With the modern state-of-the-art headquarters in place, the RNC and its NRCC began

> recruiting candidates, giving them training in modern campaign techniques, steering them to the best political consultants, supplying them with free research on their opponents, 'kick starting' their campaigns with early infusions of cash, aggressively marketing candidates to the group members of business PAC donors with money to give, and [spending] millions on a national advertising campaign, sending the message: Vote Republican, for a change.[65]

The results came in the 1980 elections. Republicans captured the presidency as well as control of the Senate and "scored a net gain of thirty-three seats" in the House of Representatives, "the second-best showing in thirty-four years."[66]

The Watergate scandal and the huge Republican losses after Watergate had forced the party to adapt to the new technology to survive. For the Democrats, however, it was the Republican landslide in the 1980 elections that provided the spur to action. "Previously, the Democratic leadership had resisted the idea of building a well-financed party apparatus. They had been able to win re-election easily enough without the help of any party organization."[67] Therefore, before the 1980 election, "Democrats had not seen much need to match what the Republicans were doing."[68] After that watershed election, which decimated the Democratic party leadership in Congress, including the head of the DCCC, Democratic leaders as well as Coelho became "afraid that with the polling, television, and mail that money could buy, Republicans might topple them from power at last."[69]

Therefore, fearing the "Republicans' money and their political 'technology' and the electoral comeback that it had won in seven years, Coelho, as head of the DCCC decided to play catch-up." Immediately upon becoming chair, "he began laying plans for a new building, a television studio, computers. Whatever the Republicans had, Coelho wanted for the Democrats."[70] And he went about it with such zeal that it created a scandal, which led to his ouster from Congress, head of the DCCC, and a jail term. But by the time the scandal was over, the DNC had a state-of-the-art national headquarters and facility.[71]

Today, both the DNC and the RNC have at their national headquarters state-of-the-art facilities.[72] And it was the building of these facilities that let them as organizations adjust and restructure themselves to meet the modern campaign technology now used by candidates. But it was not the rise of candidate-centered organizations that forced them to adjust. It was scandals and the electoral defeats that forced that to adjust, first the Republicans and then the Democrats. However, adjustment was not across the board but came in a partial and delayed and incremental fashion.

CONCLUSION

With these modern facilities, both major parties can now offer their state parties and their political candidates the same services that they must acquire from political consultants. In fact, because both parties can now offer these services, all of the major textbooks on political parties are quick to label them service or vendor organizations. Then they further declare that due to these new services, political parties have made a comeback, at least organizationally. Political parties, it is said, have once again, due to these services, become viable, meaningful, and useful to the political system. They adapted to this technology because of electoral defeats.

As a consequence of the "rebirth" and "renewal" of the major political parties because of technology, the major parties have finally become somewhat more centralized. Some degree of centralization has occurred, although state and local parties retain much autonomy. National parties now have a better relationship, and a more influential one, with their state and local organizations. The change in the practices and administration of campaign finance, as our previous chapter on party finance revealed, has enhanced this new relationship between the national parties and their lower units. Parties today have the funds that permit them to modernize our politics and to provide more integrated campaigns for public office. The cost unfortunately is high, and increasing each year. And therein lies the irony. If we reduce campaign spending, impose ceilings on contributions, eliminate "soft money," the parties will be forced to adapt again to a less financially open and generous and affluent political finance system, which will certainly have a new impact on party activities and, perhaps, on party functions.

Political party adaptation to campaign technology is not just in new services and vendors. It also has had functional and organizational consequences. Clearly parties are redirecting themselves and, the candidate-centered era notwithstanding, they have reentered the electoral equation and process in a new and quite different way. They are no longer as weak, disorganized, and nonfunctioning as they allegedly once were. The acquisition and utilization of modern campaign technology have empowered our parties very significantly.

NOTES

1. Steven J. Rosenstone, *Third Parties in America: Citizen Response to Major Party Failure* (Princeton, NJ: Princeton University Press, 1984), 121.
2. Ibid., 120.
3. Ibid., 121.
4. William Crotty, *American Parties in Decline* (Boston: Little, Brown, 1984) 73–75.
5. Herbert Simon and Frederick Stern. "The Effect of Television Upon Voting Behavior in Iowa in the 1952 Presidential Election," in Heinz Eulau, Samuel J. Eldersveld, and Morris Janowitz, eds., *Political Behavior: A Reader in Theory and Research* (Glencoe, IL: Free Press, 1956), 206.
6. Ibid., 210, footnote 5.
7. Bernard Berelson, Paul Lazarsfeld, and William McPhee, *Voting:A Study of Opinion Formation in a Presidential Campaign* (Chicago: University of Chicago Press, 1954), 134, 178, 234; and Angus Campbell, Gerald Gurin, and Warren Miller, *The Voter Decides* (Chicago: Row Peterson, 1954).
8. See Chapter 12 for the data.

9. Ceci Connoly, "Use of 'Street Money' Shows Old Ways Remain Potent," *Congressional Quarterly Weekly Report* (November 20, 1993): 3217-20.

10. James Perry, *The New Politics: The Expanding Technology of Political Manipulation* (New York: Pottes, 1968).

11. Ibid.

12. Robert Huckshorn, *Political Parties in America,* 2nd ed. (California: Brooks/Cole Publishing, 1984), 136.

13. Robert Agranoff, *The New Style in Election Campaigns* (Boston: Holbrook Press, 1972), 4.

14. Huckshorn, *Political Parties,* 131.

15. Ibid.

16. Stanley Kelley, Jr., *Professional Public Relations and Political Power* (Baltimore: John Hopkins Press, 1956), 39-66.

17. Ibid., 225-26.

18. Ibid., 226.

19. Larry Sabato, *The Rise of Political Consultants: New Ways of Winning Elections* (New York: Basic Books, 1981), xiii.

20. Ibid., 12-13.

21. Robert Agranoff, *The Management of Election Campaigns* (Boston: Holbrook Press, 1976), 1-17.

22. Sabato, *Rise,* 23.

23. Ibid., 3.

24. David Maraniss, *First In His Class: A Biography of Bill Clinton* (New York: Simon and Schuster, 1995), 330-31. See also Sabato, *Rise,* 351, on Clinton's political consultants and advisors.

25. Maraniss, *First In His Class,* 352.

26. Martin Walker, *The President We Deserve: Bill Clinton: His Rise, Falls, and Comebacks* (New York: Crown Publishers, 1996), 343-44.

27. Sabato, *Rise,* 60. See also Tom Smith, "The First Straw? A Study of the Origins of Election Polls," *Public Opinion Quarterly* 54 (Spring 1990): 31-36.

28. "Landon, 1,293,699; Roosevelt, 972,897: Final Returns in the Digest's Poll of Ten Million Voters," *Literary Digest* (October 31, 1936), 5-6; and "What Went Wrong With the Polls? None of Straw Votes Got Exactly the Right Answer — Why?" Ibid. (November 14, 1936), 7-8.

29. "Roosevelt Bags 41 States Out of 48," Ibid. (November 5, 1932), 8-9.

30. Peverill Squire, "Why the 1936 Digest Poll Failed," *Public Opinion Quarterly* 52 (Spring 1988): 125-33.

31. "Landon Keeps Digest Poll Majority," *Literary Digest* (October 24, 1936), 10.

32. "Landon, 1,293,669; Roosevelt, 972,897," (October 31, 1936), 5.

33. Edwin Amenta, Kathleen Dunleavy, & Mary Bernstein, "Stolen Thunder? Huey Long's 'Share Our Wealth,' Political Mediation, and the Second New Deal," *American Sociological Review* (October 1994): 688-89.

34. Ibid., 690-91. The results showed that Senator Long held the balance of power in the election.

35. Jacob Javits, "How I Used a Poll in Campaigning for Congress," *Public Opinion Quarterly* 11 (Summer 1947): 222-26.

36. Sabato, *Rise,* 69.

37. Ibid., 75.

38. Ibid.

39. Herbert Asher, *Polling and the Public* (Washington, DC: *Congressional Quarterly,* 1988), 95.

40. Ibid., 97-98; Sabato, *Rise,* 76.

41. Sabato, *Rise,* 77; Asher, *Polling,* 98.

42. Asher, *Polling,* 96.

43. Ibid., 99.

44. William Crotty, "Political Parties in the 1996 Election: The Party As Team or the Candidates As Superstar," in L. S. Maisel, ed. *The Parties Respond: Changes in American Parties and Campaigns,* 3rd ed. (Boulder: Westview Press, 1998), 203.

45. W. Lance Bennett, *Inside the System: Culture, Institutions and Power in American Politics* (New York: Harcourt Brace College Publishers, 1994), 506.

46. William Crotty, *The Party Game* (New York: W. H. Freeman, 1985), 83.

47. Ibid., 83-84.

48. Ibid., 84.

49. Robert Smith, *We Have No Leaders: African Americans in the Post-Civil Rights Era* (Albany: State University of New York Press, 1966), 255.

50. Ibid., 267.
51. Sabato, *Rise,* 128.
52. Ibid.
53. Smith, *No Leaders,* 3-28.
54. Keith Reeves, *Voting Hopes or Fears? White Voters, Black Candidates and Racial Realities in America* (New York: Oxford University Press, 1997), 9.
55. Kelley, *Public Relations,* 4-5.
56. Stephen J. Wayne, *The Road to the White House 1996: The Politics of Presidential Elections* (New York: St. Martin's Press, 1997), 231.
57. Ibid., 229.
58. See F. Christopher Arterton, "Campaign '92: Strategies and Tactics of the Candidates," in Gerald Pomper, ed., *The Election of 1992* (Chatham, NJ: Chatham House, 1993), 87.
59. Sabato, *Rise of Political Consultants,* 145.
60. Ibid.
61. Ibid., 286.
62. Ibid.
63. Ibid.
64. Brooks Jackson, *Honest Graft: Big Money and the American Political Process* (Washington, DC: Farragant Publishing Company, 1990), 53-54.
65. Ibid., 54.
66. Ibid.
67. Ibid., 49.
68. Ibid., 53.
69. Ibid., 54.
70. Ibid., 52-56.
71. Ibid. See "Postscript: The Unmaking of Tony Coelho," pp. 294-318.
72. Paul Herrnson, "National Party Organization at the Century's End," in L. Sandy Maisel ed., *The Parties Respond* 3rd ed., 50-80.

Chapter 15

Parties and Governance: Making Divided Government Work

Parties, as we have seen, are organizations playing a major role again today in the election of leaders — presidents, congresspersons, senators, and state and local officials. They are political action structures that mobilize money, support, and votes on behalf of their candidates for public offices. They do this in sharp competition with each other, year in and year out. But if they are to be key actors in the system, they aspire to play a significant role in presidential and legislative performance after the election. They normally seek at least three types of relevance: (1) to represent their followers, who supported them in the campaign; (2) to achieve their policy objectives outlined in the campaign; (3) to enhance their prospects for reelection, to retain power and attempt to increase it. Thus, representation, policy making, and power maximization are three key functions parties seek to achieve in the legislature.

Legislatures and parliaments are central institutions of the democratic order. Members of parliament and congressional leaders, together with presidents, prime ministers, and their cabinets are the top political decisionmakers. Parties are the major routes to, means of, and actors in access to this top decision-making authority. Historically, in many democracies the legislature was the place where parties were born, as contending groups of leaders differed over policy and sought to determine policy. Parties that came into existence outside the legislature soon fought their way into the legislature. It was the central deliberative arena, the arena of open conflict, of freedom to speak, to dissent, to protest. It was the forum for debate. It was also the authoritative lawmaker. That is why, when systems become authoritarian, legislatures are dismissed or ignored and, in the process, party conflict is proscribed and party organizations eliminated. The legislature thus must be seen as the center of democratic action, and of partisan democratic action.

The U.S. Congress plays a critical role in determining national policy. It has been referred to as "the keystone of the Washington establishment," "the center of national policy making," and "the first branch of government" (a reference to the first article of the Constitution, which deals with the powers of Congress). Woodrow Wilson placed much emphasis on it (in 1885 before he became president) by entitling his book *Congressional Government.* Congress was not always highly regarded. We have gone through periods of the so-called imperial presidency, when it is argued that Congress was clearly overshadowed by, even subservient to, the president. Today, however, it has to a large extent established itself as a major force, even during the very frequent periods of divided government, such as the current period from 1995 to the present, a period with a Democratic president and a Republican Congress. The conflict within Congress between the parties has been fierce — 227 Republicans versus 207 Democrats in the House, and 55 Republicans versus 45 Democrats in the Senate, and a Democratic president elected with less than 50 percent of the popular vote. Yet all the laws regulating American society need to be, and are somehow, passed by this legislative body. Many of these laws have major impact on our citizens because they deal with issues such as the minimum wage, health care, crime, tax cuts or increases, welfare reform, civil rights, educational subsidies, farm supports, and environmental regulations. One scholar noted that since World War II, 17,663 public laws have been passed (1945–1994), approximately 300 of them absolutely major pieces of legislation. Despite divided government, this legislative system seems to be productive. The major question we will discuss in this chapter is, What role has our party system had in this legislative performance?

THE U.S. CONGRESS: TWO CHAMBERS WITH SPECIAL FEATURES

The U.S. Congress is a legislative system with its own character and history. It is first of all a genuine bicameral body, with two houses of equivalent legislative power. Few countries have this system, preferring one with a dominant lower house or no upper house at all. In Britain, for example, the House of Lords, the upper chamber, is largely impotent and discredited as a lawmaking body. In the United States, however, the two houses share legislative power. The Senate's role in the early days, according to George Washington, was to be the saucer that cooled the hot tea of the House of Representatives. But as Polsby put it, the Senate has become a body "brewing more of the tea than they're cooling."[1] Together the two houses share governmental power with the president and the courts. The Congress has historically been a true center of politics and authority. Not only does it pass all national laws, it also spends a great deal of time overseeing the bureaucracy, approving judicial appointments, sometimes overruling judicial decisions, delegating responsibility to the state governments, and checking the exercise of presidential power. Both chambers are intensely partisan today, more so than in the recent past. Both bodies are products of the two-party system, reflecting strong competition for seats during the campaign as well as strong and sustained conflict in deliberating policy and voting on legislation. Each of the two parties in both chambers is today relatively unified on policy, and as a result the legislative parties are polarized.

The House and Senate are quite different in structure, process, and style of legislating. The Senate is very individualistic; each senator has to be taken into consideration by the Senate leadership, because any senator can at any time delay legislation. Thus, senators have considerable power if they want to assert it. The House members do not have such "veto rights" as individuals, and as a result their behavior is less individualistic and more controlled. They work more through committees to exercise influence, or through their party organizations. The House is a truly majoritarian body in that one-half plus one of the 435 members can adopt legislation; in the Senate the filibuster threat means that a "super majority," 60 percent of the 100 members, is necessary to terminate the filibuster. The Speaker of the House is a more powerful figure than the majority leader of the Senate. The Speaker has more appointive power and more power over the agenda and over floor procedures in debate over proposed legislation.

The organization of the parties in the two chambers also differs considerably. We described and discussed this party organization in detail in Chapter 6. There are certain key characteristics of party organizations in the House and Senate. The party leadership of both parties has been strengthened as a result of the reforms in the House from 1975 on, and more recently in the Senate. The floor leader and the chief whip play major roles in the House, and although they have become more important in the Senate, their influence by no means matches that of the House leadership. The party caucus (Democrats) and conference (Republicans) play a major role, particularly in the House, in the selection of the party leadership and the chairs of committees and subcommittees, but also in setting the agenda for the party. After the Republicans took over control of both the House and Senate following the 1994 election, they used their party organization, under Gingrich in the House and Lott in the Senate, as a more homogeneous party team to implement their agendas. We will discuss later the evidence of this emergence of strong party control in both chambers.

TWO CONCEPTIONS OF THE ROLE OF PARTIES IN CONGRESS

In the past, scholars have argued that many obstacles have prevented the development of strong parties in our legislatures. The traditional view was that the party organizations and party leaders could not mobilize and control the support necessary to determine or significantly influence legislative decisions. This was attributed to the decentralized and undisciplined nature of our parties, the constitutional structure of our system (bicameralism and separation of powers), and the fragmentation of the authority system within the House and Senate. Strong speakers controlled the House up to 1910, when a revolt by the members occurred. After that, it is argued, there was no central leadership control of the House, and power devolved on committee chairs. There was a variety of party officials (floor leader, whips, policy committee persons, the caucus) but none of these had real authority. The Congress was at the mercy of the president, bureaucrats, interest groups, and constituencies. As Keefe put it, in those days the party seemed "to fly apart." We had "a shared, multiple-leadership form of government."[2] Implicit in this traditional model was the contention that parties were secondary influences in policy making. It assumed that American parties could not overcome the disabilities inflicted on parties by the fragmentation, diffusion, and decentralization of the American system. The American system contrasted to the British system of centralized legislative party dominance.

Today, in examining the question of the party's relevance in congressional decision making, we face two different interpretations. On the one hand is this traditional view that since "the 1910 revolt" we have had weak party control and strong committee control. This is the "committee government" conception, actively first advanced by Woodrow Wilson in 1885. Those who support this interpretation are divided among those who say it was true up to the 1970s, when it was gradually replaced by stronger parties in Congress, and those who even after 1970 have continued to insist the committees maintained much control in Congress. The second model or conception is one arguing that party leadership has, to varying degrees, played an influential role in Congress. This we can call the "party government" conception. The supporters of this model are also divided between those who claim that from 1970 on we have had a progressive increase in party control in both houses of Congress, and those who argue that even in those earlier years the parties had much more influence than was originally thought to be true. The study by Cox and McCubbins, ***Legislative Leviathan***, is the primary advocate of this latter position.

The "committee government" theory, analyzed extensively by scholars, can be summarized fairly easily. When the Speaker lost his control ninety years ago, that power was presumably taken over by the standing committees of the House. The essential features of that committee control system were:

1. Each standing committee had autonomy, i.e., the right to handle without interference all proposed legislature in a given area (e.g., labor policy, housing, tax, budget, foreign affairs, and so on).
2. Each committee chair had unrestricted control over the committee, with indefinite tenure, with power over whether and when to meet, whether to report out proposed bills, what bills to consider, selection of staff, appointment of all subcommittees, which were controlled also by chairs responsible to the chair of the main committee, and so on.
3. The seniority system was used in the promotion of members on the committee over the years, i.e., the party members of a committee were ranked on the basis of length of service.
4. There was no control by the speaker or any party leader or party organ over committee agendas or behavior. There was, particularly, no control over the Rules Committee, which was autonomous in deciding what bills would be debated on the floor, when, for how long, and under what rules of amendment, and so on.
5. The party caucus of the majority party did select the Speaker and other House officials at the beginning of each Congress, but the caucus had no control beyond that as to agenda or procedures.
6. The lack of party discipline and absence of homogeneity in party ideology meant that voting in Congress was characterized as a cross-party "bargaining process," with party leaders having difficulty in maintaining party loyalty among members.

Thus, the House was seen in these years under this system as consisting of a set of multiple leadership power centers — the committee chairs, who often were rivals with each other, and with the Speaker and other party leaders. But the bottom line was that the committees made policy; the parties exercised no real policy-making control. Most scholars accepted this interpretation, certainly for the period up to the 1970s. One scholar who questioned this interpretation somewhat was David Truman, who suggested in his book *The Congressional Party* that behind this facade of "committee government" there might be evidence of a "congressional party."[3] It is this position which Cox and McCubbins explore carefully and systematically later. But for the most part, scholars believed as Mayhew did (1974): "No theoretical treatment of the United States Congress that posits parties as analytic units will go very far."[4]

The second conception of how Congress functioned is the "party government" model. In the analysis of most scholars, party control has emerged as a result of the reforms begun by the Democrats in the seventies. The new liberal Democrats elected to the House were upset that the conservative coalition of Republicans and southern Democrats controlled House decisions.[5] Working through the Democratic Study Group (DSG), these Democratic liberals decided to launch a drive to make Democratic leaders more responsible to the Democratic House members. A new Committee on Organization was set up by the Democratic caucus to develop proposals for reestablishing majority party control over the legislature. As a result, various reforms were adopted by the House Democrats, and these reforms revolutionized the way the House did its business. The Democratic party caucus was revived and assumed new prerogatives. All major party organs would have leaders selected by the caucus. Committee chairs would now no longer hold their positions by seniority but, at the beginning of each House, would be subjected to a secret ballot of all majority party caucus members. The decision was made to create a new body, the Steering and Policy Committee, appointed by the caucus, to make all Democratic committee assignments. This body would be chaired by the Speaker, whose power was to be strengthened in other ways also under the new rules. The Speaker was given the power to appoint the Rules Committee, subject to caucus approval, and this committee thus became the agent of the party leadership in managing the debate on the floor. The whip system was expanded, as an adjunct to the party leadership, and given more responsibility for contacting members, canvassing their positions on issues, and mobilizing their support on legislation. The Speaker could set the agenda for the House, working closely with the other leaders and the caucus. Subcommittees were also strengthened, so that they were no longer subject to control by the chairs of the main committees. Many other specific changes were made to committee power and floor procedures. These reforms, then, let to a caucus-based, policy-oriented system of party control of the House, while the Democrats were in the majority. Later, when the Republicans became the majority, they continued the same type of party control.

To summarize, the party government model focuses on, and implements, several key conditions or principles:

1. Party leadership has primacy of control over committee leadership.
2. The party caucus is the plenary body, the instrument for involving members in party discussion.
3. Decisions on legislation are no longer made primarily by cross-party coalitions and bargains but by the majority party as a result of a more cohesive and disciplined party membership.
4. The majority party's agenda, ideology, and policy preferences determine how committees function; the rules of the House and its procedures would be "consistent with majority party interests."

Thus, this model of party leadership and organizational control emerged in the late seventies under the Democrats and has survived to the present, even under the Republicans. This model was intended to, and does, bind the individual member more closely to his or her party. The legislative party is committed to ideological goals and issue positions and functions now in Congress as an organization that seeks to implement these goals and positions responsibly. When the new reforms were adopted in the House, the Democratic leadership was motivated not primarily by such lofty ideology, but by pragmatism. Through such a reformed party organization, the Democrats hoped to facilitate party control over the House and to pass legislation favorable to the Democrats. As Barbara Sinclair put it: "policy-oriented party leadership emerged in the House in the 1980s because the costs and benefits to Democratic members . . . [had] changed significantly. . . . With the Senate again under Democratic control and the Reagan presidency weakened, House Democrats perceived an opportunity to pass major legislation that had been stymied throughout the six years of the Reagan administration."[6]

Not all scholars accept the view that the "party control model" properly characterizes how the House functions in recent years. Some see the reforms as actually decentralizing the legislative process further, by giving more power to the many subcommittees. Thus, they feel it is as difficult now as before to mobilize party support for a coherent legislative program. Those who rebut this view agree that the House is still very decentralized, and that individualized members have closer ties and loyalties to their constituents in their congressional districts. Some scholars advance a modified view of the role of parties, called the concept of "conditional party government." They see party control as dependent on the homogeneity of preferences within the majority party. However, it is contended that on critical issues on which there is conflict and controversy, in recent years members are mobilized along partisan preference lines much more than before, and reflect greater party unity in response to the interests of their party articulated by and through the party organization and its leadership.[7]

This assumption of party control in the Congress has recently permeated the Senate and House Republican party organizations after they took control of Congress in the 1994 election. Even though the Senate remains very individualistic, members have also been pressing for broader involvement in policy discussions and in party decision making. Sinclair reports that Senate parties are more "homogeneous and polarized," which has meant a greater policy role for the party, its leadership working closely with individual senators: "On both sides of the aisle the party organizations and party processes have been elaborated to provide services to members and to include members in all aspects of party functioning." The image of the dominant leader (Lyndon Johnson, Bob Dole, Robert Byrd) has given way to the image of a party team to get things done.[8] The Republican Senate Conference in early 1997 changed its rules to limit chairs to three terms, require a secret ballot vote on committee chairs both in the committees and the conference, and provide for the adoption of a Senate GOP agenda in the conference by a three-fourths vote. These changes sound very similar to what happened in the seventies and eighties in the House Democratic reform of its organization. However, Sinclair reminds us that pushing through a party agenda in the Senate is much more difficult, because of the individual prerogatives of each Senator to amend, to debate, and to filibuster. While in all of the nineteenth century there were only 23 filibusters, in the 1970–1996 period alone there have been 234. Indeed, the concept of a "filibuster-centered partisan strategy" now has emerged.

The House Republican party under Speaker Newt Gingrich has also been transformed since the 1994 election. Indeed, it has been said that Gingrich operates like a "prime minister," while the Republican floor leader Richard Armey operates like the speaker.[9] It is clear that the Republicans have taken over the basic concept of party government from the Democrats and have gone further in that direction. The Republicans have made major institutional changes (cutting the number of standing committees by three and the subcommittees by at least twenty-five, as well as cutting six hundred staff employees). But their basic changes in the power of the Republican leadership mean a centralization of power in the party organization. Some of these changes reveal this clearly. They have returned a great deal of authority to committee chairs (for example, the power to name subcommittee chairs and committee staff). But these chairs are now selected by the Speaker with the consultation of the party conference. The tenure of committee chairs is definite (three terms). The Speaker can appoint persons considered to be more loyal and active, thus ignoring the seniority rule. A few first-year Republican congresspersons were already appointed committee chairs in 1995. The Republican Conference's status is enhanced also. It takes positions on key legislation, and committee chairs are obliged to adhere to conference policy as they manage bills on the floor of the House. The Republican party team thus consists of Speaker Gingrich at the center, working closely with a subordinate floor leader, a chief whip, and a caucus set up to develop policy positions and to impose these on a set of committee chairs, which in the long run serve at the pleasure of the top leadership. The Speaker's term, the Republicans have decided, is limited to eight years, but that does not interfere with his prerogatives. He in fact sought to establish himself as the national Republican party leader, pursuing his announced goal and vision of "renewing American civiliation." In summary, Koopman concludes: "Republican changes in the 104th Congress have clearly ended any vestiges of the decentralized system, and in fact done much to formalize the less explicit controls of their most recent Democratic predecessors."

PARTY VOTING AND PARTY COHESION TRENDS IN CONGRESS

For many years scholars of Congress observed a low level of "party voting" in Congress, as well as evidence of some decline in intraparty cohesion. This was presumably testimony to the weakening of parties at the legislative level, as they were allegedly becoming weaker at the organizational and electoral levels. By a "party vote" we mean a "confrontation," in a sense, of the majority of Democrats (in the House or Senate) voting against a majority of Republicans. By "party cohesion" we mean the frequency with which the Democratic (or Republican) members of the chamber vote the majority of their own party on roll-call votes.

In the past thirty to forty years, "party voting" has fluctuated. It was relatively high in the period of the Kennedy and Johnson presidencies, then declined under the Nixon and Carter presidencies (Table 15.1). But in Reagan's second term and since, there has been a steady increase in the extent of "confrontational voting," rising from 44 percent in the sixties to 64 percent in Clinton's first term. Thus, from a very low point in the early seventies of 27 percent, we have changed considerably in the pattern of legislative party conflict. In the 1995 session of the House, 72 percent of the votes were party votes. This, one must recall, was the year following the Republican victory in the 1994 election and its reorganization as the dominant party in both chambers.

It follows that party cohesion within the two parties would also result in Congress — that is the members of each party would exhibit more party loyalty in their votes. As Table 16.1 shows, there has been a steady increase in such cohesion. The low point in both parties was in 1969–74, when cohesion was at the 70 percent level. By 1995 the Republicans had jumped to 91 percent cohesion in the House, while the Democrats increased to better than 80 percent. If we look at the combined vote for both the House and Senate over the years, the same pattern emerges (although Senate cohesion has usually been lower than in the House): a low of 58 percent for both parties in 1970, rising to 70 percent or better in the eighties, and then increasing in both parties to over 80 percent on the average in the nineties. Thus, the prediction of some scholars of the inevitability of continuous decline in party voting has not been fulfilled. Both parties, in terms of the voting behavior of their members, clearly constitute majoritarian voting blocs in the Congress. As Rohde put it, "the reports of my death [the party's, that is] are greatly exaggerated."[10]

The voting behavior of members in the House is, however, only half the story. As Cox and McCubbins inform us, a key test of the revival of the party government thesis is whether or not the leadership of the two parties in Congress is unified and whether these leaders confront each other fairly regularly in roll call voting. One question is whether there is a leadership "agenda" that the party pursues, and the second question is whether there is evidence that the parties confront each other with policy alternatives, "differing agendas," and in turn are supported by their party members as these two alternative agendas of the leadership clash in Congress.[11] The measure they suggest to determine the first question (whether there is a party leadership agenda) is the extent to which the top leaders of the party (par-

TABLE 15.1 ■ Fluctuations in Party Voting and Party Cohesion in the House of Representatives, from Kennedy to Clinton (average percent)

		Party Cohesion	
Presidential Era	*Party Voting*	*Democrats*	*Republicans*
A. The Kennedy years (1961–63)	48.0	83.0	82.0
B. The Johnson years (1964–68)	44.0	78.0	80.0
C. Nixon's first term (1969–72)	31.0	71.0	74.0
D. Nixon's second term, with Ford (1973–76)	39.0	74.0	74.5
E. The Carter years (1977–80)	40.0	75.0	78.0
F. Reagan's first term (1981–84)	44.0	76.0	78.0
G. Reagan's second term (1985–88)	57.0	79.0	78.0
H. Bush's term (1989–92)	56.0	80.5	75.5
I. Clinton's first term (1993–96)	64.0	82.0	86.5
Low year	Nixon (1970, '72) = 27%	Nixon (1972) = 70%	Nixon (1969, '74) = 71%
High year	Clinton (1995) = 72%	Reagan (1987–88) = 88%	Clinton (1995) = 91%

Definitions: "Party Voting" = percentage of roll calls on which a majority of one party opposed a majority of the opposing party.

"Party Cohesion" = percentage of members of a party voting with a majority of their own party, *on party votes.*

Note: There are some small discrepancies in data on party cohesion, but we have accepted the Rohde data up to 1989 as correct, taken from Norman Ornstein, Thomas E. Mann, and Michael Malbin, *Vital Statistics on Congress, 1989–1990* (Washington, DC: Congressional Quarterly Press). From 1989 to the present we have used the *Congressional Quarterly Reports.*

SOURCES: David W. Rohde, *Parties and Leaders in the Post Reform House* (Chicago: University of Chicago Press, 1991), 15; *Congressional Quarterly Weekly Reports.*

ticularly the floor leader and the chief whip) agree on policy, as manifest in *their* roll call voting behavior. The measure they suggest using for the second question (whether there is confrontation between the leadership teams of the two parties) is the extent to which roll calls are confrontational (with the leaders of the Republicans in the chamber voting "yea" when the Democrats vote "nay," and vice versa). To test the concepts by using roll call votes is an innovative approach to the clarification of what "party voting" means, and an innovative way of operationalizing the measures to be used. What we have to study and understand is, as they state, the extent of "organized party conflict," and to do that we need to "isolate those roll calls on which the organizations are indeed active."

Summarizing the great body of data on legislative roll calls from this perspective of leadership roll call behavior provides us with new insights on the strength of parties in Congress. We present these data by presidential era (Table 15.2). Cox and McCubbins demonstrate from their data that during Franklin Roosevelt's presidency, from 1933 on, there were years of high party leadership cohesion (at times over 60 percent for the Democrats), but that from 1947 on a decline seemed to set in, particularly for the Republicans. There was a recovery, however, by the last years of Reagan's presidency (63.5 percent for the Democrats and 53.7 percent for the Republicans). They are not inclined to see a long-term erosion of party leadership cohesion, or the "party agenda" for the Democrats.[12] Other scholars have challenged this thesis, and indeed that controversy continues. What is important for us is to see the total picture as we move beyond the Reagan years to the Bush and Clinton presidencies. Our data strongly suggest (Table 15.2) that we are today much

TABLE 15.2 ■ Party Leadership Cohesion, by Party, and Cross-Party Leadership Conflict on House Roll Calls, Kennedy to Clinton (average percent)

	Party Leadership Cohesion		Cross-Party Leadership	Members Support Levels for Leaders of Own Party on Party Leadership Cohesion Roll Calls	
Presidential Era	*Democrats*	*Republicans*	*Confrontation*	*Democrats*	*Republicans*
A. The Kennedy years (1961–63)	53.0	40.5	28.0	85.0	84.0
B. The Johnson period (1964–68)	41.0	39.5	23.0	75.5	80.0
C. Nixon's first term (1969–72)	29.5	31.5	10.0	78.0	76.5
D. Nixon's second term (1973–76)	45.5	34.5	20.8	80.5	75.0
E. The Carter years (1977–80)	43.0	33.5	17.5	81.0	79.0
F. Reagan's first term (1981–84)	42.5	42.0	24.5	81.5	79.5
G. Reagan's second term (1985–88)	58.0	52.0	38.0	88.0	79.5
H. Bush's term (1989–92) *1992:*	85.0	74.0	40.0	*	*
I. Clinton's first term (1993–96)					
1995	85.0	94.0	68.0	*	*
1996	79.0	88.0	45.0	*	*

* Data not available.

Definitions: "Party Leadership Cohesion" = percentage of all roll calls on which the top leadership (floor leader and whip) agreed (called the "agenda measure")

"Cross-Party Leadership Confrontation" = percentage of all roll calls in which the top leadership (both leaders of the party) of the Republicans opposed the top leadership of the Democrats

"Members Support Levels" = percentage of party members of a party who supported their leaders on roll calls when there was party cohesion for their party's leaders

SOURCES: Gary W. Cox and Matthew D. McCubbins, *Legislative Leviathan: Party Government in the House* (Berkeley: University of California Press, 1993), 148–154; *Congressional Quarterly Weekly Reports* used for the data for selected recent years: 1992, 1995, 1996. We report here our calculations for the last year of Bush's term, and the last two years of Clinton's first term.

more in a period of party confrontation and cohesive party voting than ever before, going back to Roosevelt and the New Deal. In 1992 the Republicans already jumped to 74 percent party leadership cohesion in the House and in 1995 increased that to a remarkable level of 94 percent (88 percent in 1996), while the Democrats increased their leadership cohesion to 85 percent in 1996. Confrontational votes have also increased. As a consequence, the two parties reveal that they are engaged more than ever in organized legislative conflict — not just the result of coalition building on the floor of the House, but under the management of a policy-oriented party organization — and that policy-oriented combative legislative leadership has strong membership support. We seem to have entered a new era of legislative party organization and structured policy debate and conflict.

One must be cautious in using this empirically based model of legislative party government too exclusively in understanding Congress and policy making in the United States. Obviously, legislation is passed as a result of breaches in the roll call voting behavior, as expected by this model. Compromises have to be made. Party unity can disappear on critical roll calls from time to time. There are many examples of this. Take, for example, the vote on the final House passage of that very important welfare bill on July 31, 1996. Here was the vote:

Vote	*Republicans*	*Democrats*	*Total*
Yes	230	98	328
No	2	98	101 (one Independent)

Here we have a strong Republican majoritarian "agenda," but the Democrats could not hold their members, in the face of the president's avowal to sign the bill. There are, thus, major exceptions, but the general pattern recently seems to be toward more party voting, for leaders and members. However, there are scholars who dispute certain aspects of the Cox and McCubbins theory and argument.[13]

A major factor in the development of Democratic party cohesion was the gradual increase in loyalty of the southern Democratic wing of the party. In the 1960s and 1970s the southern Democrats, who were conservative on much of the domestic agenda, joined the northern Republicans to pass legislation. This was the conservative coalition that had irked the newly elected liberal Democrats in the mid-seventies and led them to develop their proposals for reform of the Democratic House organization. Already in the eighties during Reagan's presidency, southern Democrats defected less, and they voted more loyally with their northern colleagues. The party cohesion scores of southern Democrats reveal this clearly. They increased from a low 44 percent in 1972 to 60 percent by 1980, and then increased fairly consistently, so that by 1988 they revealed an 81 percent Democratic cohesion score.[14]

The impact of this cohesion on policy was considerable. As various scholars demonstrated, on racial issues this turnaround by southern Democrats helped in the passage of the extension of the Voting Rights Act in 1981. Republicans sought to weaken the bill, but on five such weakening amendments only a minority of the southern Democrats voted with the Republicans, and on final passage the vote was 71–6 in favor among southern Democrats. Similarly, in the 100th Congress of 1988, the Democratic effort to pass legislation to overturn the 1984 Supreme Court decision *(Grove City v. Bell)* that limited the applicability of civil rights laws to federal aid programs was opposed by Reagan and the Republicans in Congress, but Reagan's veto was overridden with the southern Democrats joining the northern Democrats 73-9.[15]

A big question in all this is, To what extent does party unity voting lead to success for the majority party in passing legislation? Is party cohesion the key factor? Or is success linked more to having a president in office of the same party as the majority in Congress? Or is success in the passage of legislation also the result of elections that result in a surge of new House (or Senate) members to Congress? It is difficult to sort out precisely what produces legislative success for a party. Yet the research done on this question tends to argue the importance of the extent of party cohesion.[16] In the period from 1955 to 1988, the data suggest the Democrats won most party unity votes under these two situations:

1. During the Kennedy-Johnson years, particularly after the 1964 election, the House Democrats had a high success rate in getting their legislation passed. This period produced moderate to high levels of Democratic party cohesion (from 75 percent to 83 percent). The success rate was 76 percent.
2. In the last two years of the administrations of Republican presidents (Eisenhower, Nixon, and Reagan), the Democrats increased their vote at the polls and their legislative cohesion increased also (in the Reagan period to 86 percent). The success rate average was 77 percent during these periods.

The Democratic legislative success rate was relatively low under President Carter (only 72 percent), attributable probably to the conflict between Carter and the Democratic liberals in the House. The Democrats' success rate was also low in the first six years of the Nixon presidency, when Democratic electoral success was relatively low (at the 57 percent level) and party cohesion was low (at the 71 percent level or lower). Thus party cohesion, linked to electoral success, did lead to a high level of Democratic victories in Congress *even under Republican presidents,* if certain conditions prevailed.

FURTHER EVIDENCE OF PARTY INFLUENCE ON CONGRESS AND ITS MEMBERS

Even in the years of weak party development, scholars noted the relevance of the party for legislation. The economic policy of the national government is one such important area. Tufte is one scholar who studied this matter and concludes that "the real force of political influence on macroeconomics performance comes in the determination of economic priorities." He then argues, "Here the ideology and platform of the political party in power dominate. . . . the ideology of political leaders shapes the substance of economic policy."[17] Indeed, his position is that one can generalize for modern democratic societies, including the United States, as follows: parties of the right (including the Republicans) favor "low rates of taxation and inflation with modest and balanced government budgets; oppose income equalization; and will trade greater unemployment for less inflation most of the time." Parties of the left, (including the Democrats) favor "income equalization and lower unemployment, larger government budgets; and will accept increased rates of inflation in order to reduce unemployment." The platforms of the national parties reveal these differences. Thus in 1976, the Democratic platform pledged "a government which will be committed to a fairer distribution of wealth, income and power." The Republican platform in 1976 pledged "less government, less spending, less inflation." In 1980 the Democratic platform promised to fight inflation but not by increasing interest rates or unemployment. The Republicans said that "our fundamental answer to the economic problem is . . . full employment without inflation through economic growth."

Tufte demonstrates that the actual employment statistics over time reveal a linkage between presidential elections and unemployment and inflation rates. These data point to the following "rules":[18]

1. Both Democrats and Republicans will reduce inflation or unemployment if there is an economic crisis and an election is approaching.
2. If there is no real crisis, the Republicans will do much better in reducing inflation than unemployment; the Democrats will do better in reducing unemployment.

Another scholar, Douglas Hibbs, has also explored this problem. He concludes that "inter-party differences in government-induced unemployment levels is 2.36 percent"— a sizable difference in national employment levels as a result of a Democratic or Republican administration. Thus, "the Kennedy-Johnson administration posture toward recession and unemployment stands in sharp contrast to Eisenhower's, . . . the basic economic priorities associated with the Eisenhower era were re-established during the Nixon and Ford administrations" and were "deliberately induced." Hibbs concludes, "The real winners of elections are perhaps best determined by examining the policy consequences of partisan change rather than simply by tallying the votes."[19]

A study of the policies of our government over the years finds that whichever party is in power for a longer or shorter period of time is crucial for the content of public policy. In an exhaustive study of laws adopted by the U.S. government from 1800 to 1968 (requiring analysis of 60,000 pieces of legislation), Benjamin Ginsberg was able to determine when the peak points in the adoption of new policies and new laws occurred. He concluded that the peak points were 1805, 1861, 1881, and 1933. These were years after major elections in which a shift in the power of the political parties occurred, called in some instances major "realigning elections." His basic interpretation is that "clusters of policy change" do come as a result of partisan change in electoral choices. He summarized as follows:

> Our findings suggest that voter alignments are, in effect, organized around substantive issues of policy and support the continued dominance in government of a party

> committed to the principal elements of the choice made by voters during critical eras. . . . Partisan alignments form the constituent bases for governments committed to the translation of the choices made by the electorate. . . . The policy-making role of the electorate is, in effect, a continuing one.[20]

Such scholarship refutes the charge that, in the days of allegedly weaker parties, congressional members were not influenced by party. Indeed, as Keefe argued in 1980, "Even though weakened, party affiliation is still the most important variable in influencing the voting behavior of members of Congress."[21]

In an early study (1969) of voting decisions in Congress, Kingdon analyzes the way in which "party" was significant. He noted that few members of the House spontaneously mentioned party leadership as the "actor" or consideration of importance to them in making up their minds on a given piece of legislation (see Table 15.3). Despite this finding that party leadership in the House appears to be unimportant because congresspersons do not emphasize party, Kingdon recognizes the differences that exist in the voting patterns of Republicans and Democrats and attempts to explain these contrasting voting patterns in Congress:

> The fact that party regularity in voting remains, despite the apparent unimportance of the leadership, leads one to search for alternative explanations. Party voting seems to begin with constituency differences; the parties have very different demographic bases and supporting coalitions. Building on constituency differences are the patterns of interaction within the House. Congressmen rely heavily on informants within their own party grouping. Common campaign experiences cement the party regularity, since Democrats and Republicans are electoral opponents both nationally and locally.[22]

On this matter of friendships and interactions in the House, Kingdon presents some very interesting evidence of the partisan nature of such contact patterns. In 75 percent of the decisions, congresspersons reported that other members of the House were important informants helping the respondent arrive at a decision on a particular issue.

The distribution of these informants by party is very revealing (see Table 15.4). The congruence in partisan associations found in these data is significant.

TABLE 15.3 ▪ Factors Spontaneously Mentioned by Congressional Members as Influencing Their Decision Making

"Actors"	*Percentage*
Party leadership	10
Constituency	37
Fellow congresspersons	40
Interest groups	31
"Administration"	25
Staff	5
Reading (of materials about the legislation)	9

SOURCE: John Kingdon, *Congressmen's Voting Decisions* (New York: Harper and Row, 1973), 18.

TABLE 15.4 ▪ Source of Guidance and Advice for Members of Congress (in percent)

	Respondent's Party		
Informant's Party	*Northern Democrat*	*Southern Democrat*	*Republican*
Northern Democrat	85	14	3
Southern Democrat	4	51	1
Republican	0	6	71
Mixed	11	29	25
N=	54	35	75

SOURCE: John Kingdon, *Congressmen's Voting Decisions* (New York: Harper and Row, 1973), 78.

They suggest that the House consists of definite partisan clusters — although southern Democrats are less predictable on this than northern Democrats and Republicans. Yet in 70 percent of the cases, members of Congress are turning to their own partisan colleagues for information and advice.

THE ROLE OF THE PARTY IN THE CONSTITUENCY

Parties can influence congresspersons through their role in the constituency or legislative district. They may do this in a variety of ways. The party organization may be very active in seeking candidates to run initially and hence also in recruiting candidates to run against a congressperson who is seen as behaving contrary to the local party organization's wishes. Another way is by giving or withholding a variety of organizational resources — manpower, money, records and files on voters, forums for presentation of the congressperson's viewpoints. Further, the organization may influence key decisionmakers in the area who are important to the success of any candidate — those who control the newspapers, radio stations, and television stations, as well as interest-group leaders and prominent financial supporters. Thus, the organization may be very important to the candidate for Congress or the incumbent member of Congress. If, in addition, he or she has come up through the ranks of the local party organization, where friends and earlier associates still remain, the relevance of the party organization may be enhanced. An early study of the relationship between members of Congress and their district party organizations reported that in half of the cases the local party played a role in either the nomination or the election process, or both. If the local organization was strong, its impact was greater.[23] In another early study (1972) of New York Congress members, the evidence of linkage was even more impressive: 57 percent were strongly supported by the district party, 74 percent had been active in that organization, and up to 75 percent secured funds and/or workers for the campaigns from that organization.[24] Since these early studies, the research by Herrnson, already discussed, reveals that candidates for Congress work more closely than ever in recent years with their district organizations, because these organizations, as a result of a closer link to state and national party committees, secure more funds as well as campaign assistance for congressional campaigns.

The legislative district must indeed be very important to members of Congress. They spend more time than ever "back home" with their constituents, providing service to those who need it and explaining in countless meetings and newsletters their, and their party's, positions on current issues. A special study of constituency service in 1980 discovered that 15 percent of the American public had directly contacted their representatives and another 20 percent had talked with others who had such communication. Further, one-fifth did remember something the incumbent had done for the district. Congress members retain district offices (60 percent have at least two) that are open five days a week, staffed by an administrative assistant.[25] Party personnel often run these offices. So the public can contact their representatives fairly easily, in Washington, D.C., or in the district. Polsby reports the staggering increase in congressional staff personnel employed by House and Senate members, from 3,500 in 1957 to 11, 500 by 1985.[26] Some 600 of these were dismissed after the Republicans won in 1994, but it is obvious that members of Congress are still well staffed for all purposes, including the servicing of constituents and maintenance of contact with the local party organization.

A major question in all this is, To what extent do the roll call votes of Congress members reflect their own independent judgment, the views of interest groups, the majority opinion of the voters in their districts, and the policy positions of party leaders "back home" in the district? Limited research has been done on this matter, particularly on the party organization's role in the constituency. The important early study (1958) by Miller and Stokes gave us important insights into the congressperson's link to his or her constituency and the probable role of the local party.[27] This study documented that Congress members are attentive to constituency opinion and behavior, and do on some issues (such as civil rights issues in 1958) vote both in apparent response to constituency opinion *and* their own perceptions of constituency opinion. On other issues this correlation was weak. Further, the public lacks information about Congress (not more than 50 percent knew which party was in the majority) and has a low interest in legislative issues (only 12 percent explicitly mentioned these issues in response to questions put to them). On the other hand, 83 percent supported their party's candidate if they knew who both candidates were (in contrast to only 60 percent who supported their own candidate if they knew only the opposition candidate). As for the candidates' explanations of their roll call voting behavior in Congress, they accorded top ranking to their "personal record and standing" with the public in the district, which they ranked more important than their position on national issues and traditional party loyalties. But it is interesting that the importance assigned to party as a "voting cue" by members of Congress was not inconsequential — 27 percent said it was "very" or "quite" important, and 22 percent "somewhat important." This, of course, is nothing like the importance of party for legislative voting in European systems.[28] But it suggests that the role of party in the constituency, for both the voter and the member of Congress, is not irrelevant.

The relation of the Congress member to the voter, as Miller and Stokes said, is not a simple bilateral one, but one complicated by the presence of all manner of intermediaries: the local party, economic interests, the news media, racial and nationality organizations, and so forth.[29] The member of Congress gets through to the voter by using these "intermediaries," and they in turn are used by the voter to arrive at a voting decision. On some issues the party may play a most important role. The Miller-Stokes study suggests that the social welfare issue at that time was an example of the dominance of party influence on the incumbent Congress member. This relationship was a "responsible-party model" relationship: the congressperson's vote on this issue was closely linked to the party position in the constituency. Today, with an increase in "party unity" and "party voting" in Congress, *and* continued constituency influence, there may be more issues on which there is high correlation between the constituency position and the vote of the member of Congress.

CONCLUDING OBSERVATIONS ON THE PARTY ROLE IN CONGRESS

It has been fashionable in years past to argue that "party" does not account for much in Congress in explaining the attitudes and behavior of U.S. Congress members. The foregoing discussions in this chapter suggest that much transformation has taken place, particularly in the last quarter century. The party leadership in both parties in Congress is stronger, the party organizations play a much morc important role than before, there is more cohesion in party votes, and there is also more confrontational voting that pits a majority of House Democrats against a majority of House Republicans (and to some extent this is also true in the Senate). Congress, which was viewed in the past as a very traditional and staid body, has responded to demands for leadership, membership involvement, and meaningful party-oriented policy making. As Polsby put it, "*Even* Congress has adapted in recent days to the complex and surprising challenges of life in our times [emphasis added]"[30]

Polsby also reminds us of the intermittent stages of innovation and retrenchment in the work of Congress as a policy-making body over the years.[31]

Innovation		*Stalemate*	
1933–37	New Deal	1937–40	New Deal reaction
1939–46	Wartime	1947–63	Conservative control
1964–68	Great Society	1969–89	Consolidation and conservative reaction

Since Polsby, we can add 1989–1998 as a new period of Democratic resurgence and in 1994 the return to Republican control of Congress. Crisis led to new policies, to institutional changes, and to new leadership.

Parties count for much more in Congress these days. And the reason is that the institutionalization of legislative parties has occurred. They are no longer empty

Political Parties and Divided Government

A major phenomenon of our system since 1960 has been divided government—twenty-four years in which one party controlled the presidency and the other party controlled at least one of the two houses of Congress. President Clinton won in 1992 and both houses were also Democratic. But in 1994 the Republicans captured both House and Senate and retained control in 1996 and 1998.

There is much controversy over the impact of divided government. Some scholars argue that it makes responsible party action impossible and even that it harms the democratic process. Above all, it is argued that it leads to deadlock, which makes it difficult to enact the policies we need. David Mayhew is one scholar who questions this. He says that somehow "the legislative process [keeps] moving along rather evenly regardless of patterns of party control," and that "all else equal, in short, unified party control contributes nothing to the volume of important enactments." Other interpretations tend to reject this, arguing that the policies we need at the national and state levels are not adopted, causing real problems for our system. We have adapted to this challenge in the past. Today the frequency of divided government, along with the bitterness of partisan conflict, may limit our capacity to adapt. While the controversy continues, more scholarly studies are necessary.

SOURCES: David Mayhew, *Divided We Govern* (New Haven: Yale University Press, 1991), 179. Gary Cox and Samuel Kernell, eds., *The Policies of Divided Government* (Boulder, CO: Westview Press, 1991).

structures that only at the beginning of the Congress make the decisions on House and Senate leadership. They are active organizations deciding not only the leaders (including the chairs of all the committees, as well as of the party), but also the legislative goals and the legislative procedures by which to achieve these goals. The key test of institutionalization, as Janda put it, is "the degree of congruence in the attitudes and behavior of party members." This has been called "party coherence."[32] Our parties in Congress, it is clear, have achieved a higher level of "coherence." Whether this moves us closer to the "responsible party model" we shall have to see, as the behavior of parties in subsequent Congresses unfolds.

These developments are perhaps all the more remarkable because of the strained conditions under which our parties have to function. Congress has two quite different chambers and the parties in each have great autonomy. We also have had a long period of divided government, the presidency controlled by one party and Congress by the other. Coordinating party leadership strategy is not easy. The public's voting behavior patterns do not help. Split-ticket voting is on the increase, and presidential coattails are short. Whereas fifty years ago (1942) the Gallup national survey reported that 58 percent of the public voted a straight party ticket, today (1996) the Roper survey reports that only 29 percent of the public claim they voted "straight."[33] Thus, candidates for Congress must run their own campaigns with their own teams of activists, working with the local party organization. Yet in Congress, remarkably, they are more than ever part of a national congressional "party."

PARTISAN BASES OF PRESIDENTIAL LEADERSHIP

The American presidency is a unique institution of governmental power, one of the most powerful offices in the world. Although the exercise of this potential power ebbs and flows with the philosophy and personality of the individual holding the office, the central place of the office in the American political system is unquestioned. The president is head of a vast army of civil servants. He directs a large White House staff of his political appointees. He is commander-in-chief of the armed forces. He appoints the judiciary with the advice and consent of the Senate. He is a legislative leader with constitutional responsibility for proposing legislation and with great opportunities for influencing the agenda and decisions of the Congress. He is the leader of the mass public in more than a symbolic sense. And in all of this he is the chieftain of his political party, presumably designated by his national party convention (or through primaries today) as its "leader" elected on a program of that party, in a campaign in which that party's organizational resources (among others) were mobilized and deployed in his election. Thus when he takes office he is the leader of his party in the government.

Many people say they would prefer to have a president "above party politics." But this is never realistically possible. The president is a "partisan," although the degrees of intensity and consistency with which he exhibits his partisanship vary from president to president. And in all aspects of his leadership his party origins, his party relationships, his party ideology, and the demands of his party on him influence greatly how he performs the variety of functions that are his when he assumes the presidency. Richard Neustadt once said, "The Presidency is a place for men [sic] of politics."[34] In its broadest sense, this means, among other things, that he is the product of a competitive party system and more or less conscious of his role in that party system.

An important aspect of the American presidency to keep in mind is that it is an institution of shared power. Because of the nature of the institutions of government in the United States, the dominance of the policy-making process by any office or body or person or collectivity of persons is difficult. That in a sense was the intent of the Founding Fathers and to a certain extent it has become reality. As Neustadt remarked, we have "a government of separate institutions *sharing* powers."[35] The president shares powers with others while also participating in the exercise of others' powers. And in order to achieve results in policy making this means "bargaining," the willingness and capability to work with others, particularly members of the House and Senate, in order to make decisions. The most capable presidents are, therefore, experts in bargaining. There are limits to command in the American system, due to the separated institutions, the limits on the control which the leaders of one institution have over another, the lack of discipline, the divided partisan control between House, Senate, and the presidency, and the absence of a parliamentary system with a cabinet collectively accountable to a legislative body. In the absence of a parliamentary system with such accountability, close linkage between executive and legislative responsibility, and party discipline, our presidential system relies on the politics of bargaining and persuasion.

We must recall that presidents are more likely than not to face a Congress in which at least one house is controlled by the opposition party. Since 1968, we have had only six years of a presidency and a Congress of the same party: Carter (1977–1980) and Clinton (1993–1994). And both those presidents were elected with a popular vote close to or under 50 percent (Carter, 50.1 percent; Clinton 43 percent). These circumstances make strong party government under a "strong president" difficult. Without a clear electoral mandate after a campaign fought on clear policy alternatives, and without the support of a sizable congressional party, the president will have difficulty functioning as a strong party president. In addition, reelection of 90 percent of incumbents to Congress these days means the president faces a rather entrenched group of legislators with which to work. Above all, one must remember that the district constituencies from which Congress members are elected are diverse and quite different from the president's constituency. Congresspersons feel linked, and accountable, to their home districts, for whom the issues may have a different meaning and saliency than for the president with his national constituency.

Theoretically the president has a variety of responsibilities as party leader, or, to put it differently, a variety of opportunities to assert himself as party leader if he chooses to do so. In a sense, he is leader of "the presidential party," his own body of supporters in the public who put him into office. This is a highly diverse group of followers, liberals and conservatives, including those who have defected to the new president's party from their previous presidential party (for example, the "Eisenhower Democrats" and the "Reagan Democrats" and the "Johnson Republicans"). A president who is a strong party leader may seek to appeal directly to this support group or to the hard-core activists within that support group.

The president is also head of the party organization. Usually he reelects the national party chair and may seek to use this organization as a sounding board for his policies, as a way to put pressure on legislators, and certainly in recent years as a central task force to raise money for Congressional candidates and for his reelection.

The president is the director of the executive branch of the government. With electoral victory comes party control over the bureaucracy. Although today under the civil service system we adopted in 1883, 80 percent of all appointments are supposed to be made under the merit system, with job qualifications taking precedence over party loyalty, a large number of positions in the federal administration are still "political" and filled by partisan appointments. The same is true for federal judgeships, although the Senate must consent to these appointments and may hold them up for long periods of time (for example, as of January 1998, the White House estimated that over eighty nominees for judgeships were being held up in the Republican Senate). Studies have revealed that every president in this century has filled close to 90 percent of judicial vacancies with representatives of his own party: Clinton, 89 percent; Bush, 93 percent; Reagan, 94 percent; Roosevelt, 96 percent; Wilson 99 percent.[36] Party turnover in elections clearly has its consequences.

Presidents vary in the extent that they want to, or try to, function as party leader, while fulfilling these various roles. The way in which presidents conceive of the office in party leadership terms influences their style and strategy in dealing with the public, the party organizations, the cabinet and administration, and the Congress. The contrasts in presidential role conceptions can be illustrated by looking at the party leadership philosophies of Kennedy and Eisenhower. Kennedy summarized his conception as follows:

> But no President, it seems to me, can escape politics. He has not only been chosen by the nation, he has been chosen by his party: if he blurs the issues and differences between the parties — if he neglects the party machinery and avoids his party's leadership — then he has not only weakened the political party as an instrument of the democratic process, he has dealt a blow to the democratic process itself.[37]

In contrast, here is Eisenhower's view:

> in the general derogatory sense you can say that, of course, I do not like politics. Now on the other hand . . . it is a fascinating business. . . . But the word "politics" as you use it, I think the answer to that one would be, no. I have no great liking for that.[38]

Obviously these two presidents see their "party" and "politics" role differently. Kennedy saw himself as the party leader; Eisenhower shrank from that image. We have had presidents who were very active party leaders, including Woodrow Wilson, Franklin Roosevelt, Lyndon Johnson, and Ronald Reagan. Others, like Eisenhower, did not emphasize this role. Some presidents enjoy the political battle and rise to the challenge of mobilizing support for their ideas by putting together a coalition and driving a good bargain. They are called by Keefe "dominant trader[s] in a trading system."[39] As "traders" they know what Congress members want and can thus parlay favors, logrolling, pork barrel, promises of future support, and patronage into votes needed to get a particular piece of legislation adopted. President Johnson was a great "trader"; President Carter was not. "I am not much of a trader," Carter confessed. "That is one of my political defects for which I have been criticized a great deal."[40]

The big question is, How successful is the president in getting his legislation passed by Congress? Specifically, (1) at what level does his own party in Congress support him on his legislative proposals, (2) to what extent does the opposition party support him, and (3) how great is the "distance" or "conflict" between the two parties in Congress as they vote for or against presidential initiatives? A look at presidential support scores over time should tell us in what direction legislative conflict over policy has moved in recent years. Are we more, or less, polarized in legislative voting in support of the president's program? Is legislative cooperation with the president on the increase or on the decrease? (See Table 15.5.)

Despite all the obstacles to presidential leadership in his relations with Congress, we find over the years that presidents have been given fairly strong support

TABLE 15.5 ▪ Presidential Support Scores for Congress: Party "Distance" in House and Senate (average percent)

	House			Senate		
Support	*Democrats*	*Republicans*	*Average Distance*	*Democrats*	*Republicans*	*Average Distance*
Under Democratic Presidents						
Johnson	65.3	44.7	20.6	55.3	47.7	7.6
Carter	62.5	38.0	24.5	66.5	46.3	20.2
Clinton	75.3	36.5	38.8	84.3	34.3	50.0
Under Republican Presidents						
Nixon	48.8	64.8	16.0	44.0	64.0	20.0
Nixon-Ford	37.3	61.5	24.2	40.5	61.8	21.3
Reagan — first term	35.8	65.5	29.7	43.8	75.8	32.0
Reagan — second term	26.0	63.0	37.0	39.0	71.5	32.5
Bush	30.0	68.8	38.8	41.5	77.0	35.5

Note: These averages are based on the percentage of the time that House or Senate members, by party, have voted on roll calls in support of legislation the President favors.

SOURCES: *Congressional Quarterly Weekly Reports*; see also Paul A. Beck, *Party Politics in America* (New York: Longman, 1996), 324.

by their own parties in Congress. In the past thirty years the support of Republicans for a Republican president ranged from a term average of 61.5 percent to 68.8 percent in the House and 62 percent to 77 percent in the Senate. Democratic legislative support of Democratic presidents has ranged from a term average 62.5 percent to 75 percent in the House and 55 percent to 84 percent in the Senate. It is interesting that President Bush's averages are highest for the Republicans, and Clinton's are highest for the Democrats.

What is particularly significant is the *trend* in support. Since 1968 (under President Johnson) the Democratic support scores in both houses have gone up, and since 1972 (under President Nixon) the Republican support scores have gone up slightly in the House and considerably in the Senate. At the same time the opposition party's support for the president has gone down, significantly. Republican support for Clinton reached a low point in 1995 — only an average of 22 percent support in the House and 29 percent in the Senate. Democratic support for Reagan in his second term dropped to 26 percent, and for Bush it was only 30 percent. As a result, the degree of *polarization* in House and Senate voting on proposed presidential legislation increased — to almost 40 percent under both Bush and Clinton (Table 15.5). This is linked to the stronger party unity voting we noted earlier, as well as to the new strategies of opposition in recent Congresses. Perhaps also it points to greater difficulties for a president today to mobilize opposition votes through presidential bargaining and persuasion techniques. Yet one senses that the American presidency is still a powerful institution. Despite the incoherence, fragmentation, and lack of discipline in our system, the president, in his capacity as party leader, can find stronger majorities in his party than ever before. The movement toward party unity in both parties is functionally related to this development.

CONCLUSIONS ON PARTY IN GOVERNANCE

The American national system of governance has changed a great deal in the past three decades. "Committee government" in Congress has been replaced more and more by "party government." The party organs and the party leadership in both houses of Congress and in both parties have assumed a more dominant role in the legislative process. The individual member has a great deal of autonomy, and leaders do not have complete sanctioning power, such as we find in parliamentary systems. We do not have 100 percent party cohesion, but we have more cohesion than before. We also have more legislative conflict and polarization in voting patterns in Congress. We have lower levels of support by opposing leaders for the president, but we do have greater support for the president by the legislative party of the president. The party thus seems to be a "bridge" helping to close the gap between the president and his party in Congress. With divided government, however, and more opposition by the second party, adoption of the president's program may become more difficult than easier. Under divided government in the past, cross-party coalitions often facilitated the passage of legislation. With the party lines drawn more firmly today, a president, despite strong party unity in his legislative party, may find it more difficult to secure action on his agenda than before. This may presage more conflict and deadlock at a time when we need bold national policy to solve our country's problems.

POSTSCRIPT: THE 1998 ELECTION AND THE IMPEACHMENT: IMPLICATIONS FOR OUR SYSTEM

All elections in a democracy are important, as we explained in Chapter 13, the chapter on the election process. The 1998 election, however, took on a special meaning, as a referendum and as a mandate. As a referendum, it was an evaluation by the public of the president's moral lapses (particularly his sexual relationship with Monica Lewinsky) and a decision about whether this should be a basis for punishing the Democrats running for Congress. The mandate issue was whether the Republicans in Congress should be encouraged to proceed with the impeachment of President Clinton. The results can be stated in the large as follows: the voters increased their support for Democrats in Congress, rather than punishing them; and their mandate was clearly communicated: no impeachment, drop the matter, and get on with the business of governing.

Throughout 1998, the Monica Lewinsky affair dominated the political life of the country.[41] In January the independent counsel, Kenneth Starr, learned of the Lewinsky matter and proceeded with his grand jury investigation. This led to the president's denial under oath of a sexual relationship. Charges against him cumulated: that he had perjured himself, tampered with witnesses, suborned others to perjure themselves, deceived the public, and "stonewalled" the grand jury. To all of this we were exposed ad nauseam during the year, the demands for impeachment becoming more insistent and the arguments against impeachment also becoming more insistent. The major allegedly impeachable charges were abuse of power, malfeasance in office, and obstruction of justice. Other criticisms, of Clinton's reckless behavior, disrespect for the high office of the presidency, lying to his family, his party, his staff, and the American public, were also cited. As the November 3 election approached, therefore, the key questions were: Should the president's party be held accountable electorally for these offenses? Would and should the public repudiate him and make his party suffer for his moral lapses and transgressions?

The 1998 election upset all predictions and baffled the prognosticators. The Republicans were expected not only to win, as normally, in the midterm election, but to "win big." Gingrich estimated they would win thirty more seats. The loss of five Republican seats was a stunning development. "We all missed it," said Larry Sabato. No *Washington Post* political writer had predicted that the Democrats would pick up seats. As a result, the election was characterized as "a devastating Republican defeat." Gingrich said "the results should sober every Republican" and then announced his retirement from the Speakership and the House. Gephardt, the Democratic House leader, called the election a "political earthquake" and Gingrich's demise an "aftershock." Leaders of the conservative right were especially dismayed and bitter. Ralph Reed, former executive director of the Christian Coalition, said "there wasn't a lot of ambiguity" about the result, admitting "we thought mistakenly that the Clinton scandal was a sufficient rationale for a Republican majority." And James Dobson, a well-known religious broadcaster, said it was "a pretty resounding defeat, especially when you have a president mired in scandal."

What did happen, and what were the results? Did they justify such a characterization? Of the 435 House seats, 92 percent were held by incumbents who ran for reelection (45 of whom were not contested at all). Of the incumbents, only 6 (1.5 percent) were defeated (the Democrats won 5 of these 6 seats). Of the 35 "open" seats, only 9 (24 percent) changed parties (the Democrats winning 5). So the Democrats did win two-thirds (10) of the 15 seats that changed parties. This resulted in a net gain of 5 seats for the Democrats. This doesn't seem to be much of a Democratic victory until one realizes that (1) the last time a party in control of the presidency also won more seats in the ensuing midterm election was in 1934 under Franklin Roosevelt; (2) the Democrats had lost 54 seats in 1994, and their control of the House, and this had been maintained with only a minor loss by the Republicans in 1996. Hence, the 1998 Democratic victory seems to be a significant reversal of the trend toward Republican control. Another gain of 6 seats and the Democrats could again capture control of the House (with the regular support of independent Congressman Sanders of Vermont).

In the Senate races, twenty-six of the twenty-nine incumbents were re-elected. The wins and losses balanced each other, so that the net result was no change: forty-five Democrats, fifty-five Republicans. But there were some reversals, such as Republican Senator D'Amato's defeat in New York, Democrat Senator Moseley-Braun's defeat in Illinois, the defeat of the staunch Republican "conservative" Senator Faircloth in North Carolina, the Republican victory of Bunning in Kentucky, and the return to the Democrats of the Senate seat in Indiana by Birch Bayh. Three states swung to the Democrats, three states to the Republicans. And the thirty-six governors' races were also almost balanced, except that the Republicans lost the governorship in Minnesota when a professional wrestler, Jesse Ventura, astounded almost everyone by getting 37 percent as a Reform party candidate in an eight-candidate race. There were, however, major gubernatorial victories, particularly in California, where Gray Davis won for the Democrats, and in the South, where the Democrats won in South Carolina and Alabama. In Florida Jeb Bush took the governorship from the Democrats, and his brother won again in Texas. Control of state legislatures changed somewhat, with enough state chambers switching to give the Democrats control of both chambers in twenty-one states compared to seventeen for the Republicans.

In terms of the magnitude of change, then, in the aggregate numbers, this election was one of minimal change. Here is the box score:

	1996	*1998*
House Seats		
Democrat	206	211
Republican	228	223
Independent	1	1
Senate Seats		
Democrat	45	45
Republican	55	55
Governorships		
Democrat	17	17
Republican	32	31
Independent	1	2

Further, the turnout was fairly low, estimated at 36 percent of eligible voters, a decline from the last off-year election. (This varied greatly, however, by state — Minnesota was high with 59 percent and Virginia low with 18 percent.) One cannot say on the basis of the national turnout that the election of 1998 excited the majority of the public enough to get them out to vote. The apathy was overwhelming, given the context for the election.

Why, then, is the election considered a major defeat for the Republicans? First, because the election is interpreted as a strong rejection of the Republican party's plans and procedures to impeach the president. The election results themselves were construed as a repudiation, particularly because the expectations were otherwise — the Republicans were supposed to win big, according to election lore (the midterm "rule") and according to their own predictions (Gingrich predicted thirty more seats). In addition, the exit poll interviews confirmed this "rejectionist" interpretation: 61 percent of over 5,000 respondents disapproved of Republican plans to impeach. Further, the election brought to Congress forty "freshmen," many of whom had gotten the word from their constituents on their views about impeachment. At the same time, the returning members of the House seemed to indicate that they were wavering in their support for impeachment. The *New York Times* (on November 19) reported that virtually all thirty-one Democrats who had voted for holding impeachment proceedings now said they would not support impeachment. And five Republicans had already openly taken the same position, claiming that "conceivably twenty Republicans" would oppose impeachment. Elections can change legislators' attitudes; in 1998 there was evidence this was certainly true. This estimate and observation by the *New York Times,* however, proved to be in error, since the Republicans in the House eventually supported the impeachment.

The election is also interpreted, secondly, as a defeat for the Republican strategic plan for appealing to voters. The Republicans predicted a big victory, and as Senator McConnell said, we should "never, never raise expectations." The Republicans then proceeded to try to win the election on one issue — the Clinton scandal. After the election, the Republicans squabbled over who was responsible for their strategic misconception. Gingrich blamed the media. The Christian Coalition said "there was no clear conservative agenda coming out of the leaders in Washington, DC." Gary Bauer, president of the Family Research Council, said, "Dozens of the Republican candidates ran for the tall grass on values issues, and the result was they demoralized their own base." Despite all the finger pointing, Republicans, whether moderates or conservatives, were all focused on the scandal, making the mistake of ignoring the other key issues. In the last weeks of the campaign, the Republican party spent millions on a television blitz, using the presidential scandal as its major issue. As the exit polls revealed, the public saw the key issues as education, the economy, social security, health care, and taxes, but only 5 percent said the issue was the Clinton/Lewinsky affair. True, 20 percent were interested in "moral standards," but they applied this broadly, to all politics.

A third aspect of the campaign interpreted as important, and which was partly responsible for the poor Republican showing, was the differential approach of the two parties to getting out the vote. The Republicans were criticized for doing a poor job in this respect. Commentators saw the contrast as between the "air attack" of the Republicans, with its heavy television emphasis, and the "ground" tactics of the Democrats, the mobilization of voters through canvassing efforts, doorbell ringing, telephoning, distributing brochures, et cetera. There is some solid basis to support this interpretation. Exit polls revealed that black turnout in the South was heavy, leading presumably to the election of Governor Roy Barnes in Georgia and to the defeat of Governor Fob James in Alabama. A special analysis in Georgia revealed that an estimated 520,000 blacks voted in Georgia, twice the 1994 total (Barnes won by 150,000 votes). The Georgia Democratic party targeted 300,000 black households and communicated with them constantly during the campaign (with ten mailings and three phone calls). While only 16 percent of blacks voted in 1994 in Georgia, 29 percent voted in 1998.[42]

The mobilization of voters by the unions was also a "ground" operation and very supportive of Democratic candidates. In 1996 the unions spent $25 million on broadcast spots, but in 1998 used most of their funds, $15 million, on a variety of "get out the vote" tactics. These included leaflets, phone banks, and four hundred coordinators working full time in congressional districts. While only 18 percent of Americans live in union households, 24 percent of the vote in 1998 came from union households. In certain states this vote was critical, as in Wisconsin, where 63 percent of union votes backed the Democratic candidate for the Senate, Russell Feingold, who was supported by only 44 percent of the voters from nonunion households. The Peter Hart Associates' study of union members found that 71 percent voted Democratic, up from 60 percent in 1994. And the key issues for union members were social security, education, and health, not impeachment. Thus, effective strategy, particularly a sound approach in selection of issues, effective use of the media, and effective technologies for mobilization of the vote were seen as possible explanations of the Democratic victory.

The election also informed us again of the group basis of party support coalitions and the voting patterns that were continuing and changing. The Republicans could not score any triumphs here. Table 15.6 illustrates the recent trends.

Scrutinizing the vote by social categories confirms that both parties maintained their coalitions rather well. Most important is that the Democratic support did not fall off at all. Compared to 1994, the Democrats held their support among women, blacks, Hispanics, and Jews, while strengthening their support somewhat among lower socioeconomic groups, voters in big cities, Catholics, and voters in the East. The Republicans won more elderly and rural support, kept their hold on the South (especially southern whites), college graduates, and Protestants. The economic prosperity of the country appears to remain a Democratic advantage. Thus, the election is valuable in telling us what did *not* happen: there was no decline of Democratic support in its major coalitional groups. Indeed, there is some evidence of expansion of Democratic support in these groups.

Elections have consequences, and 1998 was no exception. A modest revolt resulted in the Republican party of the House, leading to a change in the House Republican leadership. Here are the Republican selections:

House Leader	*Before the Revolt*	*After the Revolt*
House Speaker	Newt Gingrich	J. Dennis Hastert (IL)
Majority Leader	Richard Armey	Richard Armey (TX)
Majority Whip	Tom Delay	Tom Delay (TX)
Caucus Chair	John Boehner	J. C. Watts (OK)
Campaign Committee Chair	John Linder	Thomas Davis (VA)

Three of the five top positions changed personnel. The new Speaker, Hastert, apparently was a nearly unanimous selection by the House Republicans, supported by both the moderates and conservatives and perceived as a "low-key consensus builder."

TABLE 15.6 ▪ Voting Patterns by Social Groups (1998 compared to 1994, in percent)

Social Group	*1994*	*1996*	*1998*	*Change 1994-98*
A. Inclined to support the Democrats				
Women	53	55	53	0
Blacks	92	82	89	-3
Hispanics	61	73	63	+2
Young (18-29)	49	55	50	+1
Less educated (high school or less)	49	56	52	+3
Catholics	47	54	53	+6
Jews	77	74	79	+2
Residents of big city (population over 500,000)	72	69	82	+10
Residents of eastern United States	50	56	56	+6
Those who perceive standard of living as "getting better"	60	65	59	-1
B. Inclined to support the Republicans				
Men	58	54	54	-4
Age 60+	51	51	55	+4
College graduates	55	57	55	0
Protestants	58	57	58	0
Income over $75,000	62	60	53	-9
Those who perceive standard of living as "getting worse"	64	67	59	-5
Rural residents	57	52	62	+5
Residents of southern United States	53	55	55	+2
Whites in the South	64	64	65	+1

SOURCES: Data collected in 1998 by Voter News Service from 10,017 voters leaving 250 polling places; from 11,308 in 1994 and 16,637 in 1996. See *New York Times,* November 9,1998, A20.

Other consequences, both short-term and long-term ones, will emerge. Initially it was thought that the impeachment process might be shortened, but as partisan positions hardened, this expectation was never realized. The relevance of the election for President Clinton may also soon be apparent in whether he has more "breathing room" to pursue the critical domestic and foreign policy agenda in his last two years. How will the Republicans resolve the factional conflicts within their midst? Will they now reverse their strategy for the year 2000 and appeal more directly and broadly to minority groups, as Governor Bush of Texas advocates, using his slogan of "compassionate conservatism"? And how will all this affect the presidential sweepstakes as it now moves into high gear? Vice President Gore, many feel, is helped by the 1998 election results, but what of the other contenders on the right and left, in both parties? Has the Democratic decline in the eleven states of the Solid South really been stopped? They now have 4 of the 11 governorships (8 in 1980), 8 of the 22 U.S. Senators (16 in 1980), and they kept their hold on 54 of the 125 southern seats in the House (they had 77 in 1980). But it is interesting that with the Democratic gain of the North Carolina state legislature they have 58 percent of all state legislative seats in the South. This may be only temporary, however, according to expert observers of southern politics. There are other results to pay attention to: the fallout from the 1998 election is considerable.

What have we learned again from a study of this election? First, in an election the attentive American public does, and can, communicate rather clearly and forcefully to its leaders what its preferences for elite behavior are. Second, parties can learn how to win and how to lose elections, what strategies to avoid and what strategies have payoffs. Third, clearly personal-contact mobilization of the voters still can be employed to win elections, and grassroots party activists are important to the system. And finally, we learned again that elections really matter. They may not be landslides, but they certainly can change behavior. And in so doing they help maintain the party system while forcing it to adapt.

THE IMPEACHMENT TRIAL

In the post election period, December 1998 and early 1999, the politics of Washington, D.C., was focused primarily on the impeachment of President Clinton. After all, this was only the second time in the history of the country that an actual impeachment trial by the Senate was to take place. The House of Representatives had decided in October (by a vote of 258 to 176) to open such an inquiry. Its Judiciary Committee then met and, after much bitter deliberation and on a strict party vote, decided to report to the full House four articles of impeachment. The House then debated the question, finally accepted two charges (perjury and obstruction of justice), and on December 19 voted to impeach the president on these two articles (with a vote of 228 pro and 206 against on perjury, and 221 pro and 212 against on obstruction of justice). This was close to being a party-line vote, with the 206 Democrats voting against the charges and the 228 Republicans (with few defections) voting in favor of the impeachment charges.

The Senate convened on January 8, 1999, with Chief Justice William H. Rehnquist as presiding officer. In a very bipartisan action, the Senate had unanimously adopted a set of rules of procedure for the trial of the president. The prosecutors were thirteen House members (called "managers"), and a set of White House lawyers presented the case for the defense of the President. The Senate sat quietly for over a month hearing the case as presented by the managers and rebutted by the president's counsel. Three witnesses were called, deposed privately and videotaped (and the videotapes then presented to the full Senate). Most of these deliberations were open to the public and could be televised. Then in the week of February 7, the Senate closed its doors and the Senate debated as a jury on the case. At this time, it must be remembered, the partisan division of the Senate was fifty-five Republicans and forty-five Democrats. To convict the president, a two-thirds vote

The Founding Fathers, the Constitutional Convention, and the Definition of Impeachment

In December 1998, the House began its hearings on the impeachment of Democratic president William J. Clinton. There was much controversy then, as during most of 1998, over whether the president's scandalous behavior actually reached the standard set by the Founding Fathers at the 1787 Constitutional Convention in Philadelphia.

Article II Section Four of the Constitution states: "The President, Vice President and all Civil Officers of the United States, shall be removed from Office on Impeachment for, and Conviction of, Treason, Bribery, or other high Crimes and Misdemeanors." Before placing this section in the document, the Founding Fathers debated the standard for removal. There were delegates who wanted the president to be subservient to Congress, so they argued that the "First Magistrate" could be removed for simple "misfeasance" by Congress. This was the thinking of Gunning Bedford of Delaware, the first delegate in the Convention to utter the word "impeachment."

Delegates like Madison, who wanted the executive branch to be co-equal with the legislative and judicial ones, wanted a narrower and tougher standard. On September 8, 1787, George Mason introduced a motion that the president could be impeached for "treason, bribery or maladministration." Madison objected by noting that maladministration was "hopelessly vague." Under Madison's criticism, Mason withdrew his motion and asked that the phrase "other high crimes and misdemeanors" against the state be substituted for "maladministration." It was accepted, and that acceptance grew out of the fact that "high crimes and misdemeanors" came from British common law and had been defined in that law from the thirteenth century as Acts by the Ministers of the Crown that mislead both the Monarchy and/or Parliament on war, finance, bribery and subversion of the law. It meant serious official conduct against the state. How the House and Senate interpreted the term "impeachment" in 1998 and 1999 has been at the center of the House's inquiry and the Senate's trial of the president.

(sixty-seven votes) would be needed. On February 12 the Senate voted on the two articles of impeachment, with the following result:

Final Senate Votes

The Charges	*To Convict*	*To Acquit*
Perjury	45 (all R)	55 (10 R)
Obstruction of Justice	50 (all R)	50 (5 R)

During this period, from January 1998 to February 1999, much had gone on outside the Washington Beltway as well as inside it. Public support for the president's performance in office was high, reaching in some polls over 70 percent after the November 1998 election. Further, the public was opposed to impeachment; as late as December 15, 64 percent of the public did not want Congress to go forward with the impeachment process. A majority of the public also denounced the president's moral behavior, but clearly this did not interfere with its positive evaluation

Senators' Views on Impeachment Proceedings

This sample of the views of U.S. senators, reported in the *New York Times,* illustrates the different perspectives.

Olympia Snowe (R, Maine):

"Does the President's misconduct . . . represent such an egregious and immediate threat to the very structure of our government that the Constitution requires his removal? . . the answer, I conclude, is 'no.'"

Gordon H. Smith (R, Oregon):

President Clinton's conduct . . . has been corrosive of our culture and to our children . . . when the Chief Justice calls my name, 'Senator, how say ye?' I will say guilty."

Herb Kohl (D, Wisconsin):

"We risk opening the floodgates to more party-line impeachments if we oust a President from office for behavior that — while deplorable — isn't truly removable."

Dianne Feinstein (D, California):

"I do not believe [President Clinton's behavior] presents a clear and present danger to the functioning of our Government."

Charles E. Schumer (D, New York):

"It has shaken me that we stood at the brink of removing the president not because of a popular groundswell to remove him and not because of the magnitude of the wrongs he's committed but because conditions . . . . have made it possible for a small group of people who hate Bill Clinton and hate his policies to very cleverly . . . exploit the institutions of freedom . . . and almost succeed in undoing him."

Bob Graham (D, Florida):

"Removal of this President . . . would upset the delicate balance of power so meticulously established 212 years ago."

of his performance in office. Yet many Republican members of Congress refused to accept such strong public opinion in support of the president as relevant to their decisions. They also rejected the testimony of some of the most distinguished constitutional historians that these charges were not impeachable. Over four hundred historians signed a statement sent to members of Congress stating their arguments against impeachment. Many claimed that the Founding Fathers never intended impeachment to be employed for the removal of the president from office for personal moral lapses. But Republican leaders argued otherwise, that the president's behavior did constitute perjury and obstruction of justice and was harmful enough to the presidency and to the system generally to require his removal from office. The perspectives of the Senators varied greatly.

The question of the effects of the events of the past year, and of the impeachment process itself, widely debated, "hangs in the air." There are several facets of this question, several different types of possible effects or consequences for our system. First is the effect on the presidency itself. Senator Moynihan was worried that the removal of Clinton would "destabilize" the presidency. On the other hand, Republican Senate majority leader Trent Lott argued that the president already had been losing his credibility (for example, raising questions about the motivations of the president in the bombing of Iraq). The long-term effect on Clinton's capacity to govern was cited by others. A second "effect" anticipated by some, particularly the constitutional historians, was that impeachment would result in the presidency's being "disfigured and diminished, at the mercy as never before of the caprices of Congress." The relationship between the president and Congress under our system, some claimed, would be changed; "the balance of power between the president and Congress is at risk." Further, there are those who predict that public support for the national institutions of governance — approval of, belief in, and trust for these institutions — will fall even lower than at present. Trust in the national government has already fallen to below 30 percent, electoral turnout in 1998 already was at a 36 percent level, and the expectation is that support will decline further.

As for the partisan consequences, there is much speculation and argument about what the "fallout" will be, particularly whether the Republican party would be electorally punished in election year 2000. The Republicans allege that people will forget the next two years and not retaliate. Others refer to the vulnerability of House and Senate Republicans. As for the Senate Republicans, nineteen are up for reelection in 2000, eight of them "fresh" men or women elected in 1994, often with marginal victories. How will they fare in the 2000 election? Obviously it is too early to state with any confidence which of these effects will in fact eventuate, and what impact the public's view of the impeachment process will have. We can but wait for the future to emerge.

In the beginning of 1999, after the scandal and impeachment trial, the American party system did appear to be in a state of crisis: low voter trust, declining turnout, extreme partisan bitterness, government deadlock, and widening ideological polarization. The challenge to the system was clear: confronted with very pressing policy concerns (such as the reform of social security, health care, educational reform, and national defense) could the system right itself, recruit able leadership, adopt needed reforms, and restore public confidence in the party system as well as civility in the relations of the leaders of the two parties? The term of this Congress and the next should tell us whether our system can adapt successfully again.

NOTES

1. Quoted by Senator Fred Harris in "A Nationalized and Individualized Senate," *Extensions* (Fall 1997), Carl Albert Research and Studies Center, University of Oklahoma, Norman Oklahoma.
2. William J. Keefe, *Congress and the American People* (Englewood Cliffs, NJ: Prentice Hall, 1980), 101–5. See also Thomas E. Cronin, *The State of the Presidency* (Boston: Little, Brown, 1975), 107.
3. David Truman, *The Congressional Party* (New York: Wiley, 1959).
4. David Mayhew, *The Electoral Connection* (New Haven: Yale University Press, 1974) 27.
5. For a good description of these events, see David Rohde, *Parties and Leaders in the Post Reform House* (Chicago: University of Chicago Press, 1991), 17–34.
6. Barbara Sinclair, "The Congressional Party: Evolving Organizational, Agenda Setting and Policy Roles," In L. Sandy Maisel, *The Parties Respond* (Boulder: Westview Press, 1990), 246.
7. Rohde, *Parties and Leaders,* 173, for a discussion of this view. See also Leroy Rieselbach *Congressional Reform* (Washington, DC: CQ Press, 1986); and Burton D. Sheppard, *Rethinking Congressional Reform* (Cambridge: Schenkman, 1985).

8. For a recent discussion of Senate reforms, see Barbara Sinclair, "Unorthodox Lawmaking in the Individualist Senate," *Extensions* (Journal of the Carl Albert Congressional Research and Studies Center, 1997), 11-14.

9. For a summary of these developments, see Douglas L. Koopman, "The House of Representatives under Republican Leadership: Changes by the New Majority," paper presented at the annual meeting of the American Political Science Association, Chicago, August 30-September 3, 1995.

10. David Rohde, "'The Reports of My Death Are Greatly Exaggerated': Parties and Party Voting in the House of Representatives," in Glenn R. Parker, ed., *Changing Perspectives on Congress* (Knoxville: University of Tennessee Press, 1990), 32.

11. Gary Cox and Matthew McCubbins, *Legislative Leviathon* (Berkeley: University of California Press, 1993), 144-57.

12. Ibid., 154-55.

13. Rohde, *Parties and Leaders,* 15.

14. See the October 1997 issue of the *American Journal of Political Science* for the debate on this.

15. Rohde, *Parties and Leaders,* 58-60.

16. Ibid., 151-54. See also William Howell, Scott Adler, and Charles Cameron, "Institutional Causes of Legislative Productivity in the Post WW II Era," paper presented at the Midwest Political Science Association, Chicago, April 1997.

17. Edward R. Tufte, *Political Control of the Economy* (Princeton NJ: Princeton University Press, 1980), 71.

18. Ibid., 101-2.

19. Douglas Hibbs, "Political Parties and Macroeconomic Policy," *American Political Science Review* 71 (1977): 1486.

20. Benjamin Ginsberg, "Elections and Public Policy," *American Political Science Review* 70 (1976): 49.

21. Keefe, *Congress and the American People,* 89.

22. John Kingdon, *Congressmen's Voting Decisions* (New York: Harper and Row, 1973), 135.

23. David M. Olson, "The Electoral Relationship Between Congressmen and Their District Parties." Paper presented at annual meeting of the American Political Science Association, Washington, DC, 1977; "United States Congressmen and Their Diverse Congressional District Parties," *Legislative Studies Quarterly* 3 (1978): 239-64.

24. Barbara Greenberg, "New York Congressmen and the Local Party Organization," Ph.D. dissertation, University of Michigan, 1972.

25. Bruce Cain, John Ferejohn, and Morris Fiorina, *The Personal Vote* (Cambridge: Harvard University Press, 1987), 52-59, 60.

26. Nelson W. Polsby, "Political Change and the Character of the Contemporary Congress," in Anthony King, ed., *The New American Political System* (Washington, DC: American Enterprise Institute).

27. Warren E. Miller and Donald E. Stokes, "Party Government and the Saliency of Congress," *Public Opinion Quarterly* 26 (1962): 531-46; "Constituency Influence in Congress," *American Political Science Review* 57 (1963): 45-56.

28. See Philip E. Converse and Roy Pierce, *Political Representation in France* (Cambridge: Harvard University Press, 1986), 674.

29. Miller and Stokes, "Constituency Influence," 55.

30. Polsby, "Political Change," 46.

31. Ibid., 32.

32. Kenneth Janda, "Comparative Political Parties: Research and Theory," in Ada Finifter, ed., *Political Science: The State of the Discipline II* (Washington, DC: American Political Science Association, 1993), 173.

33. Roper Center, *The Public Perspective* 48 (October/November 1996).

34. Richard Neustadt, *Presidential Power* (New York: Wiley, 1960), 181.

35. Ibid., 33.

36. Paul A Beck, *Party Politics in America,* (New York: Longman, 1996), 350.

37. President Kennedy's speech to the National Press Club, January 14, 1960. Quoted in Robert H. Blank, *Political Parties: An Introduction* (Englewood Cliffs, NJ: Prentice Hall, 1980), 183.

38. Press Conference, May 31, 1955. *New York Times,* June 1, 1955. Quoted in Richard Neustadt, *Presidential Power,* 166.

39. Keefe, *Congress and the American People,* 108.

40. *Congressional Quarterly Weekly Report,* April 16, 1977, 691, Quoted in Keefe, *Congress,* 113.

41. The accounts of the events reported here, and much of the data, were based on news articles and reports in the *New York Times,* the *Washington Post,* and other newspapers.

42. *New York Times,* November 6, 1998, A22.

Chapter 16

The Party System and the Race Problem

The problem of racial inequality in America, and the way the parties have responded to that problem and dealt with it, has for long posed an important test for our party system. In this chapter we describe and analyze how well the parties have done. We could have focused on other minorities (Native Americans, Hispanics, and so on). Or we could have addressed any number of policy areas in the same way: health care, the environment, education, taxation, welfare, or foreign affairs. We chose the race question because of its historical significance and contemporary salience. It is a critical issue dividing us, representing a major ideological cleavage in our politics. Our objective is to discuss the state of racial inequality, how well the parties have faced the issue, and the possible strategies that might be utilized to achieve greater progress.

We have been confronted with the race problem for the entire existence of our nation. Our leaders and the public have been aware that it is a serious problem for the past two hundred years. Even though many Founding Fathers were slaveholders, some if not most of them deplored the institution of slavery. Benjamin Franklin was one early statesman who clearly felt that way. He said: "Slavery is . . . an atrocious debasement of human nature." And James Madison: "We have seen the mere distinction of color made . . . a ground of the most oppressive dominion ever exercised by man over man."[1] In the nineteenth century the French scholar de Tocqueville, observing the American scene, said "the most formidable of all the ills that threaten the future of the union is the status of the Negro."[2] The great black leader, W. E. B. DuBois, said in 1903 "the problem of the twentieth century is the problem of the color-line."[3] Forty years later the great Swedish scholar, Gunnar Myrdal, in his classic study of 1944, noted that we had not made much progress on our problem. He wrote: "The American Negro problem is a problem in the heart of the American. . . . it is there that the decisive struggle goes on . . . The moral dilemma of the American . . . the conflict between his moral valuations [and] . . . individual and group living."[4]

Shortly after Myrdal's book, the top American scholar of parties agreed that while "the American political system has accommodated itself with fair success to white immigrants . . . the Negro, however, has presented a far more difficult problem for a democratic political order. . . . from the Civil War onward, the status of the Negro was continuously an issue in American policies, an issue to which there seemed no solution."[5]

Today we find our scholars, black and white, attesting as before to the critical divide in America over the race issue. Cornel West, one of the eloquent black scholars, is an example. Recently he wrote: "Every historic effort to forge a democratic project has been undermined by two fundamental realities: poverty and paranoia. . . . Race is the most explosive issue in American life precisely because it forces us to confront the tragic facts of poverty, and paranoia, despair and distrust. . . . In short, a candid examination of *race* matters takes us to the core of the crisis of American democracy."[6]

THE HISTORICAL NARRATIVE

To achieve some perspective, it is necessary to review the evolution of the race problem and the relevance of party politics in understanding it. When we do that, we are struck by the early manifestation of the conflict over race, the passionate character of that conflict, and the incapacity — or unwillingness — of party leaders to cope with, or respond to, the problem peacefully and effectively. The history of this racial conflict is two hundred years of almost continuous, vitriolic hostility, taking many different forms, but persisting without letup.[7] The seeds of this hostility were sown in the post-Revolutionary War era. There were then strange anomalies in our society. Of 750,000 Negroes, 59,000 were "free" in 1790, while many of the top revolutionary leaders were slaveholders. And the North and South were already divided on the status of Negroes. There were many more slaves in the South, with no political or economic rights. In some, mostly northern, states (New York, Pennsylvania, Maryland, North Carolina) Negroes had the right to vote, and the parties actually competed for their votes in local elections. In the South, as John Hope Franklin describes the status of blacks, "a veritable world was erected around the black man and he found it necessary to develop his own life and his own institutions." Negroes were not allowed to go to public schools, their mobility was severely restricted, and they had no right of assembly. Communication was limited. "Mixing with Negroes" was a crime in certain areas.

In the North Negroes might be called "quasi-free," for there also they were subjected to restrictions. And as more of them fled to the North, tension developed there also, over jobs, schools, hiring conditions, et cetera. In the thirty years before the Civil War, riots broke out in many places in the North. Yet the one big difference, as Franklin points out, was that in the North the blacks to a certain extent had the law on their side; in the South, there were no legal protections. While intra-sectional tensions were contained, even though superficially, by 1861 inter-sectional hostility broke the union. Eleven slave-holding states seceded (only four did not: Missouri, Kentucky, Delaware, and Maryland). The party system had failed to preserve the union, indeed had contributed to its dissolution. And during this period of war, the Negroes suffered. They sought to enlist in the North but were basically excluded until after Lincoln's Emancipation Proclamation in 1863. Those who were fugitives to the North were poorly treated under an unclear policy in the early days, and finally put in northern interment camps. In the South the

Confederate forces, as the war dragged on, sought to mobilize Negroes, if their slave-holding masters would allow it. When they did serve, they fought well.

Reconstruction in the post–Civil War period could have been a period of integrating Negroes into our society, but it really resulted finally in an assault on them. For the first years blacks had some power as free men. Many white southerners were disfranchised and blacks had the ballot, at least theoretically. They appeared at southern state constitutional conventions, helped write new constitutions, and were elected to state legislatures (in South Carolina they had majority legislative control for a time). But in actuality their power was limited. And they sent few representatives to Congress (three in 1869, five in 1870, and seven in 1873–75). Their "empowerment" subsided fast, and whites actually controlled these states. By 1877 the opportunity for black power was over when President Hayes withdrew federal troops from the South.

What happened then was a return to heavy-handed white dominance while the two parties fought for control, and as Franklin puts it, so far as blacks were concerned we witnessed the "losing of the peace." He concludes that the two parties "share the guilt for their utter failure to establish peace between the sections and the races."[7] With the return of Democratic party control in the South in the last quarter of the nineteenth century, we saw the bitter backlash against the Negroes. They were almost totally disfranchised. Various ingenious techniques were used: polling places were relocated far away from where Negroes lived, ballots were not uniform and were distributed by the parties (resulting in "stuffing"), the overcounting of ballots, the nomination by whites of several Negroes for an office in order to split the block vote, gerrymandering, poll taxes, suffrage disqualification for petty larceny, the confusing "office block ballot" used to reduce straight party voting, and the limiting of blacks to 2½ minutes in the voting booth. By 1889, one observer concluded, "the Negro as a political force has been dropped out of serious consideration." Middle-class and lower-class whites had banded together against the Negroes. Once the Negroes were potentially eliminated, the two white classes fought each other.

There was some hope that the Populist party in 1892 would provide an opportunity for recovery of some Negro influence. Liberal agrarian whites and blacks did join together and secure some victories, taking over at least one state legislature, North Carolina, in 1894. But by 1896 the Populist threat was over. The whites continued to perfect their black disfranchisement techniques, among them a tough literacy test such as that of 1895 in South Carolina (the ability to read and write any section of the Constitution, or understand it when read aloud to them). Plus the addition of the "grandfather clause" (adding to the list of eligible voters anyone whose father or grandfather could vote on January 1, 1867).

Thus, Democratic party politicians in the South engaged in an assault on black political rights and opportunities instead of facilitating the entrance of blacks into the political system, which was supposed to be happening since the Emancipation Proclamation of 1863. After the turn of the century, black leaders sought to organize and fight for their rights. In 1905 a group of them led, by the famous writer, editor, and scholar W. E. B. DuBois, met at Niagara Falls to develop a strategy of resistance and reform. They attacked the "blasphemy" of "stealing the black man's ballot." Their actions led to the founding of the NAACP in 1910, whose membership grew fast throughout the country, and whose publication *Crisis* sold 100,000 copies by 1918. In the meantime the Supreme Court began to come to the defense of Negro political rights. In 1915 the Court struck down the "grandfather clause" and in 1917, 1921, and 1923 returned other verdicts dealing with such issues as Negro jury rights and the unconstitutionality of "white primaries." The Court did not always rule for the Negroes, but often decisions were monumental, as in 1954 *(Brown v. The Board of Education of Topeka)* when it revised its 1896 decision and ruled that separate education was not equal education.

What did the parties do to deal with racial inequality in the first part of this century? President Theodore Roosevelt at first attracted some black support when he decided to run as a "Bull Moose Progressive" in 1912 against his own Republican party. But blacks discovered at the convention of his party that he was not responsive to their interests, and they withdrew. The Democrats won in 1912 with Woodrow Wilson, who had assured blacks that he would treat them fairly. But after his election he mostly ignored them and pursued if anything, a discriminatory policy, approving separate cafeterias for federal employees as well as separate restroom facilities. He also did not facilitate their consideration by the Civil Service for federal employment. Thus, despite all black efforts, we were still operating in a two-party era in which both parties, if not anti-Negro, were certainly not pro-Negro. And then World War I engulfed the nation's politics. Blacks sought to enlist and many thousands served. But they again ran into a discriminatory response to their efforts to secure officer training for blacks (only a few hundred finally were commissioned). Nevertheless, the blacks served the country well in the war.

Returning after the peace, they were hopeful that the black race as well as the world was now "safe for democracy." They discovered instead the same old racial hostility, in renewed, violent form. The Ku Klux Klan was revived, committed to the "supremacy of the white race." Attempts by blacks to assert their rights and reach out for new opportunities were met by floggings, hangings, and burnings. Some Negro soldiers were lynched. In 1919 we experienced what came to be known as "the red summer," with over twenty-five race riots, particularly in large cities like Detroit and Chicago. Where were the parties? They made one effort in 1921 to pass a bill outlawing lynching. It passed 230–119 in the House but was blocked in the Senate by a filibuster of southern senators.

The Negroes began to lose confidence in the Republican party in 1924 when John Davis, the Democratic candidate, and Robert La Follette, the Progressive candidate, promised to eliminate discrimination on the basis of race or creed. The Republicans won handily in 1924 and in 1928 again, after a strong Republican campaign in the South where the party organizations were called "lily white" because prominent black leaders in the Republican party were shunted aside. "By 1928," Franklin observed, "Negroes were beginning to learn to vote for candidates who were not Republicans." It was no wonder. The racial hostility was pervasive. It was still possible in 1930 for a candidate for the Supreme Court, John J. Parker, to say that "participation of the Negro in politics is a source of evil and danger to both races." Top leaders in the country were still unresponsive to the race problem in America as the Great Depression hit us in 1929.

Thus, seventy years after the "emancipation" of blacks, we were still living in "the dark ages" when the Franklin Roosevelt Democrats took control of the country in January 1933. The depression had brought great economic destruction, and blacks had been hit particularly hard. Up to 40 percent of blacks were on relief (three to four times that of whites). In some cities it was much higher — 65 percent in Atlanta, 80 percent in Norfolk, for example. Hoover and the Republicans did nothing in particular for the blacks. What would the new president and the Democrats do?

FDR did not immediately attract blacks (only 23 percent voted for him and the Democrats in 1932). But his and Eleanor's actions in the months, and years, after his inauguration in 1933 convinced many of them to switch parties. Arthur Mitchell, a former Republican, was the first Negro Democrat elected to Congress since Reconstruction (in 1934), much to the concern of his southern colleagues. Eleanor was openly friendly to blacks, inviting the National Council of Negro Women to tea at the White House, to the consternation of many people. President Roosevelt consulted with top Negro men and actually had what was called a "black cabinet" of advisors. Many were given positions in departments or agencies connected to the recovery program. The new federal "alphabet" agencies helped Negroes in many ways: AAA (giving grants to farmers), FSA (helping farmers purchase farms), NYA and CCC (providing relief for Negro youths and helping them learn trades), FPHA (constructing low-cost housing), PWA (employing blacks to build hospitals and other public buildings), WPA (assisting black writers and artists), et cetera. Although discrimination continued, there was much progress. Negro employment in the federal government increased from 50,000 in 1933 to 200,000 (although these were mostly menial jobs), by the end of 1946. Blacks also ran successfully for state legislative positions and were appointed to judgeships. The pressure also was put on firms refusing to employ Negroes, and this type of picketing of firms was declared legal by the Supreme Court in 1937. The Fair Labor Standards Act of 1938 set a minimum wage (twenty-five cents an hour) and a forty-hour work week. The CIO left the AFL in 1935 and announced its policy and intention to organize employees in all urban plants regardless of race — a real breakthrough in racial discrimination by unions.

Thus there was some progress on a variety of fronts, and FDR's executive orders and decisions were responsible for much of this progress.[8] The parties in Congress, however, were impotent. There were 150 civil rights bills introduced in Congress between 1937 and 1946, focused particularly on lynching, the poll tax, and fair employment. None of these passed. And Roosevelt was blamed because he did not come out in support of these bills, "fearing southern opposition to his entire economic recovery program." He did issue executive orders on fair employment practices and set up the FEPC to eliminate discrimination in defense areas and, later, in other federal contracts. Roosevelt's record in toto was thus evaluated, by some, as relatively limited and weak. Yet Roosevelt broke the commitment of blacks to the Republicans and, millions of blacks became "Roosevelt Democrats."

During World War II a half million blacks served overseas, and many labored in the home industries to provide the products for war. The government tried to improve the morale of the blacks in the armed forces and at home, but tension continued and riots broke out from time to time.

The parties took different positions on rights issues, as manifest in their party platforms in 1944. The Democrats had a bland statement that racial and religious minorities had the "right to live, develop and vote." The Republicans specified in more detail their demands: a congressional investigation on mistreatment of blacks, federal legislation, setting up a Federal Employment Practices Commission, a constitutional amendment against the poll tax, legislation to make lynching a federal crime. It took until 1948, however, for the Democrats to take a strong position on civil rights. There was a fight that year at the Democratic convention in Philadelphia, led by northern liberals. Truman had become president and had pressed for more action, appointing a committee on civil rights that recommended a variety of actions. He also ordered the end to discrimination in recruitment into the armed forces. Truman spoke before the NAACP in favor of reforms and argued that "the national government must show the way." He sent a civil rights bill to Congress in February 1948, the first since Reconstruction; the southern Democrats were stunned. The national Democratic convention then met and, after a battle on the floor, adopted a platform calling for legislation on equal voting rights and employment opportunities, among other proposals — the first time the Democrats had taken a collective party position for racial equality! Some southern delegates, known as Dixiecrats, walked out and organized the States Rights party, which won four southern states in the election.

Now we entered a period when both parties took stands in favor of civil rights, but the Democrats were more moderate. The Democratic platforms of 1952 and 1956 for the most part repeated 1948, but the 1956 one pandered more to states rights. Also Adlai Stevenson's wishes in 1952 resulted in a more conciliatory platform, since he was clearly concerned about splitting the party and losing the Democratic South. The Republicans continued strongly supportive of civil rights. The Supreme Court decision in 1954 *(Brown v. Board of Education)* that outlawed segregation in the schools was endorsed by the Republicans in 1956 (but only mentioned by the Democrats). And under President Eisenhower, civil rights acts were passed in 1957 and 1960 (the first since 1875) that, although weak, did create a Commission on Civil Rights and gave the Justice Department some power to ensure that eligible blacks could register and vote. Further, Eisenhower sent federal troops to Little Rock in 1957 to enforce school desegregation. So some action by the Eisenhower Republicans followed the actions by the Truman Democrats. In a sense, blacks and Southerners were both in a state of uncertainty as to which party would best further their racial justice and equality concerns. We were in a period of admittedly ambivalent two-party inclinations toward more racial justice. One could call it a "window of bipartisan opportunity" for civil rights.

Protests and demonstrations — sit-ins, bus boycotts, freedom marches, and brutal riots — led up to the 1960 election of Kennedy. It was as though the whites were organizing a "last-ditch stand." In the 1960 Democratic convention, a strong civil rights plank was adopted that included a permanent FEPC (Fair Employment Practices Commission), continuation of the Civil Rights Commission, more power to the Justice Department, and implementation of school desegregation. The Republicans also adopted a strong platform, after Nixon and Rockefeller worked out an agreement. So there was really no basic confrontation on racial policy in the platforms. Kennedy, however, did not press for civil rights action at once after his election. Not until November 1962 did he by executive directive put an end to racial discrimination in federally supported public housing. Blacks had voted for him, but one-third also supported Nixon (40 percent had voted earlier for Eisenhower). Blacks were upset because many of them felt their votes had made Kennedy a winner in key states (Illinois, where the Kennedy margin of victory was 9,000, and Michigan, where the margin was 67,000). They claimed that 250,000 blacks supported Kennedy in each state. Kennedy did set up a Committee on Equal Opportunity in Housing and appointed many blacks, particularly to judgeships (Thurgood Marshall to the circuit court, for example), and Robert Weaver to a top federal housing finance position (which subsequently led to his becoming the first black in the cabinet, under Johnson). In February 1963 Kennedy submitted a preliminary, and on June 19 a comprehensive, civil rights bill to Congress, with an eloquent speech of support calling it a moral issue and concluding with this question: "Are we to say to the world and . . . to each other that this is a land of the free except for Negroes; that we have no second-class citizens except Negroes; that we have no class or caste system except . . . Negroes?" Tragically, Kennedy's assassination then intervened.

President Johnson announced at once that he would fight for comprehensive civil rights laws: "We have talked for one hundred years. . . . it is time now . . . to write it in the books of law." The 1964 Civil Rights Act was pushed through Congress with a 70 percent favorable vote, but 90 percent of southern Democrats in the house voted against it. This act dealt with public facilities, school segregation, job discrimination by employers and unions, and the right to vote. With this act and the Voting Rights Act of 1965, the national government asserted its authority, as well as challenged all those seeking to perpetuate racial hostility and block the rights of Negroes.

The 1964 election was a turning point in our electoral politics. (See Figure 16.1, pages 362-363.) It brought about realignment of the parties and of groups on the race question. The Republican candidate, Goldwater, who had voted against the 1964 act, argued that "race relations . . . [are] best handled by the people directly concerned." And the Republican platform that year backed away from its explicit and strong support of civil rights that had characterized its earlier platforms. Goldwater won only six states, five in the Deep South, by large margins (87 percent in Mississippi and 70 percent in Alabama, for example). He lost 90 percent of the black vote, however. And thus we had a reversal of parties on the race question. Johnson pursued his agenda on civil rights after the election with the new Voting Rights Act, which gave the attorney general the power to appoint federal examiners to supervise voter registration in states with a literacy test and where fewer than 50 percent of working-age citizens cast ballots in the 1964 election.

Mostly these were in the South. This act produced a staggering increase in black registration — from 7 percent in Mississippi in 1964 to 68 percent in 1970; from 23 percent in Alabama to 64 percent, for example. But it was clear that the Democrats had lost their control of the South. White southerners turned against the Democrats. And their representatives in Congress continued the battle, defeating Johnson's 1966 bill to prohibit discrimination in housing (on a Senate filibuster). He succeeded, narrowly however, to defeat the filibuster in 1968.

After Johnson, the Democratic party continued its liberal approach to the race question with Humphrey and McGovern. Its platforms enthused about the 1965 and 1964 acts and committed the party to future efforts. Carter continued this support in 1976, receiving, however, only 46 percent of the white southern vote, attesting to how regional politics had changed. The Republican party under Nixon and Ford followed the southern strategy of Goldwater, even though neither of them was as racially conservative as Goldwater. Nixon, when elected, moved immediately to his own racial agenda: nominated southern conservatives to the Court, forced the resignation of the Civil Rights Commission chair, opposed the extension of the Voting Rights Act, sought to postpone desegregation of schools, also requested a delay in school busing. Nixon had promised these actions to southern leaders, and he delivered. Reagan followed after Nixon and Goldwater. "I am opposed . . . to discrimination," he said, but he also opposed the Civil Rights Act, the Voting Rights Act, and the Fair Housing Act. Thus Reagan was "a chief apostle of contemporary racial conservatism, breathing new life into the Republican southern strategy."[9] He received the lowest black vote of any Republican candidate.

The year 1964 was thus a significant time in the confrontation between the two parties on race, a dividing and transforming moment in our history. It led to progressive legislation, after bitter battles in Congress and under a Democratic president, and then to a slowdown in supportive civil rights implementation under the Republicans: Goldwater, Nixon, Reagan, and finally Bush, who in 1988 won with an attack on Dukakis that had racial undertones, exemplified by the Willie Horton ad. Bush did nothing particular to deal with the race problem. Clinton's victory in 1992 produced much more expectation and empathy for black leaders and the black cause, resulting in more appointments of blacks to leadership positions, such as the selection of Ron Brown, the chairman of the Democratic National Committee, for the commerce secretary post. Under Clinton there was more administration support of blacks and at least discussion of what policies might be helpful to blacks. But explicit efforts to reduce racial hostility and to lead to more social opportunity and justice seemed not a priority. Some scholars argued that the Clinton administration de-emphasized racial concerns. And a divided government after 1994 certainly did not enhance the probabilities of national governmental action facilitative of black interests. Indeed, as Clinton moved more to the center after 1994, the policies of the liberal Democratic left were put on hold. Liberal Democrats were upset with his joining the Republicans in the adoption of the new welfare bill, with all its possible negative consequences for the black poor. The new issue entering the spotlight was affirmative action, triggered by the adoption

FIGURE 16.1 ▪ Race and the 1964 Realignment

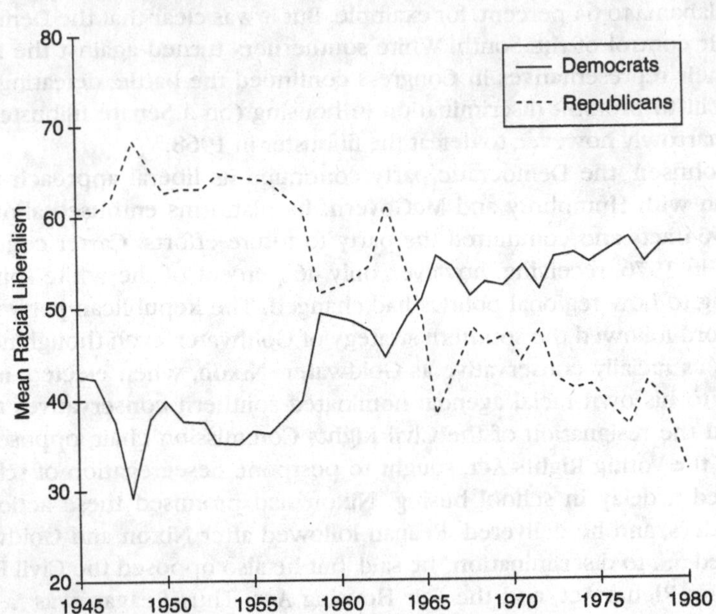

FIGURE A Senate Racial Voting Scales by Party, 1945–1980.

Source: Compiled by Carmines and Stimson from analyses of annual roll-call votes of the U.S. Congress, 1945–1980.

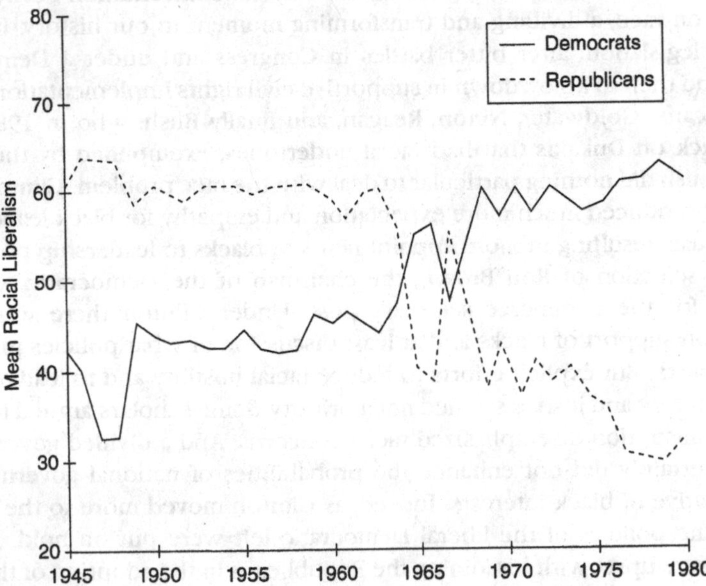

FIGURE B House Racial Voting Scales by Party, 1945–1980.

Source: Compiled by Carmines and Stimson from analyses of annual roll-call votes of the U.S. Congress, 1945–1980.

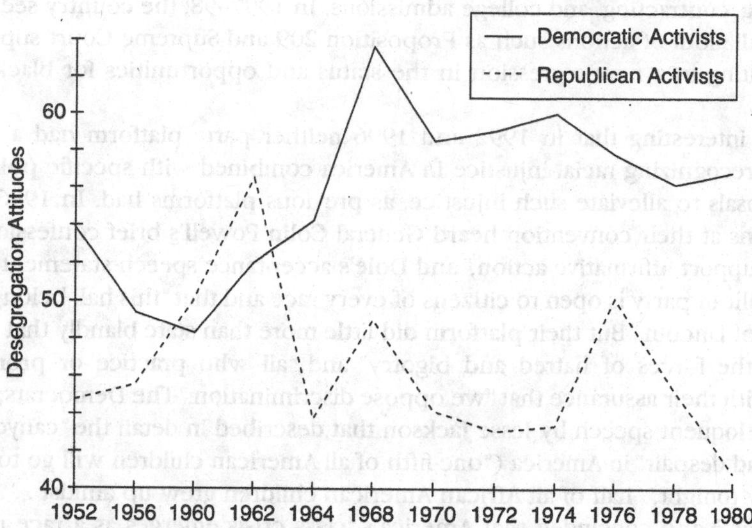

FIGURE C Desegregation Attitudes of Party Activists Weighted by Degree of Activity, 1952–1980.

Source: SRC/CPS National Election Study Series, 1952–1980.

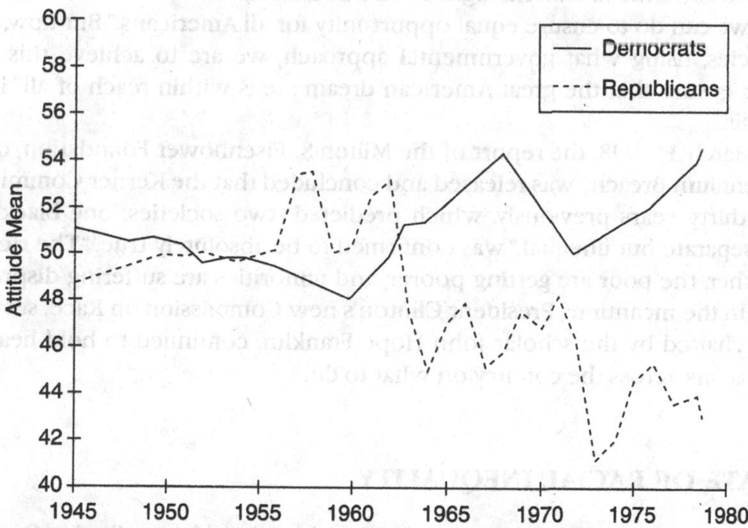

FIGURE D Racial Liberalism of Democratic and Republican Party Identifiers.

Source: SRC/CPS National Election Study Series, 1956–1980, and Louis Harris and Associates, Study #1285.

SOURCE: Edward G. Carmines and James A. Stimson, *Issue Evolution: Race and the Transformation of American Politics* (Princeton: Princeton University Press, 1989), 63, 64, 105, 168.

in California of Proposition 209 in 1996, ending race and gender preferences in state hiring, contracting, and college admissions. In 1997-98, the country seemed ready to fall afoul of actions such as Proposition 209 and Supreme Court support of it, resulting in a new regression in the status and opportunities for blacks in America.

It is interesting that in 1992 and 1996 neither party platform had a plan explicitly recognizing racial injustice in America combined with specific policies and proposals to alleviate such injustice, as previous platforms had. In 1996 the Republicans at their convention heard General Colin Powell's brief confession, "I strongly support affirmative action," and Dole's acceptance speech statement that the Republican party is open to citizens of every race and that "this hall belongs to the party of Lincoln." But their platform did little more than state blandly that they rejected "the forces of hatred and bigotry" and "all who practice or promote racism," with their assurance that "we oppose discrimination." The Democrats, too, heard an eloquent speech by Jesse Jackson that described in detail the "canyon of welfare and despair" in America ("one-fifth of all American children will go to bed in poverty tonight," "half of all African American children grew up amidst . . . broken dreams") and concluded that America's "class crisis emerges as a race problem." But the platform itself says nothing explicitly about racial injustice and the specific approaches the Democrats would now use to work on the race problem. "We must renew our effort to stamp out discrimination and hatred of every kind. . . . we continue to lead the fight to end discrimination. . . . We know there is still more we can do to ensure equal opportunity for all Americans." But how, with what policies, using what governmental approach, we are to achieve this "mission . . . to ensure that the great American dream . . . is within reach of all" is left up in the air.

On March 1, 1998, the report of the Milton S. Eisenhower Foundation, called "The Millennium Breach," was released and concluded that the Kerner Commission report of thirty years previously, which predicted "two societies: one black, one white — separate but unequal," was confirmed to be absolutely true. "The rich are getting richer, the poor are getting poorer, and minorities are suffering disproportionately." In the meantime, President Clinton's new Commission on Race, set up in 1997 and chaired by the scholar John Hope Franklin, continued to hold hearings and discussions across the country on what to do.

THE STATE OF RACIAL INEQUALITY

Many scholars today write of the continuing crisis of racial inequality in often very pessimistic and alarming language. Cornel West, in his *Race Matters*, says this is "one of the most frightening moments" in our history.[10] Danziger and Gottschalk, in *America Unequal*, begin their analysis with a chapter titled "The Diminishing American Dream." Michael Dawson wrote a disillusioning piece pointing out how the Clinton campaign distanced itself from the black community and how some scholars have recently "assaulted virtually the entire moral, theoretical, political and policy edifice that is the legacy of the Civil Rights Movement."[11] Sniderman and his colleagues agree that "racial optimism has shrivelled."[12]

A number of conditions and developments have led to these negative evaluations of the state of racial inequality today in America. One major basis for pessimism is the increasing pattern of economic disparity between whites and blacks. We review some of the data on that subject first, as a beginning of the answer to the question, "Where do we stand today on the race problem?"

It is quite clear from the most recent studies that as the gap between rich and poor in America has widened, so has the gap between whites and blacks. As Oliver and Shapiro state the problem at the outset of their book, there is a paradox here. Even as blacks have improved in economic status in the aggregate, they have fallen farther behind whites. "There is a disturbing break," they say, "in the link of achievement and rewards." There are "persistent and vast wealth discrepancies among blacks and white with similar achievements and credentials" and this poses a "daunting social policy dilemma."[13] This is not a haphazard phenomenon but "structured over many generations through the same barriers that have hampered blacks throughout their history . . . slavery, Jim Crow, so-called de jure discrimination, and institutionalized racism."

How bad are these discrepancies? Oliver and Shapiro look not only at income but also at "wealth" (net financial worth and net financial assets). In 1988 blacks had a median income that was 62 percent of white income, but only 8 percent of white median net worth and zero percent of white median net financial assets:

Race	*Median Income*	*Median Net Worth*	*Median Net Financial Assets*
Whites	$25,384	$43,800	$6,999
Blacks	$15,630	$3,700	—
Ratio	.62	.08	0.00

And the gap has been increasing. In 1967 the difference in mean net worth for the families of blacks and whites was $16,000; by 1988 it was $72,000.[14]

A similar analysis emerges from the February 1998 report of the Council of Economic Advisors to the president, including a chapter on the extent of income inequality for racial and ethnic groups. Although black earnings from 1965 to 1975 narrowed the gap, since 1975 no progress has been made. The poverty rate of blacks in 1997, the report says, is three times that of whites, and the median black family had one-tenth as much wealth at its disposal as whites.[15]

The poverty rates, if examined in detail, tell a grim story, as Danziger and Gottschalk's data reveal:[16]

Percent of children living in poverty in:	*1973*	*1991*
White families	8.1	11.6
Black families	39.4	42.4
Hispanic families	26.5	35.5
All children	14.0	19.7

Percent of male high school graduates 25-34 with earnings below the poverty level:[17]

	1969	*1979*	*1989*
Whites	11	18	27
Blacks	22	40	51

As Fred Harris, co-chair of the report released by the Milton S. Eisenhower Foundation on March 1, 1998, says: "People need to become aware that things are getting worse again."

There are many other studies and reports in the 1997-1998 period that reinforce this evidence of white-black disparities. The thirteenth annual survey of U.S. cities on hunger and homelessness, released December 1997, reported an increase in requests for emergency shelter in 59 percent of the cities; in some it was extreme (San Antonio, 35 percent increase; Detroit, 20 percent, for example). Requests for food assistance were on the increase in 86 percent of the cities (22 percent of the requests were unmet). Ten thousand to fifty thousand are homeless in Detroit each day. Richard Douglass, who has studied homelessness at length, observed that "we have a high-tech economy that's largely irrelevant to the poor and unskilled. This economy can't bail them out." It is alleged that the situation began to deteriorate ten years ago when young, unskilled adults found they were unemployable. The new welfare policy of the government is identified by most cities as partly responsible for these increases.[18] In June 1997 the UN Development Program, using data from its economists on worldwide conditions, reported that 50 million Americans now live below the poverty line, an increase of 3 percent over the past two decades. The gap between the average income in the United States and the poorest 20 percent of the population is twice that found in countries like the Netherlands and Japan.[19]

Another report, by the U.S. Conference of Mayors (February 1998) on housing discrimination, points out the discrepancy in home ownership in the United States (whites, 71.3 percent; blacks, 43.6 percent; Hispanics, 41.7 percent). There is continuing evidence of racial discrimination in approving mortgages, based on a Federal Reserve Board study with the Joint Center for Housing Studies at Harvard. Minority households, they report, are "much more likely" to be rejected for mortgages.[20] The National Fair Housing Authority also reported in 1997 that "housing discrimination is more subtle and sophisticated than ever before." Of 2,000 complaints received in 1996, 36 percent were for racial discrimination against blacks.[21]

Finally, the evidence continues to mount on the disparity in educational performance by black and white children. A study by the United Negro College Fund, chaired by William Gray III, reported a grim picture and an "uphill climb." The study measured the performance of white and black students from preschool through twelfth grade. While there was a slight gap among preschoolers, the gap widens staggeringly early in grade school. By grade four, 76 percent of black students read below "the basic level," compared to 28 percent of white students. By twelfth grade the gap is 66 percent "below basic level" for blacks and still 28 percent for whites.[22]

Have we arrived at the point in the United States where we have actually legitimized racial economic and education inequality? Certainly, if one goal of "the American dream" is to provide the chance to all blacks to live above the poverty line and achieve a decent standard of living, as well as have an opportunity for some upward economic mobility, we seem farther away from it today than twenty years ago. How can this be? Why are blacks mired in such gross economic inequality? Is our political system, including our party system, the reason we have failed to respond to this need?

THEORIES OF THE CAUSES OF RACIAL INJUSTICE AND INEQUALITY

One could argue that economic forces and developments are largely responsible for neglecting the problems of blacks. The United States as a global economic power must continue to cooperate effectively in the world. In order to do so, it has to facilitate the interests of corporations and the business elites. They seek raw materials, new markets, new and cheaper labor. The role of the government is seen as doing what is necessary to increase the productivity of the economy, in the context of increasing world competition since the collapse of Communism. One can argue that the welfare and progress of blacks is not then central to business decisions. "Trapped as they are in high-cost areas of the inner cities but with low levels of skill, many blacks cannot compete with new sources of labor in Mexico and farther afield." Those who hold economic power, operating under the laws of the market, see no responsibility for the disadvantaged and unskilled laborers who are unable to contribute. Economic survival and productivity, then, based on increasing profits excludes most blacks. They are seen as irrelevant.[23]

Yet one must ask, Does this have to be the primary rationale for, and is it the defense of, the increasing black economic inequality in America? Why has not this same logic applied to other Western industrialized nations whose economies have also expanded and become "global," but who also have been able to do a much better job of taking care of their disadvantaged and poor? Their poverty rate is lower, their "safety net" larger and more secure. Perhaps the economic argument is not the real, primary explanation for the American race problem.

One could also advance cultural explanations, but with what credibility? Gunnar Myrdal argued in 1944 that our American creed of liberty, equality, rule of law, consent of the governed, et cetera was at odds with democratic practice, particularly in the treatment of blacks. It was, he said, the most important problem our system faced, its central dilemma. But he was optimistic. Our political cultural orientations, our egalitarian principles would surely lead to racial equality. After all, we all believe it: "We hold these truths to be self evident, that all men are created equal." He was wrong. But perhaps it was another aspect of our culture, ingrained white hostility toward blacks that in the end would dictate our unwillingness to accept blacks as equals. Indeed we had, perhaps, two cultures — a white racist culture and a culture pressing for integration, for equality, the great "melting pot"—and the two would forever clash. The dream that economic progress would be followed by social and cultural changes, including value liberalization, and that this would lead to a general conversion to belief in racial and ethnic equality, which would lead to economic justice, seems quite empty. Whatever socialization has taken place in our society in the past thirty years has certainly not led to a basic value change, for the majority of Americans, that would lead to the acceptance of racial equality, not on just the abstract level, but the level of actual practice, actual policy, actual behavior reflecting meaningful racial tolerance.

What is left to us, then, in terms of grand theory is the theory of the political system, our system of democratic pluralism and particularly the functioning of the two-party system, to which we must turn in order to understand the conditions for our success and failures in dealing with the race question. Race is a political problem; it can only be solved adequately by political action if it is to be solved at all. Economic progress has not led to a solution, nor has educational advancement, nor has the abstract belief in equality. We return to the basic query, then: What are the political conditions responsible for these disparities in black-white income, wealth, health, education, welfare, housing, and general social well-being? What has gone right with our political system when we made progress; what has gone wrong with our political system when we declined to the terribly unjust state of racial inequality that today grows worse each year? Has the party system failed us? Let us look at the pluses and minuses, the accomplishments and disappointments.

PROGRESS IN THE PAST: WHAT CAN WE LEARN?

We should admit our great victories in the past under this party system. We have not always neglected "the Negro problem." The period preceding the Civil War saw the mounting of a major campaign to end slavery, beginning at the time of the Revolution in certain northern states. But the bitter confrontation had no peaceful resolution. The Emancipation Proclamation in 1863 was a unilateral Republican party declaration imposed on the South. Similarly, after the Civil War the Negroes achieved considerable opportunity under Reconstruction for involvement in the politics of the South, winning legislative office in many states and actual control in certain counties. This was reform by coercion and lasted only to 1877, when federal troops were withdrawn from the south. Whatever major advances under the two-party system occurred were the product of irreconcilable confrontation, war, and coercion.

The second period of the two-party approach to racial problems lasted essentially from the 1870s to 1933 and can be called the period of "two-party indifference" to the racial question. As our narrative indicates, the Democrats returned white supremacy to the South, and in the North neither party pushed the black cause. The Populists in the 1890s did attract blacks with their program and some areas of the South, such as North Carolina, joined with them and the remnants of the old Republican party to briefly, in 1894, take control of the legislature. But all this collapsed with the demise of the agrarian revolt in 1896, and blacks were consigned by all parties again to a disfranchised status. There was no interest in the Negro problem by the Republicans, Teddy Roosevelt's Progressives, or the Wilson Democrats. The minimal interest these leaders had was soon dissipated. And so the leaders of both parties did nothing until the Democrats under FDR came to power in 1933.

This ushered in the third two-party period from 1933 to 1964, which might be broken into two periods: (a) the "incipient democratic enlightenment" and (b) the "bipartisan window of opportunity." During the Roosevelt period, the Republicans and the Southerners in Congress would not do anything, but the Democrats by executive decree did a lot. Their actions were described earlier: appointments of blacks and the establishment of the National Recovery Program's "alphabet agencies" to provided economic assistance to blacks. The Supreme Court assisted by declaring legal the picketing of firms engaging in employment discrimination. At Roosevelt's death, President Truman became even more an advocate of black rights. He spoke to the NAACP in favor of action to deal with racial inequality and took the basic position that the parties must recognize the responsibility of the federal government to act.

In the 1944–1948 period we see the first signs of a possible "bipartisan window of opportunity." The Republicans adopted in their platform a series of explicit proposals for action on the poll tax, lynching, employment commission, et cetera. The Democrats as a party were not yet ready to go this far and had a mild, ambiguous plank. But the stage was set for the famous battle in the Democratic Convention of 1948. The liberal Democrats won a classic platform fight on civil rights that resulted in some southern delegates walking out of the convention. During the rest of this period, 1948–1964, both parties were supportive of civil rights in all of their national platforms including 1960, the Republicans slightly more explicit in their support. Both Truman and Eisenhower took action in favor of civil rights, and two early civil rights bills were passed under the Republican president, the acts of 1957 and 1960. Eisenhower also backed up the 1954 Supreme Court decision against segregation by sending troops to Little Rock in 1957. President Kennedy was slow to act after his election but finally did so in 1962, ending discrimination in federal public housing by executive order. And in 1963 he sent his civil rights bill to Congress.

Thus for a period of almost twenty years, 1944–1964, we had the "bipartisan window." Blacks had apparently recognized this and had continued considerable support for Republicans in presidential elections — 40 percent for Eisenhower, over 30 percent for Nixon. All this ended in 1964. With Goldwater's nomination and his rejection of the Civil Rights Act of 1964, thus playing for the "white southern vote," the big switch was on. We entered the period of "the transformation of racial politics." This turned out to be a long period for Republicans, who opposed or were indifferent to the civil rights of blacks from Nixon to Dole. The Democrats under Johnson pushed through the two important pieces of civil rights legislation in 1964 and 1965 and pressed for other legislation, particularly on housing. After Johnson, the Democratic candidates were civil rights believers (indeed, Dukakis was berated as a "liberal"), but by the time of Clinton's victory in 1992 with less than a majority of the vote, the Democratic party's obvious latent empathy for blacks and their problems did not translate into much of an active program. Indeed, one can say that to some extent the two-party system was in a state of relative indifference and inaction again after 1980.

What do we learn from this analysis of the two-party cycles in terms of their race policies? We are reminded that the parties, one or the other or both, initiated those significant actions, executive or legislative, that did occur. They gave us the Emancipation Proclamation, the critical constitutional amendments, the appointments of blacks to important positions, the memorable civil rights acts, and the great variety of presidential directives altering the behavior of bureaucrats toward blacks in the government. We also are reminded, however, that there have been long periods of two-party indifference and neglect of the race problem from 1877 to 1933, and from 1980 to today. They remind us, finally, that the parties can act, can support the cause of racial justice. But often to the parties racial justice was an issue of bitter party conflict rather than bipartisan cooperation. When both parties strongly supported policies to improve the status of blacks, cooperation didn't last. It foundered on the rocks of extremism and the calculations of electoral advantage.

BLACK ACCESSION TO, AND PARTICIPATION IN, PARTY POLITICS: THE 1960s AND BEYOND

The earliest national surveys of political behavior reported a low level of black political involvement and interest. In the 1948 and 1952 elections only 33 to 36 percent said they voted in presidential elections, compared to 66 percent to 79 percent of whites. And blacks had a much lower sense of political efficacy and of a citizen's duty to vote.[24] In the eleven states of the Solid South, on the average only 32 percent of eligible blacks were registered. There had gradually been some improvement during the New Deal period in the status of blacks, and FDR's recognition of black leaders was significant, but at the mass level there was not much change in political involvement. All this changed radically after the passage of the 1964 civil rights act and the 1965 Voting Rights Act, combined with the civil rights movement, Democratic party action, and deputizing of federal registrars in the South. By 1964 in the southern states, black registration reached an average of 44 percent and this jumped to an average of 68 percent by 1970.[25] By the 1968 and 1972 national elections, voting participation reported by blacks in presidential elections increased to over 60 percent. And it remained over 60 percent in subsequent elections despite some fluctuations, and some decline in the eighties and nineties.[26]

One must not forget that the actions by the civil rights movement, beginning in the forties and reaching a peak in the early sixties, played a major role in developing pressure and mobilizing support for these changes. Called "the most important social movement in the United States in this century,"[27] this coalition of a wide variety of liberal action groups used many different tactics to improve the status of blacks: boycotts, sit-ins, judicial action, freedom rides, economic warfare, marches, voter registration drives. Its participants, as we know full well, suffered incredibly for their efforts, at the hands of police, local leaders, the Ku Klux Klan, and southern citizens, often being beaten as well as imprisoned. A systematic study of the 2,000 or more individual actions documents what happened. Its targets were white racism, police brutality, legal inequality, segregation of accommodations, fair housing, transportation and services, and eventually the aim of winning some political access and power for African Americans.[28] One of the developments during this period was the increase in political party contact which also increased voter turnout, of both whites and blacks.[29] As a result of these efforts, black turnout increased 35 percent from 1952–56 to 69 percent by the 1980s. Rosenstone and Hansen attribute this to a set of five key "causes":

1. The change in voter registration laws and their administration during this period;
2. An increase in black education (20 percent more completed high school during this period) and many got better jobs;
3. An increase in personal contact by parties and civil rights activists;
4. Increased feelings of personal efficacy, as a result of the mobilization efforts in (3);
5. A change in the party orientations of blacks, so more of them began to commit to, and vote, for the Democratic party.

They concluded: "the rise in black turnout was dramatic," with the largest part of the increase coming as the result of elimination of legal barriers.[30]

The transformation that occurred in the party loyalties and party voting from 1964 on is, of course, very significant. This was a modest restructuring of the party system. This 1964 transformation has been documented in detail by Carmines and Stimson.[31] We have updated the basic evidence for that monumental change in Table 16.1. The phenomenal swing of blacks to the Democrats from 1952 to 1964 (1968) of over thirty points in party identification was accompanied by a mass desertion of blacks from the Republican party. Many blacks remained "independent" or "apolitical," however. The confusion in voting among the few blacks with a Republican orientation is suggested in Table 16.2. While Democratic identifiers voted heavily Democratic, Republican identifiers first voted heavily Democratic also, but then were more likely not to vote at all. One should note also that in 1996 there was some decline in African American loyalty to the Democrats and an increase in nonvoting among Democratic identifiers. What is particularly notable, however, is the high percentage of blacks declaring a "strong" identification with

TABLE 16.1 ▪ The Change in Party Identification of African Americans 1952–1996 (in percent)

Party Identification	*1952*	*1964*	*1968*	*1980*	*1988*	*1992*	*1996*
Democrat (strong or weak)	52	72	85	72	61	64	66
Independent (all three categories)	18	15	11	19	29	28	30
Republican (strong or weak)	13	7	2	5	7	5	3
Apolitical/don't know	18	6	3	4	3	3	2
N (cases)	171	159	149	187	369	319	206

SOURCE: National Election Studies.

TABLE 16.2 ▪ Change in Presidential Vote of African Americans by Party Identification (in percent)

Party Identification	*1952*	*1964*	*1968*	*1980*	*1988*	*1992*	*1996*
A. *Democratic*							
Voted Democratic	38	70	68	70	60	70	56
Voted Republican	3	0	2	3	4	1	0
Nonvoters	59	30	30	27	35	27	43
B. *Republican*							
Voted Democratic	12	73	25	8	17	38	31
Voted Republican	20	0	25	17	9	25	6
Nonvoters	68	27	50	75	74	37	63

Note: The small number of cases of blacks who were Republican identifiers indicates that caution should be used in accepting these "absolutist" percentages. For example, in 1996 the *N* for Democratic identifiers was 163, for Republican identifiers only 16.

SOURCE: National Election Studies.

the Democrats in these early years, compared to the norm for American citizens. And this has continued at only a slightly lower level in the nineties:

Percentage of African Americans with Strong Democratic Affiliation

	1960	*1964*	*1968*	*1972*	*1992*	*1996*
Total Sample	21	26	20	15	18	19
African Americans	24	51	56	36	41	44

When they committed to the Democratic party in 1964, they did it with real conviction. Even in 1992 (and 1996) they were the most loyal subgroup in the party. These data thus document the extent to which, in terms of party commitment and voting, this was a period when blacks really entered the mainstream of electoral politics, attached themselves to one party, the Democrats, and stayed committed. There was some later evidence of wavering in loyalty, possibly because they became aware that they were not achieving as much as they had hoped by this one-party affiliation.

Further evidence of the increasing involvement of blacks with parties and campaigns emerges from the political activism data. Using national survey responses to specific questions on participation, Rosenstone and Hansen found that blacks on some items were not too far behind whites in their level of activity. For the entire 1952–1988 period they found these comparisons:[32]

Political Activity	*Blacks (percent)*	*Whites (percent)*
Tried to influence how others voted	22	27
Contributed money to a party or candidate	4	10
Attended a political meeting or rally	7	8
Worked for a party or candidate	4	5
Attended a public meeting or a town or school affair	12	18
Wrote a letter to a member of Congress or Senator	5	16
Signed a petition	18	38

When we break down the data by time periods, we find a surge in such activism in the 1964–1968 period, declining only slightly after that. Another indication of blacks' involvement is their selection as delegates to the national conventions of the two parties. With the reforms introduced by the Democrats, the percentage of black delegates increased from 5 percent in 1968 to 15 percent in 1972, and since then the percentage has increased slightly, to 21 percent in 1996. The percentage of blacks who achieved delegate status at the Republican convention was never high (2 percent in 1968, 5 percent in 1992, and 2.6 percent in 1996). This party differential attests again to the completely different role African Americans play in the two parties. It is revealed also in the striking contrasts between the compositions of the two national party committees — there was only one black on the Republican National Committee in 1998 (out of 165 members), but eighty-eight (out of approximately four hundred) on the Democratic National Committee. If one adds to this the number of blacks in the U.S. Congress, one begins to get a measure of the black representation in national party and legislative institutions. There are thirty-seven blacks in the House and one in the Senate, which is about 7 percent of both houses. This is more than double what it was twenty-five years ago when the number was only sixteen (or 3 percent). Yet clearly our system is not very congenial to upward black mobility to the national level. Shortly after the Voting Rights Act was passed in 1965, the number of blacks elected to state legislatures in the eleven states of the Solid South increased from one hundred (in 1965) to fourteen hundred (in 1974), probably close to one-third.[33] But penetrating the system at the national level proved more difficult for blacks; in fact, it never really occurred in significant magnitude. That is not to say that blacks in Congress had no influence. Through the congressional Black Caucus they mobilized support for legislation, working often with non-black members of Congress. Yet the caucus had neither the resources nor voting power to exert the type of leverage necessary to compete effectively in the game of party politics in Congress.

Blacks were very successful in certain major cities, winning the mayor's office *and* electing a majority of blacks to the city council. Atlanta, Detroit, Gary, and Washington, DC, are examples of such victories, and in other cities blacks were successful in electing black mayors: Chicago (Washington), Los Angeles (Bradley), New York City (Dinkins). Even in the South some progress in winning office occurred after the sixties, but very little before that time. Matthews and Prothro observed in their 1966 study that "negro office holding is so rare in the South that this is not a very satisfying measure of negro political leadership" and argued that informal leadership was more important.[34] Yet even today, the percentage of African Americans holding seats on city councils is relatively low: 3 percent in small cities, 7 percent in medium-sized cities, 19 percent in large cities (based on a 1989 study of nine hundred council members in 243 cities).[35]

One must conclude from this review of black political participation that the record is mixed. In 1964–1968 the system, in a sense, opened up and blacks responded. They went to the polling booths, they joined the Democratic party with a strong commitment, they became active in campaigns, and in between elections they ran for office and were for a time increasingly successful. Then a decline, born perhaps of disillusionment, began to set in, and their strength of Democratic identification declined somewhat, as did their activism. As Katherine Tate put it in 1994: "There are signs that Black expectations have declined and increasing numbers of Blacks are dropping out of local politics." [36] Blacks' trust in government, at the 70 percent level in the 1964–1968 period, dropped sharply (as it did for whites also) to the 19 percent level in 1972, and only recovered slightly in the eighties to 27 percent. In 1992 it was 24 percent (almost equivalent to that of whites). Miniscule advances have been made in securing Democratic seats in Congress and in positions in the Democratic party organization. But blacks have been virtually excluded by the Republicans.

Some scholars have embraced the easy optimism that after the 1964–1965 civil rights and voting rights legislation blacks would gradually, through incremental linear progress, improve their status and quality of life in American society, that "incorporation" would gradually take place. If we look about us, we note that "the rising tide" of black progress didn't continue to rise, except for a minority of blacks. Political empowerment has not happened. And today, for all our improvement in black political participation, the striking disparities in black economic and social status (which we described at length earlier) still exist. A basic problem may be that, for all the improvement in political participation, blacks still do not have political influence. As Carmichael and Hamilton put it long ago, "the basic goal is not 'welfare colonialism'. . . but the inclusion of black people at all levels of decision-making. We do not seek to be mere recipients *from* the decision-making process, but participants *in* it."[37] And this means participation that results in influence over outcomes.

The Electoral and Party Involvement of Other Minorities

African Americans are not the only minority group whose participation in elections and party politics is significant for our system. We should not forget the roles of Latin Americans, Asian Americans and Native Americans. They were admitted to citizenship and given the right to vote at different periods: Mexicans were given the vote in 1870, Puerto Ricans in 1917, Cubans in 1959, Native Americans in 1924, Chinese in 1943, Japanese in 1952.

The partisan behavior of these groups has varied over time. Latinos generally are more Democratic (57 percent) than Republican (25 percent). In the 1996 election, 72 percent of Latinos supported Clinton, 21 percent Dole, and 6 percent Perot. The Asian American vote was split fairly evenly between the two major parties in 1996. But the turnout at elections is low for Latinos (30 percent) compared to African Americans (54 percent); Asian Americans also have a low turnout rate –27 percent. The voting behavior of Latino subgroups varies: in Mexican congressional districts 65 percent vote Democratic, in Puerto Rican districts the vote is 88 percent Democratic, while Cubans are more likely to be strongly Republican in the few districts where they are a majority.

A comparison of strength of party identification reveals the following contrasts in our recent NES studies:

Party Identification	*Mexicans*	*Puerto Ricans*	*Cubans*
Strong Democrats	31	37	14
Weak Democrats	34	33	11
Independents	12	12	6
Weak Republicans	18	11	21
Strong Republicans	4	7	48

The percentage of party activism, however, does not differ for these minority groups very much from whites and African Americans — about 8 percent in presidential elections. It is the low voting turnout of Latinos that is perhaps more significant. Rodney Hero explains this as follows: "Latino political participation is lower than that of other groups, a pattern that seems to hold even when accounting for socioeconomic status. This may reflect some aspects of Latin culture. . . . it may also be traced to complex voter registration requirements and difficulty in understanding U.S. politics. . . ."

SOURCE: Rodney E. Hero, *Latinos and the U.S. Political System: Two-Tiered Pluralism* (Philadelphia: Temple University Press, 1992). See also F. Chris Garcia, ed., *Latinos and the Political System* (Notre Dame: University of Notre Dame Press, 1988) and Rodolfo O. de la Garza, ed., *Ignored Voices: Public Opinion and the Latino Community* (Austin: Center for Mexican American Studies, 1987).

THE PLURALIST AND TWO-PARTY MODELS: DO THEY FACILITATE BLACK PROGRESS?

We come to the basic question: Is the American political institutional system one through which blacks can work to further their welfare in this society? That forces us to think through the nature of our system — the pluralist model and the two-party model. The pluralist model conceptualized by Robert Dahl in 1967, and later others, on the surface suggests optimism for black progress. Once a group achieves its political rights in our system, gains access to the political arena, theoretically it should move ahead. It should achieve group political status, and the welfare of the members of the group should be improved; in short, a group becomes "empowered." But political opportunity may not lead to political relevance, political participation may not lead to political influence, in the sense of a decisional role over policy making in the areas of black social welfare and racial justice. Why is this the reality for blacks and similar minority groups?

The essence of the pluralist model is the existence of a set of intermediate groups linking citizens to their government, groups which compete for power and for the distribution of economic and social resources in our system. Our society has a consensus on our basic values (such as equality, liberty, justice, government by consent of the governed, and so on), and as part of this creed we believe that governmental decision making should reflect the best representation of group interests. It is difficult to achieve governmental action because of what we call "the multiple centers of power" in our system (the two houses of Congress, the courts, the presidency, as well as the states). For interests to be successful in their struggle, they must survive several gauntlets because of this "checks and balances" system. There are several key requisites for success under such a pluralist system:

1. considerable group self-consciousness and internal unity, i.e., agreement among members on goals and strategies;
2. an effective political organization of its own, with an involved and active membership;
3. effective external leadership, i.e., leaders who represent the group's interests in the government and in the society;
4. alliances or bargaining relationships with other groups, a willingness to compromise, under a strategy of meaningful "trade-offs" and "payoffs";
5. in lieu of alliances, if they are not attainable, sustained and massive mobilization of the public from time to time to maintain saliency and to exert pressure on governmental authorities;
6. participation in electoral politics so as to demonstrate political "clout" while reserving the option of electoral defection or withdrawal;
7. development of important public support, a public sympathetic to the group's objectives so that a favorable "climate of opinion" prevails;
8. avoidance of radical separatism, a decision to work outside the pluralist model, which may be perceived as threatening the consensus.

This pluralist model with these requisites, if one accepts it, has clear implications. It sees politicians as brokers of group interests and demands at any given

The Balance-of-Power Theory: The Role of Minority Voters

NAACP publicist Henry Lee Moon set forth in his book in 1948 the basic strategic thinking and the raison d'être for minority voting — that in close elections their vote could determine the outcome in presidential elections. In that tome Moon argued his theory as follows: "the Negro's political influence in national elections derives not so much from its numerical strength as from its strategic diffusion in the balance-of-power and marginal states whose electoral votes are generally considered vital to the winning candidate." However when Nixon won twice, 1968 and 1972, and Reagan twice, 1980 and 1984, and Bush once, 1988, Walters challenged the balance-of-power theory and declared that the reason the theory was invalid was that African Americans after the 1964 election had become not an independent electoral force tied to no one party but a dependent one with an almost complete alliance with the Democratic party.

Smith, building upon the theoretical and empirical insights of Walters, says: "in 1992 the party that [African Americans] have loyally supported the last twenty-five years decided that it could only recapture the presidency by running a campaign based on a strategy of ignoring them and their policy concerns while pandering to racist and antiblack sentiments among whites." As Smith sees it, although the African American vote might have become central to the Democratic party's recent presidential victories, a sort of balance of power for the party, it has helped them very little. He concludes that because of this "dependency on the Democratic Party, blacks have been unable to develop the requisite pressures to get the system to respond to its most pressing post-civil rights era demand: the need for employment in the context of some kind of overall program of internal ghetto reconstruction and development."

Using his own analysis of the party system and hearing the insights of Walters and Smith, Morris, in an innovative assessment of the American party system, suggests that race is no longer relevant to either of the two parties. Each major party, in Morris's opinion, decided in the post-civil rights era to distance their organization from this issue. The implication here is that older theories seeking to explain minority party behavior are no longer viable. If this is true, a major reassessment of the strategic direction for minorities is necessary, as this chapter suggests.

SOURCES: Ronald Walters, *Black Presidential Politics in America: A Strategic Approach* (Albany: State University of New York Press, 1988).

Robert C. Smith, *We Have No Leaders: African Americans in the Post-Civil Rights Era* (Albany: State University of New York Press, 1996), 274, 280.

Lorenzo Morris, "Race and the Rise and Fall of the Two-Party System," in Lorenzo Morris, ed., *The Social and Political Implications of the 1984 Jesse Jackson Presidential Campaign* (New York: Praeger Publishers, 1990), 65-90.

Henry Lee Moon, *Balance-Of-Power: The Negro Vote* (Garden City, NY: Doubleday, 1948), 198.

point in time. Theoretically the state is seen by some as neutral in the entire process, although, of course, the current power structure (currently the white middle class–upper class) has its own group interests to serve. It is biased in favor of those now in power and their values. This is a model that implicitly prefers stability, not radical change; it prefers incrementalism. Theoretically, if the system works as it should, governmental actions are a balance of major group interests, and the outcomes will fairly represent a compromise of the most important of these interests, thus maintaining the democratic order.

How does the two-party system fit into this pluralist model? In actuality it overlays the pluralist conception, and while it imposes constraints, it should be instrumental for it. But the party system in a sense jerks us back to the reality that the system of party politics affects the pluralist system's processes a certain way, different in the United States from multiparty systems. Above all, it restricts the paths for groups seeking power.

One must visualize each of the two parties as a coalition of interest groups, with some groups overlapping both parties, other groups distinctively, if not exclusively, pro-Democratic or pro-Republican. There are "100 percent" groups, "75 percent" groups and "50 percent" groups, thus named for the proportion of their commitments to one or both parties. Parties in a two-party system strategically seek to mobilize group interests over a fairly wide spectrum, because of the "imperfect mobilization of social groups," in order to have enough votes to win. Interest groups also may seek to penetrate both parties in order to maximize the probability of attention to their demands. A party may espouse the cause of one major interest if it is large enough and capable of mobilizing enough voters to be a significant factor in the voting calculus. Normally both parties play to "the median voter," so to speak, a little to the left of center or right of center ideologically. If an interest group becomes attached almost exclusively to one party (as the blacks to the Democratic party, or the Christian right to the Republican party), it faces the prospect of being ignored or only minimally responded to, unless it is part of a larger alliance or can mobilize extreme and massive pressure outside the party system, or it finds its elite representatives in the party (the president and in Congress) sympathetic and committed to the group's needs.

The alternatives, thus, for a group such as African Americans, as it confronts the two-party system, are:

1. become part of a large intraparty alliance within the Democratic party, ideally without losing its separate identity; *or*
2. strengthen the group's organization and leadership, and mount a massive campaign for political action mobilizing the public in support of its program (like the civil rights movement of the sixties); *or*
3. develop a bipartisan strategy, including the creation of a cross-party alliance of groups sympathetic to the group's needs; *and*
4. work to elect a set of representatives in government, ideally in both parties, who are aware of the group's needs, want to take action, and *do* take action by formulating policies in furtherance of these objectives.

The alternatives of working within the parties without alliances with other groups, or launching a third party, or threatening reduction of support for the Democrats, do not seem to be very viable. Finally, of course, we could argue that we should continue in the future with the same laissez-faire attitude and perspective that seems to characterize the position of the parties and the government today, leaving the cities and the counties and the states to develop their own programs, relying on their own limited resources, with no significant new national policies or programs and no new infusion of national funds to deal with the problem of race at lower levels of the system.

ALTERNATIVE SCENARIOS: WHAT ARE THE PROSPECTS?

If we reflect on the long history of the two-party system's engagement with the race problem, there must be both positive and negative observations. The balance sheet contains both. On the positive side it is clear that there has been considerable progress in the improvement of the status of blacks. A minority of blacks are much better off in economic terms, and the poverty rate is not as high as it was fifty years ago. Blacks have gained the franchise and are voters, party identifiers, strong party identifiers, political activists, and successful candidates for public office. In the past they demonstrated great skill in the mass political action movement, and when mobilized they were a significant political force. Many of their leaders in office are articulate and able, and in addition they have scholarly protagonists who speak eloquently for the problems of the black race, whether a Jesse Jackson, a Michael Dawson, or a Cornel West.

The negative observations one must make, however, are very disturbing. The plain fact is that economic inequalities between whites and blacks are large, and they are certainly not decreasing. The debate that goes on, among conservatives particularly, tends to de-emphasize the findings on economic disparities. They claim the fight against discrimination has been won. The problems are with the black lower-class family structure and behavior, they argue. But the Danziger and Gottschalk analysis reveals that is not true. And the 1998 Council of Economic Advisors Report concurs.

Many blacks are disillusioned about their lack of progress. Voter turnout is down, as is strong party affiliation. Blacks are essentially shunned by the Republican party. They are a one-party group. And a rather disunited, divided group — at both leadership and followership levels. Even the Democratic party is not assertive about fighting for new policies to ameliorate racial injustice. Its national party platform is in a state of ambiguity, if not actual denial, about its vision for blacks. And white support in the population is lower today than ten to twenty years ago. True, blacks have, and in the past have had, mayors in quite a few cities, but it is difficult to see much evidence that these cities, sometimes controlled by black leadership, have been able to cope with their problems and improve the economic condition of blacks in their city limits, in the absence of action by government at higher levels. So today there is a more hostile environment, there is dissension in the internal politics of the black community, and no evidence of action by white or black leaders, Democratic or Republican, on the crucial questions of black economic and social welfare. All those black votes, and activity, and party loyalty produced some progress up to the seventies but in recent years have failed to be converted into public policy — that is, the type of public policy we need to deal with the race problem today. As one scholar put it, black "optimism has shriveled."

Faced with these plain facts, particularly that no progress on reducing racial economic inequality is occurring, what are the political options for blacks? First is the option of working harder through the Democratic party, increasing political activity, running more black (or sympathetic) candidates for office, mobilizing a larger vote, developing a stronger presence at caucuses and conventions (from local to national), and working closely in alliance with white liberal groups in the party in support of a liberal agenda. Research has shown that there are sociopolitical groupings within the Democratic party with strong correlations in affect, attitudes, and policy preferences with blacks — groups such as union supporters, feminists, welfare liberals, civil rights leaders, and environmentalists. Miller finds a fairly high intergroup correlation (.33) for these Democratic groups, just as he finds a relatively high correlation in Republican party groups such as big business, moral majority, and right-wing conservatives (.37)[38] The potential for a strong liberal-black coalition exists, but such proximity of affect and attitudinal reciprocity does not necessarily translate into proposals and policies that blacks advocate. Merely having a liberal Black Caucus in Congress is not enough. It takes House and Senate and presidential leadership with a vision and a program and a commitment, bolstered by pressure from below.

Within the Democratic party at the middle leadership level there is strong support for a policy of government aid to blacks (Table 16.3). Fully three-fourths of the Democratic county and state chairs and national convention delegates feel that the government has the responsibility to help blacks economically and socially. And such support has been increasing since the eighties. So there is a sort of "hard core" potential organizational support in the Democratic party for such a black agenda. However, we notice in Table 16.3 also that the Democratic *followers* in the public have been declining in support for blacks on this question — from 41 percent pro in 1984 to 33 percent in 1992. This produces a big gap, then, between elite and mass opinions in the Democratic party today. It is necessary not only to activate liberal leaders in the party, but to turn around, and educate, the Democratic party's followers on this question. That requires leadership at the top as well as at the bottom, both white and black.

The persistent efforts by Jesse Jackson as a candidate for the Democratic presidential nomination in 1984 and 1988 illustrate the uphill battle of a black leader in the Democratic party. Jackson in 1984 was the first black leader to seriously seek the presidency, and his eloquence and appeal made him a real threat to Mondale, the eventual winner in the Democratic nomination battle. Jackson tried to create his "rainbow coalition," including Hispanics, Native Americans, white liberals, nuclear freeze advocates, environmentalists, and Jews. He was not very suc-

TABLE 16.3 ■ Opinions of Party Leaders and the Public on the Role of Government in Providing Aid to Blacks (in percent)

	Democrats			Republicans			Change 1984–92	
Opinion	*1984*	*1988*	*1992*	*1984*	*1988*	*1992*	*D*	*R*
Leaders								
Government should aid	57	66	75	12	25	27	+18	+15
Blacks help themselves	22	18	9	66	55	51	−13	−15
Followers (identifiers)								
Government should aid	41	37	33	21	16	12	−8	−9
Blacks help themselves	29	40	41	48	59	63	+12	+15
Difference — Leaders and Followers (government aid)	16	29	42	9	9	15	—	—

Question: In your opinion, should the government provide aid to blacks to improve their economic and social position, or should blacks help themselves?

Note: "Leaders" include a sample of county party chairs, delegates of the national party convention, state party chairs, and members of the national party committees. The number of cases was over 1,000 for each party in 1984, nearly 600 in 1988, and a total of close to 900 in 1992.

SOURCE: National Election Studies, plus special elite studies by several political science scholars in each year. See particularly William Crotty, John S. Jackson III, and Melissa Miller, "Political Activists Over Time: 'Working Elites' in the Party System." Manuscript prepared for conference in Ann Arbor, Michigan, in 1997. To be published by University of Michigan Press in 1999.

cessful in putting this coalition together, but his challenge was dynamic and serious, evangelical in style, and his threat of a possible withdrawal of black support forced Mondale to respond cautiously. The basic issue of the Jackson candidacy was the need to use the resources of the nation to redress black economic injustice — by redistribution of income from rich to poor, an increase in the corporate share of taxes, a graduated income tax, a 20 percent reduction in the defense budget, and the adoption of new proposals to help the nation's poor. He unfortunately alienated the Jews and adopted foreign policy positions perceived as too extreme.[39] In the end, he received 77 percent of the black vote cast in the primary but only 5 percent of the white vote, and he won 12 percent of the delegate votes at the convention. (He actually won the Louisiana primary with 43 percent of the vote.)

In 1988 Jackson tried again and did much better. He appealed to the white liberals as well as the blacks. He did very well in the early primaries, winning six states and the District of Columbia. He won in southern states like Georgia, Louisiana, Mississippi, Alabama, and Virginia due primarily to the black vote. His victory in Michigan was a big surprise and kept him in contention until New York's primary in mid-April, which Dukakis won, and in which the "populist alliance" of Jackson lost badly. However, his policy agenda, similar to 1984, was explicit: "the incentive must be to reinvest in America, retrain our workers, reindustrialize our nation." Jackson's image grew more positive during the campaign.[40] His vote in the primaries increased to 29 percent, which was also the percentage of delegate votes he received in the convention. He really mobilized the black vote in the primaries, getting 92 percent of them to vote for him; but only 12 percent of the whites supported Jackson. In a sense these two Jackson candidacies illustrate the problem of a black-led coalition in American national politics. However, there are cities where such a coalition has been more successful.

A second option is, obviously, a *bipartisan* black strategy. Indeed, there are those who see this as the only hopeful scenario. The argument is that today, unless the leaders of both parties join to endorse a policy of redressing black injustice, it will not happen. The strategy is suggested that some blacks begin to consider joining the Republican party and voting for, and working with, Republicans in order to demonstrate that there is a base of mass black support available in both parties. Some writers go so far as to recommend that blacks begin to "colonize" the Republican party.

Realistically, and empirically, can this happen? If we look again at Table 16.3 we can refresh our minds on the level of support in the Republican party for blacks. What percentage of Republican leaders believe the government should take action to provide economic and social aid to blacks? Interestingly, there has been a small increase recently at the Republican leader level — from 12 percent (1984) to 27 percent (1992) who favor government aid to blacks. At the Republican follower level, however, the reverse is true: 21 percent supported government aid in 1984 but only 12 percent in 1992. The result is a fairly homogeneous Republican party opposed to using government for this purpose. On this empirical basis it is difficult to have much confidence in the present Republican party as a partner in an alliance to improve the economic status of blacks. The basic conclusion must be that the parties are polarized at the elite level, and the level of public support for doing something on the black problem is increasingly negative.

In the past our progress has been either through one party or the other, if any action took place at all. The Republicans emancipated the slaves and imposed reconstruction on the South, providing for the empowerment of blacks during a brief period that lasted only to 1877. In 1933 the Democrats became the dominant party, and they did what they could by executive directive, under FDR and Truman. Yet there was a bipartisan period from 1948 to 1964 during which both parties took action on civil rights. If we could resurrect that bipartisan scenario, there would be hope. But it seems highly implausible today. Perhaps there are cities whose experience could be instructive, such as the history of Atlanta in the postwar period, particularly under mayors Maynard Jackson and Andrew Young in the seventies and eighties.[41]

In the last analysis, perhaps, returning to the mass mobilization strategy and tactics of the earlier civil rights movement is the scenario needed to resocialize people to the problem of race, activate a mass-level coalition of liberals, and force the hand of the national elites. One of the most careful assessments of the civil rights movement from 1948 to 1976 is found in the study by Rosenstone and

Hansen.[42] To them that movement was "the most important social movement in the United States in this century." They describe carefully the character, extensiveness, and persistence of that movement, not only in the South where it began, but nationally from 1960 to 1970. It was a massive mobilization of blacks and whites fighting against restrictive and discriminatory voter registration laws in the beginning, and then expanding objectives to incorporating blacks into the American political system, and then directing attention to policies needed to improve the economic opportunities of blacks, their educational level, and their social involvement with whites.

The effects were striking. In fifteen years there was a 35-percentage-point increase in black voting participation. Blacks became strong members of the Democratic party, active in electoral politics: attending political meetings, contributing money to the party and to candidates, working in campaigns, running for office, getting seats in state legislatures and city councils, and even being accepted, though belatedly, as candidates for Congress. The political attitudes and perspectives of blacks changed. They developed a sense of political efficacy. The system opened much more to them, providing for more opportunities. They had access to better education, and the percentage completing high school increased to 61 percent in a fifteen-year period. Many got better jobs, reducing gradually the wide gap in economic status. The percent of blacks who earned incomes putting them in the lowest third of families decreased from the sixties to the seventies by 15 percent (from 71 percent to 56 percent). This was, thus, a powerful movement with powerful effects. In the face of the bigotry of the past, it was what this democracy needed to begin to redress extreme and unacceptable racial injustice.

While the effects of the civil rights movement were certainly enduring, and changed the status of blacks in the society, the expectations of continued political, economic, and social development in that direction were unfulfilled. After about 1975 progress stagnated, as the empirical research on inequalities has voluminously demonstrated.

The prospects for mobilizing the American public today to accept and work for a new civil rights movement, in order to continue the work of the past, do not look good. We have seen the evidence of public attitudes toward proposals that the government should adopt new policies to provide economic and social aid to blacks. The support has certainly diminished. Further research on the extent and depth of racial animosity adds to this concern. In a recent national study of racial resentment among white Americans, Kinder and Sanders found that such negative feelings are held by 60 percent to 67 percent of respondents. Here are some examples of their findings:[43]

Specific Statements Used in the National Surveys	Agree (in percent) *Strongly*	*Somewhat*	*Total Agree*
1. "Irish, Italian, Jewish, and many other minorities overcame prejudice and worked their way up. Blacks should do the same without any special favors."	32.9	33.7	66.6

Specific Statements Used in the National Surveys	Agree (in percent) *Strongly*	*Somewhat*	*Total Agree*
2. "Most blacks who receive money from welfare programs could get along without it if they tried."	25.4	35.3	59.7
3. "It's really a matter of some people not trying hard enough; if blacks would only try harder they could be just as well off as whites."	*22.4*	*36.9*	*59.3*
	Disagree (in percent) *Strongly*	*Somewhat*	*Total Disagree*
4. "Over the past few years, blacks have gotten less than they deserve."	20.5	*38.0*	*58.5*

Over the period 1986–1992 these levels of resentment were stable. Further, the authors demonstrate that such attitudes correlate high with the positions of these respondents on policy positions dealing with employment, school desegregation, preferential hiring, and college admissions. Such levels of prejudice make democratic interaction, and joint efforts to support changes in the environment and to press for new legislation, very difficult and unlikely. In the 1996 National Election Study the following statement was put to respondents: "We have gone too far in pushing equal rights." In total 68 percent agreed, 82 percent of the Republicans and 57 percent of the Democrats. At the public level it is clear that only a minority in either party is interested in action on egalitarian issues. We are in a period of considerable racial confrontation and tension. Yet there is a strong minority willing to pursue this cause.

Where do we go from here? Do we accept the present, in effect legitimizing inequality, or do we develop a new strategy? Charles Tilly, in his new book *Durable Inequality,* raises the specter of probable inaction, of two-party neglect of the problem, and concludes with an interesting observation and an exhortation:

> Widespread categorical inequality threatens democratic institutions twice: by giving members of powerful categories incentives and means to exclude others from full benefits, and by providing visible markers for inclusion and exclusion.
>
> Here is a chance to work seriously toward reducing durable inequality. . . . In the midst of twenty-first century abundance, we should leave no place for bitter confrontations of the tall and the short, the fat and the thin, the overfed and the hungry.[44]

Our democratic society needs to mobilize to face the challenge of racial inequality — ideally through the party system, perhaps with some outside assistance.

NOTES

1. Thomas West, *Vindicating the Founders,* (New York: Rowman and Littlefield, 1997), 5.
2. Alexis de Tocqueville, *Democracy in America,* 1835. Quoted in Donald Kinder and Lynn Sanders, *Divided by Color* (Chicago: University of Chicago Press, 1996), 11.
3. W. E. B. DuBois, *The Souls of Black Folk* (New York: Vintage, 1903), 1.
4. Gunnar Myrdal, *An American Dilemma: The Negro Problem and Modern Democracy* (New York: Harper and Row, 1944), 1.
5. V. O. Key, Jr., *Political Parties and Pressure Groups* 3rd ed. (New York: Crowell, 1954), 132.
6. Cornel West, *Race Matters* (New York: Random House, 1994), 155-56.
7. We have relied a great deal on the historical scholarship of John Hope Franklin, *From Slavery to Freedom* (New York: Knopf, 1974), for many facts and insights presented in this summary.
8. Edward G. Carmines and James A. Stimson, *Issue Evolution: Race and the Transformation of American Politics* (Princeton, NJ: Princeton University Press, 1989). Their historical account is relied on greatly here, 27-58.
9. Ibid.
10. West, *Race Matters,* 155.
11. Michael C. Dawson, "Black Power in 1996 and the Demonization of African Americans," *PS* 29 (September 1996): 456-61.
12. Paul M. Sniderman et al., "A Test of Alternative Interpretations of the Contemporary Politics of Race," paper presented at 1997 meeting of the Midwest Political Science Association, Chicago, Illinois, 4.
13. Melvin L. Oliver and Thomas M. Shapiro, *Black Wealth/White Wealth* (New York: Routledge, 1997), 12-13.
14. Oliver and Shapiro, *Black Wealth,* 86, 99-100.
15. *Detroit Free Press,* March 24, 1998. See also *New Republic,* February 1998.
16. Sheldon Danziger and Peter Gottschalk, *America Unequal* (Cambridge: Harvard University Press, 1995), 90.
17. Ibid., 86.
18. *Detroit Free Press,* December 16, 1997.
19. *Ann Arbor News,* June 12, 1997.
20. *Ann Arbor News,* February 23, 1998.
21. *Detroit Free Press,* May 22, 1997.
22. *Detroit Free Press,* June 5, 1997.
23. For an elaboration of this theory, see Stephen Burman, *The Black Progress Question* (Thousand Oaks, CA: Sage Publications, 1995), 213-16. See also Danziger and Gottschalk, *America Unequal,* chap. 5, "Why Poverty Remains High." They disagree with those who blame the black family structure as the main reason, and also develop a theory on the economic changes that have accompanied demographic shifts.
24. Angus Campbell, Gerald Gurin, Warren E. Miller, *The Voter Decides* (Evanston, IL: Row, Peterson, 1954), 10, 191-92, 197.
25. Carmines and Stimson, *Issue Evolution,* 49.
26. Rosenstone and Hansen present data suggesting that by 1968/72 the turnout was 67 percent (increase of 35 percent over the fifties.) They also report a decline of about 13 percent 1972 to 1988, to 54 percent. However, NES data on the reported vote do not agree with this level of decrease, suggesting a 66 percent turnout in 1964-1968, 65 percent 1972-1976, 64 percent 1980-1988, and 67 percent again in 1992. These authors, however, question the calculation of the turnout in a lengthy footnote. See Steven J. Rosenstone and John Hansen, *Mobilization, Participation, and Democracy in America,* (New York: Macmillan, 1993), 58, fn, 220-23.
27. Rosenstone and Hansen, *Mobilization, Participation, and Democracy in America,* 188.
28. Rosenstone and Hansen provide an excellent review of this period of civil rights action in 188-96. They cite the valuable study of Doug McAdam on the Civil Rights Movement, 1982, for evidence of the volume and types of actions.
29. Ibid., 193.
30. Ibid., 221-22.
31. Carmines and Stimson, *Issue Evolution.*
32. Rosenstone and Hansen, *Mobilization, Participation, and Democracy in America,* 44.
33. Robert Putnam, *The Comparative Study of Political Elites* (Englewood Cliffs, NJ: Prentice Hall, 1976), 171.
34. Donald Matthews and James Prothro, *Negroes and the New Southern Politics* (New York: Harcourt Brace, 1966), 177.

35. James H. Svara, *A Survey of America's City Councils* (Washington, DC: National League of Cities, 1991).

36. Katherine Tate, *From Protest to Politics* (Harvard: Russell Sage Foundation, 1994), 179.

37. Stokely Carmichael and Charles V. Hamilton, *Black Power* (New York: Vintage Books, 1967), 183.

38. Arthur H. Miller and Bruce E. Gronbeck, eds., *Presidential Campaigns and American Self Images* (Boulder, CO: Westview, 1994), 202-5.

39. For a description of the 1984 campaign, see Pomper, "The 1984 Election," 1985, 16-23, 45-48.

40. For a description of the 1988 Campaign, see Pomper, "The 1988 Election," 1989, 37-50.

41. See the analysis by Burman, *The Black Progress Question*, especially Chapter 7, "The Illusion of Progress? Race and Politics in Atlanta, Georgia." Burman is not convinced that blacks made as much progress as some claim, nor that they really shared power.

42. Rosenstone and Hansen, *Mobilization, Participation, and Democracy in America*, 188-210, 219-27.

43. Donald Kinder and Lynn M. Sanders, *Divided By Color* (Chicago: University of Chicago Press, 1996), 107.

44. Charles Tilly, *Durable Inequality* (Berkeley: University of California Press, 1998), 245-46.

CHAPTER 17

Change and Adaptation in the American Party System

Ours is a party system that has shown signs of considerable change. In recent years it continues to exhibit such signs. It appears to be a system constantly adapting to the new developments and challenges in our society. As we have demonstrated in this book, much has happened in the organization and behavior of our two-party system. In this final chapter we want to answer four questions: what are the major changes that have occurred in the past fifty years in the ways our parties work; are these changes transitory or apparently permanent; what seem to be the conditions and factors that caused these changes; have these changes altered the basic character of our party system and, if so, how? We have argued that our party system is in a state of dynamic equilibrium, that significant changes have occurred but that the two-party system is still viable, still basically the same type of two-party system. Is this true? We need to review the evidence carefully. Is it the same type of two-party system of the past, or one with a different character?

We need to have a theory of party system adaptation by which to interpret the changes in the parties. Parties are complex collectivities of leaders, activists, and followers, with a diversity of interests and orientations, that seek to gain power, or hold on to power, by being alert to the changing problems and needs in their societies and then responding to these changes. They are institutions which must change, and do change; they are not stagnant, not irrevocably fixed structures. Parties that are successful learn to adapt effectively. There may be times when adaptation is minimal, perhaps because of a presumed lack of need to adapt, internal resistance by entrenched supporters, or rigid commitments to key clientele groups. Yet the basic proposition holds: adaptation is necessary, and if it is strategically well conceived it will lead to electoral success; if it is strategically misconceived it can lead to party decline, even failure. There are a variety of ways in which a party may have to adapt, i.e., in terms of which adaptation probably has to take place. We have discussed these at length in this book. They include the following:

1. The type of leadership selected by a party as its candidates for public office: personal backgrounds, policy positions, experience, record of performance, et cetera. Particularly in the United States, with our presumably "candidate-centered politics," this is so, but it is also true elsewhere. Those who are presented for public office should fulfill certain credential needs

in the society, reflecting public expectations, tolerances, and preferences. One can certainly find examples of failure of the parties to meet this adaptation requirement. Our parties have sometimes selected ineffective leadership.

2. The organization form and process which the party adopts, whether very democratic and decentralized internally, or hierarchically controlled from the top down, or an in-between "stratarchical" structure with considerable antonomy in decision making at all levels of the party. It all depends on what a political party culture is used to and will tolerate, and which form will best maximize the most efficient decision making, while also contributing to the most successful mobilization of mass support in elections.
3. The social base of support for the party, or the character of its social group coalition. The party in the United States is always confronted with the question of which sectors of the population it *can* get support from, and which it *must* get support from, if it is to be successful. How narrow or broad should the coalition be, and should it be modified from time to time as new groups enter the electorate, or as the opposition party seeks to invade the party's social territory? The party must solve the dilemma of looking for support from new groups while not relinquishing the support of its core clienteles.
4. The ideological line and direction of the party in the context of public opinion shifts. As Downs said long ago, while a party "fashions an ideology which it believes will attract the greatest number of votes," there must also be "continual readjustment of ideologies within each party."[1] While each party has primary ideological positions, supported by its voters, it has to be able to modify its ideologies if conditions internal or external to the party threaten electoral success.
5. The strategies and tactics used in campaigns that will activate those loyalists, floating voters, and apathetics to ensure victory. Parties have to develop messages and effective techniques to communicate these messages. In this they have to be alert to their opposition's campaign approaches, new technologies, and the types of campaign appeals most likely to penetrate to their potential supporters. Should the campaign emphasize one or two key issues (and which ones), and how should it "sell" the candidates' personalities? Should it be a "stand-pat" or "laid-back" campaign, or one in which the party goes on the attack? A party's strategic decisions in any campaign, and over time, need to represent an adaptation to the realities of change at both public and elite levels.
6. The party activists at the middle and lower levels of the organization, the "working elites" at the base of the system, upon which the party relies heavily. If the party organization is to be healthy and viable, it must be concerned with the level of satisfaction, commitment, and efficiency of these lower-level activists — in a sense, their "morale" and their performance. Since these activists come and go, the organization is in a constant state of flux. The organization must be involved constantly in recruiting

new workers and in responding to new demands for involvement, for upward mobility of activists, for giving activists a chance to have decisional roles and a feeling of genuine satisfaction and accomplishment while performing important organizational tasks.

These six types of party adaptation illustrate how we must think about parties as they respond to their changing environment. There are others that could be identified. The point is that parties constantly face challenges and are confronted with dilemmas, and there are alternative scenarios by which to deal with these challenges. One must recognize, of course, that there is always the possibility that a party may not be able to adapt successfully. One study dealing with such cases, *When Parties Fail,* explains the causes of party demise in democratic systems.[2] It has happened often in Europe and for third parties in the United States. Some parties are absorbed by others. Or the programs of parties become irrelevant, or the organization deteriorates, or the supporters disappear gradually over time or shift their allegiances. In this book we argue that our two major parties have been adaptive enough to survive. The American society and economy have certainly changed, as well as the political culture and governmental institutions within which our parties work. The parties, however, have been able to react positively and creatively enough to maintain themselves, but not by "standing pat." At least we can say they have responded creatively enough to survive, although scholars may argue heatedly as to whether, despite their survival, our parties have responded enough to meet the needs of the American people (as the previous chapter illustrates).

The conceptual model that underlies our thinking about party adaptation, and that informs our discussion about it, is fairly simple (Figure 17.1).

FIGURE 17.1 ■ Conceptual Model of Party Adaptation

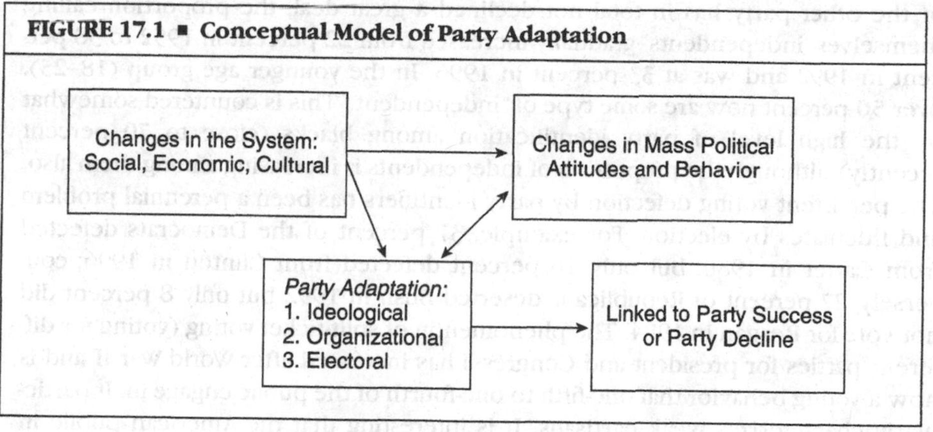

WHAT MAJOR CHANGES HAVE OCCURRED IN THE AMERICAN PARTY SYSTEM?

In focusing on the major changes in party politics since World War II, we should begin by stating what has *not* changed very much, if at all. We still have two-party dominance of public offices from the presidency to the counties. Rarely are independents or third-party candidates successful (there is one independent congressman, from Vermont, and he usually votes with the Democrats). The capacity of the two parties to win despite third-party efforts is impressive. The major third-party challenges since World War I were in 1924, 1948, 1968, 1980, and 1992–96. Our two parties consist of broad coalitional teams we have had for some time, even though the groups within these large coalitions have changed. The parties are evenly matched in their control of the presidency: from the year 1960 to 2000 we will have had twenty years under the presidency of each party. The Congress is another matter, but even there the Republicans have demonstrated their capacity to take over, in the 1994–96 elections. The loyalty of the public to the two parties is still high, over 60 percent identifying with the Democrats or Republicans since 1952. In 1996 27 percent identified as Republicans, the same figure as in 1952; the Democrats declined — 47 percent in 1952 and 39 percent in 1996. Hence the two-party system seems fairly deeply anchored in public support despite assertions to the contrary.

Changes in Mass Support

Major changes have occurred, however, as we have demonstrated empirically in this book. First, as we have seen, there have been changes in the *patterns* of mass support for the parties. While the percentage of adults declaring allegiance for one or the other party has in total not declined a great deal, the proportion calling themselves "independents" gradually increased from 22 percent in 1952 to 38 percent in 1992 and was at 33 percent in 1996. In the younger age group (18–25), over 50 percent now are some type of "independent." This is countered somewhat by the high level of party identification among blacks (close to 70 percent recently), although the proportion of independents is increasing among them also. The persistent voting defection by party identifiers has been a perennial problem and fluctuates by election. For example, 37 percent of the Democrats defected from Carter in 1980, but only 16 percent defected from Clinton in 1996; conversely, 27 percent of Republicans deserted Bush in 1992 but only 8 percent did not vote for Reagan in 1984. The phenomenon of split-ticket voting (voting for different parties for president and Congress) has increased after World War II and is now a voting behavior that one-fifth to one-fourth of the public engage in. It particularly characterizes weak partisans. It is interesting that the American public in interviews conveys the impression that it "typically" splits its ticket (in 1996 71 percent responded this way in a Roper survey), when in any actual election the proportion is much smaller.[3]

The level of attitudinal support for the parties also suggests less strong "affect" for the parties — more neutrality in expressing evaluations of the parties and less-positive evaluations, as Wattenberg's analysis has revealed. He also documents the striking decline in presidential candidate popularity from 1952 to 1988, among party supporters as well as among opponents.[4]

Finally, of course, one must remember the decline in voter turnout in elections in the United States. In presidential elections we have declined from a 64 percent turnout of those theoretically eligible to vote in 1960 to 48.8 percent in 1996. Yet for certain groups in the population, turnout increased — blacks and women, for example. Other types of political activity have not decreased significantly, if at all.[5] But the public is more apathetic and less supportive of parties now on election day.

Reflecting on the meaning of these changes in public support for the parties, one must conclude that party allegiance is probably more open and less rigid today. The new generation, the post–New Deal generation, as Miller and Shanks insist, is more educated and less tied to traditional, familial, party loyalties. A hard core of strong identifiers still exists, but even these are not immune to the appeals of the other party. We have had in the electorate the "Reagan Democrats," people who said they were Democrats but voted for Reagan (over 20 percent in the eighties). Some of them stayed on to vote for Bush or Perot. We also have the "Clinton Republicans" (perhaps 15 percent today). The parties continue to be vulnerable to successful appeals by the competition. Above all, we must note that Congress and the president are elected now by a minority of eligible voters, and that parties today have a narrow popular base numerically. Some groups of citizens, then, may be left out of the representation process.

Coalitional Changes

A second type of change we have experienced is in the group character of the party coalitions. In the past forty years we have undergone a major group realignment in our party system. The Republicans and Democrats as interest-group coalitions quite different today from the coalitions in the early postwar period of the fifties. The evidence that this is so has emerged from the research of many scholars. One must remember that the Democratic New Deal coalition, forged in 1932 and continued long after the war, consisted of a South that had been overwhelmingly Democratic since the Civil War plus certain groups in the big cities of the North, particularly ethnic groups, union members, and, eventually blacks. The Republicans were left with a coalition of white-collar and non-union whites, found mainly in the rural and suburban North but also including the Midwest and much of the West. These groups did not always support the two parties at the same level of support, since our politics has always seen some fluctuation from election to election in group loyalties to the parties and no perfect mobilization of any group by either the Republicans or Democrats. Thus, for example, Catholics, who normally have been on balance pro-Democratic in their voting, have shown periods of major shifts in party support: after strong support for Eisenhower, a 36 percent shift to the Democrats 1956-60, and then again a 39 percent shift away from the Democrats in 1964–1972. Similarly, union members, who are usually pro-Democratic but often do not openly endorse Democratic candidates, revealed wide swings in their support for the parties: 30 percent pro-Democratic in 1956-64, 40 percent pro-Republican 1964-72, and 21 percent pro-Democratic 1972-76. Such shifts can and do occur because of the vagaries of particular election campaigns and presidential personalities. The key question is whether a *sustained* shift occurred that is durable enough to reveal that the group-coalitional basis of the two major parties is changing fundamentally.

Such a sustained shift did occur between 1960 and 1968 and involved two key groups, southern whites and all blacks. The Democratic party under presidents Roosevelt, Truman, and Kennedy had gradually attracted support from the blacks, even though President Eisenhower and the Republican party had also supported civil rights legislation. When President Johnson successfully pushed through the Civil Rights Act of 1964, over the strong objection of the southern, mostly Democratic, Congress members, the challenge over racial policy came to a head. The election of 1964 became a crisis point, a realigning point in American politics. In the 1964 election campaign the Republican candidate, Barry Goldwater, denounced the Civil Rights Act (which he had voted against) and appealed to voters in the South to support his candidacy. He lost the election but won in five major southern states. Thus was precipitated a major realignment in the group support for the parties. As a result, the proportion of Democratic support among whites in the South declined from 78 percent in 1952 to 57 percent in 1960 to 47 percent in 1972, to finally 34 percent for Clinton in 1992 and 36 percent in 1996. In the meantime the blacks, South and North, who had remained divided even in the Eisenhower elections, although gradually revealing a pro-Democratic shift, eventually shifted almost completely to support of the Democratic party: 96 percent in 1968, 93 percent in 1980, and over 80 percent since that time. In 1992, 7 percent of the blacks supported Perot (but only 10 percent supported Bush); in 1996, 4 percent supported Perot (and 12 percent supported Dole). Thus occurred what Miller and Shanks refer to as a "virtual revolution" in the group realignment of our parties, a transformation that really affected primarily only one region of our country, the South. But since that region was so central to the success of the Democratic party, it was a change of "massive proportions" and significant implications.[6] Miller and Shanks argue that there was essentially no change in the North from 1952 to 1980 in the group support for the parties, except for the continued shift of the blacks to the Democrats. They do see a "limited national realignment" under Reagan 1980-88.[7] But the major upheaval occurred in the sixties. It produced a sustained change in the coalitional character of the parties.

The factors responsible for this striking shift and reversal of special-interest-group loyalty to the parties are many. The increasing pressure from the twenties and thirties from black and white liberals to do something about racial political inequality culminated in the massive civil rights movement of the fifties and sixties, beginning in the South and spreading to the North. Both parties sought initially to respond to it, but conservatives in Congress, southern Democrats and northern Republicans, obstinately refused to yield. Eventually the Democratic party assumed the leadership to pass the 1964 Civil Rights Act and in 1965 the Voting Rights Act. The Republican leadership refused to endorse such efforts to give blacks the vote and to eliminate efforts to disfranchise blacks in the South. The 1960 Republican platform was pro-reform; the 1964 Republican platform took a diametrically opposite position. Thus public demand for change, plus massive liberal mobilization efforts, demonstrated the need for action and this led to the decision by the Northern Democratic leadership in Congress and President Johnson to reverse the historic neutrality and opposition position of Democrats. They adapted to this pressure for change. The Republicans responded by reversing themselves also and, calculating they could now win the South, essentially deserted the cause of blacks, arguing it was not a problem that could be solved nationally. Thus, elite action at the national level in the face of mass pressure led to a Democratic ideological, organizational, and electoral adaptation that triggered a counter-response by the Republican elite in Congress that was eventually accepted by Republican activists and identifiers. Hence, the 1964 realignment in black and southern white patterns of party support. It had tremendous impact on the electoral strategy of both parties, continues to the present. Jimmy Carter and Bill Clinton, the Democratic southern governors, were to get only 35 percent of the southern white vote in 1980, 1992, and 1996.

Ideological Shifts?

In the early postwar period, the structure of public attitudes was apparently well balanced, as 1956 survey data revealed: 31 percent "liberal" or "leftist," 41 percent "centrists," and 28 percent "conservative" or "rightist." The bulk of the public was in the middle. In the late fifties and sixties, public opinion became more preoccupied with liberal issues (e.g., in 1963 52 percent of the public said the race issue was the most important problem we faced), as well as more "ideological" (increasing from 18 percent in 1956 to over 40 percent in the 1964–1972 period). After the Nixon victory over McGovern in 1972, there appeared a change in the structure of public attitudes. Already by 1973 we find the following: 33 percent "leftist" or "liberal," only 27 percent moderate, and 40 percent "conservative" or "rightist."[8] Later shifts based on response to policy questions, as well as self declared statements of ideological position (liberal, moderate, conservative), have tended to suggest a continuation of this drift toward conservative. For example, by 1988 45 percent of the public reported being "conservatives" and only 23 percent "liberals." This set of findings suggested that ideological groups in the public were wider apart ideologically, more polarized.[9] Using a somewhat different approach for the analysis of 1992, Miller and Shanks present data for a "summary index of policy-related opinions," which reported that 45 percent of all voters could be considered weak or strong conservatives and 34 percent weak or strong liberals.[10]

There continues to be considerable debate over how much change in ideological direction has actually occurred. Wattenberg, for example, saw no real shift toward conservatism during the Reagan period.[11] Others who have studied issues and ideology in relation to voting are not necessarily convinced of the importance of this shift; yet some scholars see ideological shift as significant. There is no question that these shifts have been larger at certain times than at others and that the public may take more liberal positions on some issues and more conservative positions on other issues.

The interesting question here is, how have the parties responded to these public opinion or ideological shifts? If we utilize the theory of Anthony Downs, we would expect the parties to exhibit shifts in ideology also, by identifiers and activists, because theoretically parties need to adapt to public opinion change and follow public opinion in such a way as to maximize their voting power. If citizens tend to shift to the right, the parties must calculate carefully their response. While the inclination may be to shift decisively and noticeably to the right also, the parties must realize that many voters still are left of center. And the "party of the left" has a "hard core" of supporters on the left whom it cannot desert. Likewise, if the "party of the right" deserts its moderates in the middle, it may jeopardize its electoral success. Thus the options are not clear-cut, and success is certainly not easily calculable.

What have American parties done when faced with changes in opinion direction? In the postwar period, the parties responded in different ways. For the identifiers, here are the data for the period 1956–1972:

	Democratic (in percent)		Republican (in percent)	
Ideology	*1956*	*1972*	*1956*	*1972*
"Left"	37	39	25	23
Center	41	26	41	30
"Right"	24	35	35	47

The Republicans clearly moved to the "right," while the Democrats increased slightly their proportion of identifiers who were "left" and to a greater extent those who were conservative. Furthermore, the activists of the two parties responded even more differently during this period of a shift to the "right." The Republican activists remained conservative while the Democratic activists moved more to the left (20 percent were on the extreme left in 1956 compared to 43 percent in 1972)[12] Of course, this was in the period of liberal reform in the Democratic party, so this increase in left activism is understandable. Nevertheless, these party shifts may have contributed in the end to the Democratic defeats in 1968 and 1972.

After 1972, the party identifiers in each party exhibited continuing patterns of change, if we look at their ideological self-placement (whether they called themselves "liberals," "moderates," or "conservatives") (Table 17.1). The Democrats by 1992 were 44 percent "liberal" (compared to 10 percent of the Republicans), while the Republicans became 64 percent "conservative" (compared to 24 percent of the democrats). So the parties over the twenty-year period from 1972 to 1992 were moving in opposite ideological directions and were more polarized than ever. In 1972 the conservative difference was 24 percent; by 1992 it was 40 percent. As public opinion was gradually drifting to conservatism, the Democratic supporters were continuing to believe in liberal and moderate positions; the Republican supporters, on the other hand, were following more the conservative line. At the same time, their presidential candidates, Dukakis and Clinton, Bush and Dole were by no means taking extreme "left" or "right" positions. In fact, their supporters on the

TABLE 17.1 ▪ Ideological Self-Placement of Partisans, 1972–1992 (in percent)

	Democrats				Republicans			
Self-Classification	*1972*	*1976*	*1980*	*1992*	*1972*	*1976*	*1980*	*1992*
Liberal	33	37	38	44	13	10	10	10
Moderate	41	39	34	33	37	30	23	27
Conservative	26	24	28	24	50	60	67	64

SOURCE: National Election Studies.

extreme liberal and extreme "right" wings of the party were complaining that their party leaders were not listening to them.

In summarizing ideological trends for the parties over the past forty years, we can focus on three major types of change. The first focus is on the differential patterns of ideological response of the parties to the drift toward conservatism. Table 17.2 captures the essence of these variations: the Republicans consistently and very decidedly more conservative and less liberal, the Democrats shifting to more liberalism as well as maintaining a 25 percent cadre of conservative believers. The second focus is on the resultant increasing ideological polarization of the loyal party supporters. This is presented simply in Table 17.3. The basic findings are very striking: in terms of liberal trends, the party identifiers have widened the gap from 12 percent in the 1950s to 36 percent in the 1990s; in terms of conservative trends, the gap is even wider, 41 percent recently, compared to 11 percent in the 1950s. There is indeed much less ideological consensus *across parties* today than earlier.

The third focus is on the *intraparty* patterns that have emerged. Using available data from 1984 and 1992, we see in Table 17.4 the internal party homogeneity

TABLE 17.2 ▪ Summary of Long-Term Trends in Ideology by Party in the Postwar Period (in percent)

Political Ideology	*1950s*	*1970s*	*1990s*	*Percent Change*
A. *Liberals*				
Public	31	33	34	+3
Democratic Identifiers	37	33	44	+7
Republican Identifiers	25	13	10	−15
B. *Conservative*				
Public	28	40	45	+17
Democratic Identifiers	24	26	24	0
Republican Identifiers	35	50	64	+29

SOURCES: For the 1950s and 1970s public data, see Norman Nie, Sidney Verba, and John Petrocik, *The Changing American Voter* (Cambridge: Harvard University Press, 1976), 143, Based on an attitude scale in five issue areas (see pages 23–24); the later public dates are based on the working of Warren Miller and J. Merrill Shanks, *The New American Voter* (Cambridge: Harvard University Press, 1996), 353; The date for the 1950s and 1970s for the identifiers is found in Nie, Verba, and Petrocik, 1976, 199. The 1990s data are from the NES for 1992, based on ideological self-placement responses. The measurements for "ideology" vary somewhat for certain years, by study, but the overall trend is basically the same.

TABLE 17.3 ■ Increasing Polarization in Ideology for Party Followers, 1950s to 1990s

	Percentage Liberals			Percentage Conservatives		
Party Loyalists	*1952*	*1972*	*1992*	*1952*	*1972*	*1992*
Democrats	37	33	34	24	26	24
Republicans	25	13	8	35	50	65
Difference	12	20	36	11	24	41

SOURCES: See Table 17.2.

TABLE 17.4 ■ Leaders and Followers in the Parties: How Much Intraparty Ideological Congruence? (1984 and 1992)

			"Distance" between Activists and Identifiers	
	1984	*1992*	*1984*	*1992*
Republican (percent conservative)				
Activists	68	72	3	7
Identifiers	65	65	—	—
Public	41	44	—	—
Democrats (percent liberal)				
Activists	43	56	4	12
Identifiers	39	44	—	—
Public	26	34	—	—

SOURCE: The data for identifiers and public are from National Election Studies. The Activists data come from a study by William Crotty, John S. Jackson III, and Melissa Miller, "Political Activists Over Time: 'Working Elites' in the Party System," paper prepared for Ann Arbor Conference, Summer 1997 (to be published by University of Michigan Press).

on ideology, even as the activists in both parties move toward the extremes — the activist Republicans continually more conservative and the activist Democrats continually more liberal. Both parties have moved further from the public in 1992 compared to 1984; that distance is considerable. But what counts also of course is the relation between activists and loyal followers — here we find high congruence.

One must visualize our parties ideologically, then, as consisting of fairly congruent ideological structures, very distinctive in basic beliefs, in a society that has been perceptibly moving in a more conservative direction.

ORGANIZATIONAL DEVELOPMENTS

We have spent much of this book describing the changes in party organization in the United States during the past four decades. Clearly much has been happening, not necessarily in the transference of control over the party from one level to another level (i.e., from the local to the national, or vice versa), as is the focus of the discussion of organizational change in Europe recently. Ours has been a system in which the parties at all levels share power; it has never been a system with strong, dominant, national control. And that remains true today. Rather, what has been occurring is a change in how the party organizations operate, what activities they have taken over, and how they function somewhat differently today from previously. The basic observation is that virtually all branches and centers of the Republican and Democratic parties have been strengthened in recent years. There are many specific ways in which this strengthening is manifest.

In quick, summary form we can review the types of organizational changes that are most important, and that we have described at length in earlier chapters. At the top, the national committees and their staffs have taken on more responsibilities. The staff sizes are ten times larger than in the seventies. They have yearly budgets much larger than before to support these large staffs. They recently have each raised over $100 million, which they spent in the national presidential campaigns and which they also share with the state and local party units. They engage in a variety of campaign activities, including recruitment of congressional and local candidates, planning and managing campaigns, preparing media materials, working with political action committees, and keeping in close touch with and providing assistance to local party campaign organizations. They are more closely associated with state organizations and congressional campaigns throughout the country than previously. The national committees also continue to do all the planning for the national conventions as previously. And the Democratic National Committee in particular has control over (i.e., specifies the rules in detail for) the delegate selection process, by caucus or by presidential primary, as in the past.

The state organizations also are better funded and staffed today. In recent years they have raised millions of dollars on their own to run their campaigns. Studies of the county organizations reveal that they too are relatively more active than previously. And at the base of the system — in the wards and precincts and clubs — numerous studies reveal that activity today is as high as or higher than in the past, with many thousands of campaign activists distributing literature, preparing media publicity, canvassing for votes, calling people at their homes, and holding rallies and parties in order to mobilize the vote.

The evidence points to an increase, not a decline, in party organizations' activity at all levels. This activity is better financed than ever and involves a more diverse set of activists than before (women and minorities particularly), people who are deeply committed and motivated, many of whom are experienced and well trained in campaign work and whose efforts, studies show, have a significant impact on the vote. The activists seem more ideological and issue oriented than previously, while also admitting that they enjoy the social rewards of party activity. We find, thus, continuous social renewal of the organizations, increasing interest in policy goals, and high morale. Thus, from top to bottom American party organizations exhibit strong signs of expanded involvement in party work, greater integration of national, state, and local party units, and effective vote mobilization efforts. American party organizations today are better led and are engaged in more party-type functions than they were thirty years ago.

What factors have been responsible for these party organizational changes? In part, of course, as with other organizations it is a natural consequence of trial-and-error experience and maturity. To a great extent it is due to better leadership, from Washington down to the states, counties, and cities. We have had able and dynamic national party chairs who had a vision of reforming and activating the organization, as well as presidents who sometimes take an interest in what the party organization does. And the parties seem to respond and adapt to each other. When the Republicans began to reform their national organization in the post-Watergate period of the late seventies, the Democrats began to realize they were falling behind, and so they rushed to mobilize more money and institute reforms. Similarly, after the Democrats reformed their party organization in Congress, particularly in the House of Representatives, the Republicans finally also strengthened their party caucus and the role of the House leadership in the policy process. So some of the adaptation that occurred is due to reciprocal interparty actions. No doubt ideological change, in the country and in the party, may play a role. When a new set of liberal Democratic congresspersons, the "New Breed Democrats," were elected between 1974 and 1978, the result was strong pressure for organizational reform in the House. And when in 1994 a new set of seventy-one Republican first-termers came in under Gingrich's leadership, they completely reorganized the Republican party in the House. Further, the much more ideological character of our politics was manifest in Congress and contributed to more cohesive party organizations in both House and Senate. The financing of party campaigns recently also has been responsible for organizational change, as has the technological advances available to the organizations and the technological changes in communications. The party organizations have more money and have had to develop organizational approaches to spending this money most effectively. The bottom line in all this, of course, is the competitive drive for votes. The campaigns today use more high-priced media and vote mobilization approaches than ever. The parties, in playing a more central role in the management of these events at all levels of the system, have to innovate organizationally. Studies show that this appears to be exactly what they have done — innovated adaptively. Thus, various external conditions (such as ideological and social developments in society), as well as internal factors, competitive pressures, new leadership, and ample resources, have been responsible for the new developments and directions in our party structures.

AS PARTIES CHANGE, WHAT TYPE OF PARTY SYSTEM HAS EMERGED?

If we reflect on the meaning of these changes and adaptations by the parties in the last four decades, the first point to be made is that our parties in a sense have defied the evaluations that they are moribund, deteriorating, and decomposing. If that were ever so, the parties attest today to the opposite: they are alive, more vital than ever, and innovative. Any decline in loyalty has not destabilized the party system. The realignment that has occurred, such as the 1964 transformation, has not really disrupted the two-party system. That does not mean that there are no signs of concern about mass involvement or commitment, or even elite involvement or commitment. But the parties have proven their capacity to reform despite problems at the public and elite levels. The central observation is that they adapted to many of the pressures and challenges of the society.

If we were to evaluate the consequences of these many changes for the character of the party system, there are a few observations that may be significant. Our system now seems to be more competitive than before, with closely fought primaries and elections, and small majorities and much party voting in Congress. It seems also to be more ideological, in the sense both that issues are a more motivating force in party activity, and that the political beliefs of party identifiers, leaders, and members of Congress seem to be more ideology based and also more polarized than previously. The organizations are apparently more combative. Second, we have more of an organization-based and organization-centered politics in the legislature and outside the legislature. The system is more integrated organizationally in terms of the relationships between the national, state, and local levels of the party. However, there is still a great deal of "stratarchy," and there is no centralized control. Yet the different levels of our massive party structures share ideas, personnel, technologies, money, and strategies. While there is still much local autonomy, there is also much national help to party organizations in the states and localities. This may lead to a more coherent and cooperative cross-level politics.

Third, there is still much volatility and uncertainty in the system. Voters if anything are more "independent," more free to split their tickets, to defect and transfer their party affections, temporarily or for a longer term. In a sense it is a more sophisticated electorate, more individualities. Yet while the voting public will withhold loyalty and will give considerable support to a Perot, it will also eventually come back to the two-party system. The parties thus are continually being tested and challenged to perform; they have certain "hard-core" sets of supporters, but these do not assure victory by themselves, and some may defect. There is a constant body of "floating" voters. The parties in our two-party systems are continually under scrutiny. Voting behavior is very much based on perceptions of party performance, which influences the behavior of the activists, the organization leaders, and the party's legislative elites. Before we become overly enthusiastic about party changes and adaptations, we may well ask whether to these characteristics of the "modernized" American party system we can add the label "more responsive," and "more responsible." The twin concerns many scholars have are, first, that the system cannot mobilize even 50 percent of the eligible American adults to register and vote. This means that in a sense it is a minority-based two-party system that it is "elitist"— the vast body of apathetics, who are a majority, have withdrawn from participating in this system. The second concern is, then, how representative of the American public is the two-party system? Above all, to what extent does it deal effectively with the needs of the political "underclass," and, perhaps more important, to what extent does it deal effectively with the social and economic welfare of those at the lower reaches of American society? This is the question we dealt with at length in the discussion of the race problem in Chapter 16. It is a question to which we must continually revert. Our party system is reformed and revitalized, apparently healthier and more dynamic than it has been for a long time. But is it more humane?

These observations about the changes in our party system view the parties as part of the broader political process and the changes as part of a continual adaptation of parties to the changes in our society and in public expectations about how the parties should function. Thus, we have not specifically discussed at length the more narrow concepts of partisan "realignment" or "dealignment." For a long time during the postwar period, scholars were preoccupied with analyses and speculation about whether, and in what respects, party system "realignment" was taking place. Many scholars, as we, have moved away from this approach, primarily because it does not deal with the broader phenomenon of system change.[13] Occasionally we discuss "realignment" in connection with certain particular developments that occurred at certain key points in our party history. The election of 1964, for example, we see as a critical turning point in our party history that led to a realignment in the partisan support of white southerners (from Democratic to Republican) and of blacks (from Republican to Democratic). Primarily, however, we have focused more broadly on party system adaptation. But it is well to keep in mind what the long history of "party realignment" in the nineteenth and twentieth centuries was all about. Knowledge of this history will contribute to our understanding of the American party system's historical development. This scholarship on realignment also alerted us to the key aspects of change in the system, which we need to focus on. Therefore, the following discussion of realignment is included because of its value in identifying certain aspects of partisan change that are part of the overall picture of party adaptation as we have discussed it thus far.

THE THEORY OF REALIGNMENT

Realignment means different things to different people, although all who discuss the phenomenon see it as a change in the basic pattern of public support underlying the party system. For some, realignment is a *sudden* happening that occurs at the time of a "critical election" (such as 1860, 1896, or 1932) in which a minority party becomes the majority party and remains dominant for a considerable period of time. For others, realignment is a *gradual* change in party support called secular realignment. Again, for some writers realignment means electoral replacement (older voters of party A replaced by new voters supporting party B). Realignment also means to some primarily a change in the party loyalties and voting behavior of the public, while for still others realignment means much more: a change in the organic structure of the party process and in the citizen's relationship to that process, with significant consequences for the political system.

In the views of the most prominent writers on the subject, the nature of realignment can be described as a series of changes taking place in a society.[14] These views can be summarized as follows.

During any era of party control (such as the period after World War I when the Republicans were the dominant party and the Democrats were in the minority), social and economic changes occur that may be the result of specific crises (such as war, economic recession, or racial conflict) or gradual changes in the nature of the population (economic well-being, technological improvements, changes in population movements, and changes in the public's attitudes toward issues). New groups may appear along with these social and economic changes, making new demands on society.

New political issues appear as a result of these crises or "social maladjustments," or because of the gradual changes occurring in society. These issues will replace the older controversies and will relate to the role of government in the society or economy (such as whether the government should provide more social welfare, do more to end discrimination against blacks in education, withdraw from Vietnam, or pardon all deserters). Certain of these issues may trigger confrontation between or within the parties and result in ideological polarization. Issues become more important to politics than ever before at these times, both for political leaders and for citizens. Further, interest in politics presumably is intensified during these periods of issue conflict, often resulting in the involvement of citizens previously disinterested.

The response of the political parties to these new issues, or to a single triggering issue, determines the extent and nature of party realignment. Since these new issues usually cut across both Democratic and Republican parties, the leaders and activists (and finally the voter support groups) have to take positions on the issues and somehow resolve the conflict The response of the parties may take the following forms:

1. There is no significant response, or at least a very slow reaction to the crisis, and the opponents of change, or those who want to straddle the new issues, win out in both parties. (Some writers would argue that this was the scenario of the Democrats and Whigs before the Civil War in their failure to deal with the slavery issue.) A major realignment occurs through the replacement of both old parties, if the crisis continues.
2. One major party, either the Republican or Democratic, responds affirmatively, with the "new issue adherents" who seek more progressive policies winning out in the control of the party. The other major party does not respond affirmatively, continuing to straddle the issues. (Perhaps the party crisis of 1928–32 during the depression fits this scenario.) A major realignment occurs and one party attracts the progressives, the other the conservatives.
3. Both major parties respond to the new issue politics by developing "progressive" new-issue positions — that is, the internal struggle within the parties is won in both cases by the "progressive" new-issue adherents. (After Hoover's defeat, the Republican party was changed considerably, so that by the 1950s both parties were more "liberal" and "internationalist.") The old party system survives, although changes occur in both parties.

4. A new third party appears on the scene, organized on the basis of the new issue controversy. But this new party is eventually absorbed, because one, or both, of the older major parties responds affirmatively to the crisis, thus making it difficult for the newly formed third party to maintain itself. (The Populist party threat in 1892 is a prime example of this.)
5. The new third party may withstand efforts for its absorption into one of the major parties, and thus a new type of party system is born, which includes a viable third party as a meaningful, continuous threat to the older major parties (as in Canada) or even a replacement of one of the major parties (as the Liberal Party in Britain was replaced as a major party by the Labour Party after World War I).

Depending on which scenario takes place at these times of national crisis, the party system may be modified considerably, transformed into something different, or maintained as it was. But in each case, even if it is maintained in basic form as it previously was, political realignment of some type occurs in society. The following types of changes, according to realignment theory, are likely to occur.

1. There will be a shift, sudden or gradual, by groups or "blocs" of voters away from support for the party to which they have been loyal — not by all the individuals in a particular group or "bloc," but by a significant proportion, large enough to indicate that the coalitional character of the parties is changing.
2. There will be a loosening of individual citizen ties of loyalty for and identification with parties, depending of course on the citizen's perception of how satisfactory his or her party's response is to the new-issue politics.
3. Issues will play an important role in the citizen's support for the parties during this realignment period. "Issue polarization" develops within and between the parties. The citizen's own position on the issues is sharpened and then is related to that of the parties and their candidates. The vote is then more closely linked to issue positions.
4. During realignment, citizen interest in politics and voting participation will increase. This is not necessarily so for all sections of the country or for all groups, however. Participation increases for those groups most concerned about the new issues.

These phenomena should be looked for in seeking to discover whether a "realignment" has occurred or is occurring. Theoretically, this is what happens in a society like ours when there is a basic change in the party system. As V. O. Key said, a "critical election," one in which a "sharp and durable realignment between parties" occurs, is one "in which more or less profound readjustments occur in the relations of power within the community."[15]

It is generally agreed that we did have major party realignments linked to the "critical elections" of 1828, 1860, 1896, and 1932. These were spaced at fairly regular thirty-two-to-thirty-six-year intervals, a fact that to some scholars is very important. The basic question after World War II was whether the United States entered a new realigning era, leading to a national transformation of the party system, and if so, when. Burnham wrote in 1970:

> The political parties are progressively losing their hold upon the electorate. A new breed of independent seems to be emerging. . . . This may point toward the progressive dissolution of the parties as action intermediaries in electoral choice and other politically relevant acts. It may also be indicative of the production of a mass base for independent political movements of ideological tone and considerable long-term staying power. . . . There is considerable evidence that this country is now in a realigning sequence and that we are enroute to a sixth party system.[16]

And right after the 1980 election David S. Broder suggested that we might well be heading into a realigning era. He said:

> Sifting through the returns of [the] election [of 1980] makes it clear that for the first time in a generation it is sensible to ask whether we might be entering a new political era — an era of Republican dominance. . . . To find an earlier example of the kind of basic realignment of party strength some observers think may have begun this month, you have to go all the way back to the Roosevelt-Democratic victories of 1932 and 1936 . . . something like the reverse of that occurred this month.[17]

Other scholars are less alarmist. After examining the evidence in their study, *The Changing American Voter,* Nie and his associates conclude that although there is some evidence of changes that "clearly point to a realignment . . . all in all, we see little prospect for the emergence of a new party system from the disarray of the present system."[18]

Similarly, James Sundquist, after a careful analysis of realignment, concluded in 1973 that a radical transformation of our system was not clearly in evidence: "In the long run, the prospect may well be for a further gradual decomposition of the two-party system. But there is at least as much reason to believe that in the shorter run the headlong march toward decomposition . . . will be checked and even reversed. The New Deal party system will be reinvigorated, and most of those who ceased identifying with one or the other major party . . . will reidentify."[19]

If we review the basic indicators that presumably in the past resulted in a major realignment, and look carefully at the evidence since World War II, it is difficult to come to the conclusion that we did have a *major national* realignment which resulted in a basic restructuring of our system sometime in the sixties or later. As we noted earlier, there was some decline in party loyalty and in strength of party identification. More people wanted to call themselves independents, particularly those in the younger age cohorts. But it certainly was not a "dissolution" of the parties in terms of public support for them. There were indeed voting behavior changes, but voting in America is always volatile. Issues certainly did become more important as a factor explaining the vote, but this was a phenomenon not clearly linked to realignment, except in 1964. There were shifts in group support for the parties, as we demonstrated earlier, an expected result of the loose character of our party coalitions. If, as V. O. Key said, realignment occurs when the public becomes more interested in politics, then for most of the postwar period we do not find such higher public interest. In fact, there was a continuous decline in voting interest and turnout after 1960. If anything, we had a "demobilization" rather than a realignment.

In reality, based on the evidence, it is difficult to see any national party realignment that restructured our party system. The one exception one might try to argue was in the 1964–68 period. As we reviewed this period in Chapter 16, we saw the intense conflict in racial politics between and within the parties during this period. A series of developments occurred that enhanced the probabilities of realignment: Goldwater's conservative campaign in the South, following the passage of the 1964 Civil Rights Act, in which he reversed the Republican party's pro–civil rights stance; the passage of the 1965 Voting Rights Act in the Johnson period, which paralyzed Congress further; the Vietnam protests leading to Johnson's decision not to run for reelection; George Wallace's candidacy in 1968 (getting 13 percent of the vote); and finally the election of Nixon, also on a conservative platform critical of civil rights legislation. This period of intense ideological controversy, reversal of party issue positions, a third-party threat, and the emergence of new party leadership certainly suggest that a major realignment might be occurring. In our previous analysis we call this a partial system alteration, a limited and partial realignment, not a national realignment leading to a basic restructuring of the party system. As Miller and Shanks have demonstrated, the patterns of party identification and voting in the North were largely not affected; it was the South where the major change in party alignments occurred. Nationally there was no catastrophic erosion or reversal, or exchange of party loyalties. The party loyalty pattern nationally was:

Party	*1952 (in percent)*	*1968 (in percent)*
Democratic identifiers	47	45
Republican identifiers	27	24
Total percent *strong* identifiers	35	30

But as we demonstrated earlier, the party loyalties of southern whites and of blacks, North and South, *were* essentially reversed. So a *regional* party loyalty "revolution" occurred, but there was no fundamental *national* realignment.

In the last three presidential elections there were certain developments that might suggest considerable partisan change, including the big success of a third party in 1992 (19 percent of the vote), the defeat of the Republicans in 1992, and then the stunning Republican takeover of Congress in 1994. Yet if one examines the patterns in party identification, the change overall was negligible:

Party	*1988 (in percent)*	*1992 (in percent)*	*1996 (in percent)*
Democratic identifiers	35	35	39
Republican identifiers	28	25	27
Total percent strong identifiers	31	29	31

Similarly, if we examine the voting behavior of certain key groups of voters, there are few large shifts:

Percent of Group Voting Democratic	*1988*	*1992*	*1996*
1. Race: Blacks	86	83	84
2. Sex: Women	49	45	54
3. Age:			
18–29	47	43	53
60+	49	50	48
4. Religion:			
Catholic	47	44	53
White Protestant	33	33	36
Jewish	64	80	78
5. Family income:			
Low (under $15,000)	62	58	59
High (over $50,000)	37	39	44
6. Community type:			
Big cities	62	58	68
Suburbs	42	41	47
Rural areas	44	39	44
7. Whites in the South	32	34	36

There are a few fairly large changes — for example, with Jews, women, youth. But certainly these are within the vote-swing patterns of the past and do not suggest major reversals. In fact, such data suggest a high level of behavioral regularity and stability. Thus, although shifting currents can be found internal to the party from one election to the next, recently we see no persistent, decisive modification of the party coalitions since 1964–68. In fact, it is interesting that the "Reagan Democrats," who defected in 1980 from the Democratic party in fairly sizable numbers (at least 25 percent of Democratic identifiers), were beginning to return to the Democratic party by 1988; by 1992, only a few remained to support Bush.[20] In sum we have indeed had important, if not "critical," elections in the past thirty years, and the parties' fortunes have ebbed and flowed during this period. But we have not seen a major partisan realignment of a national character. We have, however, seen many changes in how parties are led, how they have been organized, and how they function, which attest to significant evolution and adaptation of our party system.

CRITICISMS OF OUR PARTY SYSTEM: IS MORE REFORM NECESSARY?

In evaluating our party system, finally, it is well to keep in mind the criticisms of it as well as the proposed reforms, in the past and recently. Many scholars and political leaders have attacked our system from time to time. Indeed, one must remember that at the very beginning, in the 1790s, great statesmen like Washington, Madison, and Jefferson considered the parties to be "divisive and dangerous."

Ranney, in his history of party reform, reminds us that Madison, who he says was "the cofounder of the world's first modern political party," initially was very anti-party when he wrote the *Federalist Papers,* which argued that our new Constitution would protect us from the evil effects of parties.[21] So our parties came into existence at a time of great concern about the "mischiefs of faction," and throughout the nineteenth century, at both the elite and mass levels of our society, there continued to be repeated pressures for reform. As Ranney suggests, we have from the beginning been very "ambivalent" about parties.

The criticisms of our parties and the party system have been varied. As organizations they have been seen as too fragmented, decentralized, and incohesive, consisting of loose aggregations of factional subgroups and autonomous units of leadership and action. At the level of government, they have been described, for good or for bad, as undisciplined sets of representatives who do not work and vote as a collective group of party policymakers but as individualists running their own campaigns and acting in Congress on the basic of personal preferences rather than as part of a party team. The conduct of campaigns has been heavily criticized for being uncoordinated, poorly designed strategically, very improvisational, even not very rational — and above all, too expensive. The leadership recruitment process has always been a favorite target of attack, for being either poorly controlled by the parties or, to the contrary, too heavily under the control of "backroom party bosses." Recently, leadership selection is seen as too haphazard, not linked to party organizational responsibility, and requiring such large funds that it has become very elitist. On the one hand the parties, it is said, have at the top no centralized organs of leadership policy making, and at the bottom no effective mobilization of the votes by activist local organizations. Finally, these parties have over the years been decried as not programmatic and not responsive enough to public opinion. Indeed, one continual criticism is that they have poor communication lines to the public. They thus are alleged to be failures as good linkage structures between the government and the people.

Historically, we have seen many reforms of our parties, and the most important of these have been described in some detail in the pages of this book. We have certainly not been irrevocably locked into one system with no movement toward change. Many scholars have documented these reforms. Ranney argued that in the postwar period we had three types of reformers: the regulators (who wanted the parties to be subjected to the law and not permitted to function as private associations), the "representative party structures" reformers (who worked hard to make the party organizations, particularly the national conventions, more reflective of the characteristics of the party rank and file), and the "responsible party" advocates (who wanted more programmatic, well-organized, and disciplined parties).[22] All three types of reformers have been active in the last thirty to forty years. And much has changed, as our previous chapters have indicated. But not all objectives of the reformers have been realized. In fact, the "regulators" have lost out somewhat in recent years because the courts are giving back to the parties the right to make certain decisions, especially in the area of leadership nomination. The "representation" reformers accomplished a great deal in both parties, as the makeup of our national conventions now reveals, but unfortunately that whole series of reforms seems less significant these days with the decline of the convention as the arena for presidential selection. And the "responsible party reformers" have made some progress, perhaps, while still pressing for more. We will discuss this type of reform a little later.

The reform process and reform objectives from 1970 on have included several new directions in which party change has been moving, changes that go beyond the types of reform Ranney was preoccupied with. First, we have seen great effort in both parties to strengthen the organization, at all levels, with able, often salaried, personnel who have enough funds to engage more effectively in a variety of activities and functions. They work more directly, efficiently, and strategically in campaigns, with skilled activists performing a variety of functions — from media publicity to registering voters, planning rallies, and canvassing for votes. Second, our reforms have led to greater party cohesion in Congress, with the party caucus and leadership organs having prominent roles in deciding legislative procedure as well as policy. Third, there is more party integration of the national, state, and local levels of the party, leading to more cooperation and mutual support. Fourth, we have seen more emphasis on the local party, and the maintenance of these local activists' structures for mobilization of the vote.

There are two specific reform areas preoccupying a great deal of our time and concern these days that provoke continued debate and for which we have not yet achieved a satisfactory resolution. These are campaign finance reform and presidential primary reform. Both of these have been on the reform agenda for almost a century. In the 1880s we began worrying about the way money was raised for party campaigns and, later, how it was disclosed and spent. We are still deeply involved in this controversy today, despite the post-Watergate reform efforts. Particularly with the developments in the 1996 election, this reform challenge is directly before us and neither party appears to be willing to propose serious reform, as the situation in the present Congress makes all too clear. The alternatives for campaign reform were discussed in detail in Chapter 13.

The other major reform problem on our agenda is the presidential primary system. The parties continue to tinker with the details of the procedures for these primaries (and caucuses), their timing, particularly "front loading," their financing, their "closed" or "open" natures, the candidate procedures for access to the ballot in each state, the "vote threshold" for candidate success in a state, and the selection of party organization delegates outside primaries (the "super delegate rule"). More basically, there are those who are concerned about the effects of this leadership recruitment system for the American system. This debate is presented in detail in Chapter 11.

Obviously, despite all the party reform that has occurred in our system, there is much still to be concerned about if we are to maintain and strengthen our party system as a viable, responsive, and effective part of the political process. When we talk about party reform, we in political science circles usually return to the question what should we desire in a party system, what *model* of party system should we work toward? And that leads us invariably to the "responsible party model" developed by a committee of the American Political Science Association in 1950.[23] This was an ideal-type model to which scholars of our party system today still refer, either as a model to which we should aspire, or as a model that is really inappropriate for the American system. So it is a model that has evoked a great deal of advocacy, as well as strong opposition, over the years.

The APSA committee, in presenting the "responsible party model," took the basic position that our party system was inadequate "in sustaining well-considered programs and providing broad public support for them," that "with growing public cynicism . . . the nation may eventually witness the disintegration of the two major parties" and the possible emergence of "extremist parties." Further, the authors argued that we needed a party system that is "democratic, responsible and effective." This meant (1) "parties which bring forth programs to which they commit themselves"; (2) parties with "sufficient internal cohesion to carry out these programs"; and (3) the basic concept of the "responsibility of both parties to the general public, as enforced in election," as well as "responsibility of party leaders to the party membership." Thus, *program, cohesion,* and *accountability* were primary objectives of the responsible party reformers in 1950.

Some specific reforms were recommended by the committee.

A. *Organizational Changes*
   1. A national party council of fifty members with power to adopt platforms, recommend candidates for Congress, and discipline state and local parties deserting the national platform;
   2. Professional staffs for the national committee;
   3. National committees more representative of party strength;
   4. Smaller national conventions, but more frequently held (every two years).

B. *Platform Writing and Implementing Process*
   1. Members of Congress to be active in writing the platform;
   2. The party council to adopt and interpret it;
   3. State party platform to be adopted *after* national platform is adopted;
   4. Platform is binding on all party office-holders;
   5. Members of Congress to be deprived of committee chair positions if they opposed the party program.

C. *Congressional Parties*
   1. All Senate and House leadership positions to be consolidated in one leadership committee;
   2. The caucus of party members to be more important, to meet more frequently, and the decisions of caucus to be binding.

D. *Nominations*
   1. Closed primaries;
   2. More pre-primary conventions;
   3. A national presidential primary.

These were the reform proposals of the 1950s. Most of these recommendations have not been adopted in the specific form proposed. But some of them have, notably: professional staffs for the national committee, more representation in national party committees, a strengthened congressional party caucus, more closed primaries. But the basic concept underlying the idea of the "responsible party model" has not emerged yet. The idea of cohesive, program-oriented, disciplined, centralized parties, with (1) candidates selected because of their commitment to the party program, (2) party workers recruited on the basis of their commitment to the program, and (3) voters urged to support parties and candidates in programmatic terms, is a model that by no means has been implemented. Yet we have moved closer with our better organized, better integrated party organizations and with our congressional parties more ideologically cohesive. We still have what used to be called a pragmatic model, but with some modifications in the direction of more coherence and discipline.

The basic outlines of our pragmatic party system with ideological overtones remain. It is one in which the parties are still decentralized and pluralized, containing factions and subgroups often in considerable conflict over policy. These are fairly open structures, with great diversity in the types of people who are activists and who run for public office. Protest and dissent occurs often and is tolerated, and the insurgents remain within the party. There is still much autonomy, stratarchy, and variation in party program and performance by state and locality. Ours are not consensual, hierarchic, or controlled parties.

In any efforts at major party reform one must remember, first, that our pragmatic model has emerged after a long history of political development, after much experimentation with a variety of parties and party organizational systems. It is part of our political culture and constrained by many elements of that culture: our populism, our disinterest in strict discipline, our acceptance of duopolistic competition, our dispersion of power, our structural openness and tolerance of protest, our eclecticism in ideology, and our capacity for volatile electoral change. As our party system has evolved, particularly since the Civil War, we have come to appreciate certain virtues of that system. By and large it has been a stable system, maintaining an equilibrium of power between two major parties, strongly competitive but nonpolarized and nonextremist — a dynamic and open system, yet stable. Ours is a system that has demonstrated considerable consensus, but not a stagnant consensus for the most part. While competitive in orientation, it has also been integrative in function. And it has permitted us to get things done — to produce great leaders, great laws, an affluent economy, a society as free and just as any.

What, then, do we say about reform? Clearly some reform actions are necessary, and any system, however successful, must adapt to changing conditions in its society. If there is evidence, as there is, of a decline in public support for the parties — in the vitality of the organizations, and in their functional relevance — then we should seriously consider change and reform. But we should not delude ourselves into thinking that we can move very close to the responsible party model, although we can work toward certain of its proposals.

Certain proposals for further reform have been made throughout this book. In the larger sense, as we view the state of this party system, it is clear that we need ***more coherent policy leadership*** at all levels of the party. We are doing much better in this regard today in Congress, but our party organizations outside the legislature need to be more policy conscious. We also need organizations that are *more deliberative on questions of policy and leadership.* We need to involve citizens interested in politics in discussions of the issues of the day, and then forward the results of such discussions to the public leaders of the party. We should not spend all our time and energy on battles over strategy, personality conflicts, planning social occasions for the party, et cetera. We need a better educated and articulate party activist cadre. We also need *better means of communication* between the party and the public. The images that citizens have of politics are less positive and more derogatory than they should be. What parties communicate and how they communicate to the public is more important than ever. Finally, *the mobilization efforts of the local party* organizations are often unreliable and unsustained. We know how important such work is, from the research that has been done. Effective mobilization not only gets out the vote, but it educates citizens and involves them in a large array of participatory actions. These are good for the political process generally and also functional to party success.

Many improvements in the American party process are still possible and necessary. We know that our organizations often operate at only 50 percent or less of their effective potential. While our parties in recent years have been revived and revitalized, there is much opportunity and need for moving beyond this 50 percent threshold. But in all this our major principles and objectives must be kept firmly in mind. Primarily we seek to make our parties more basic and useful instruments for the popular control of government, and more concerned with finding humane solutions to our very serious social problems. Our parties have been the instruments for monumental achievements in the past. They continue to be major instruments for democratic government today. With necessary reforms, we can make them even better structures, central to the governmental process and to the lives of American citizens. One hundred years ago an English scholar of our system, James Bryce, said, "In America the great moving forces are the parties."[24] If we work hard enough at it, American parties will still be "the great moving forces" of our system.

POSTSCRIPT: THE VALUE OF STUDYING AMERICAN PARTIES

From a very broad intellectual perspective, one might ask what a case study like this, of the party system of one country, in depth, and over a long period of time, contributes. It certainly does not settle the issue which still today divides scholars over whether the two-party system of the U.S. type is more effective, or less effective, than the British system or multiparty systems elsewhere in the world. That controversy is particularly current today as Eastern European systems move toward democratization and the choice of party system confronts the elites and the publics of those countries. The criteria for evaluating party systems vary somewhat, depending on who is doing the evaluating. But usually the emphasis is on certain types of performance and outcomes for the system, such as meaningful citizen participation, governmental stability, maintenance of public order, effective representation of social groups, responsiveness of elites to mass needs and problems, and increasing acceptance of the political values of democracy.

When we reflect on the American two-party system and its functioning today and in the recent past, certain conclusions emerge. It is a system of parties that has emerged in the United States over a long time and has developed deep roots in our society. It is firmly embedded in the minds of most Americans as a system which, although periodically challenged, performs acceptably. It is a rather durable party system contributing to both changes in governmental control and continuity of the governmental system. It seems to be considered a system that functions effectively enough, despite lapses and crises (indeed, particularly during crises), to achieve most of the goals articulated by party elites, despite long periods of divided government, and party turnover, and electoral volatility.

The basic observation our book documents again and again is that this party system, as any other party system, can survive and perform effectively if it engages in strategic adaptation to the changes in the economic and social environment and the threats of those who seek to change this party system for another. In the face of all the dire predictions of scholars and journalists that our parties were "decomposing" and in a state of "demise," the American two-party system has survived. Why? For many reasons. Among them would be the creative effort of national, state, and local party leaders, along with the organizational resiliency and vitality of the two parties throughout the country, their linkages to the public and most of the major groups at the mass level, and the development of loyal cadres of supporters and mobilization of these supporters at election time. We can generalize that any party system in a democracy can do the same, thus maximizing its capacity for survival.

Every party system has its weaknesses, however, and ours is no exception. One problem is the evolving "culture of electoral apathy." Further, we worry over the failure of the two parties to adequately address the needs of the "underclass"—the blacks, the ethnic, the poor and disadvantaged. We have discussed one aspect of this problem in this book in detail. Another concern is the constant prospect of legislative-executive deadlock, although research reveals that much can and has been accomplished despite divided government when the party leaderships decide to do so. There are other problems, but essentially the American party system has over the years performed at a high level. It has become part of our cultural heritage. Even as it changes, and as we want to change it more, it maintains its central, functional role in the American system of governance. This imposes constraints on change and channels such change to fit the norms and virtues of our particular democratic society.

NOTES

1. Anthony Downs, *An Economic Theory of Democracy* (New York: Harper Brothers, 1957), 100, 109–11.

2. Kay Lawson and Peter H. Merkel, eds., *When Parties Fail* (Princeton NJ: Princeton University Press, 1988).

3. Roper Center for Public Opinion Research, *America at the Polls 1996* (Storrs, CT: Roper Center, 1997), 191.

4. Martin P. Wattenberg, *The Rise of Candidate Centered Politics* (Cambridge: Harvard University Press, 1991), 38, 44, 74.

5. Russell J. Dalton, *Citizen Politics,* (Chatham, NJ: Chatham House, 1996), 49.

6. For a complete discussion of the 1964 realignment, See Warren E. Miller and J. Merrill Shanks, *The New American Voter* (Cambridge: Harvard University Press, 1996), 140–46; Norman H. Nie, Sidney Verba, and John R. Petrocik, *The Changing American Voter* (Cambridge: Harvard University Press, 1976), chap. 13; Edward G. Carmines and James A. Stimson, *Issue Evolution: Race and the Transformation of American Politics* (Princeton, NJ: Princeton University Press, 1989), 37–58.

7. Miller and Shanks, *New Voter,* 149–50.

8. Nie, Verba, and Petrocik, *Changing Voter,* 143, 102, 115.

9. See Martin Wattenberg, *Rise,* 96.

10. Ibid., 353.

11. Ibid., 101.

12. Nie, Verba, and Petrocik, *Changing Voter,* 203. Based on an issue position scale for five issue areas: size of government, welfare, integration, welfare for blacks, and the Cold War.

13. See, for example, Byron E. Shafer, ed., *The End of Realignment?* (Madison: University of Wisconsin Press, 1991).

14. This description is drawn largely from Burnham and Sundquist's treatments of the subject: Walter Burnham, *Critical Elections and the Mainsprings of American Politics* (New York: Norton, 1970), 130–35; James Sundquist, *Dynamics of the Party System* (Washington, DC: Brookings Institution, 1973). See also John R. Petrocik, *Party Coalitions: Realignments and the Decline of the New Deal Party System* (Chicago: University of Chicago Press, 1981).

15. V. O. Key, Jr., "A Theory of Critical Elections" *Journal of Politics* 17 (1955).

16. Burnham, *Critical Elections,* 130–31. 135.

17. *Ann Arbor News,* November 20, 1980.

18. Nie, Verba, and Petrocik, *Changing Voter,* 241, 354.

19. James Sundquist, *Dynamics of the Party System* (Washington, DC: Brookings Institution, 1973), 373.

20. Julio Borquez, "The Life and Times of the Reagan Democrats: Defection of Democratic Party Identifiers in Recent Presidential Elections," doctoral dissertation, University of Michigan, 1998.

21. Austin Ranney, *Curing the Mischiefs of Faction: Party Reform in America* (Berkeley: University of California Press, 1975), 24–25.

22. Ranney, *Curing the Mischiefs of Faction,* 42–47.

23. "Toward A More Responsible Two-Party System" (Report of the Committee of Political Parties), *American Political Science Review* (Supplement) 44 (1950).

24. James Bryce, *The American Commonwealth* (New York: Macmillan, 1916), 5.

Index